Living with the California coast

Gary Griggs and Lauret Savoy, editors

Sponsored by the National Audubon Society™

Duke University Press   Durham, North Carolina   1985

THE NATIONAL AUDUBON SOCIETY
AND ITS MISSION

In the late 1800s, forward thinking people became concerned over the slaughter of plumed birds for the millinery trade. They gathered together in groups to protest, calling themselves Audubon societies after the famous painter and naturalist John James Audubon. In 1905, thirty-five state Audubon groups incorporated as the National Association of Audubon Societies for the Protection of Wild Birds and Animals, since shortened to National Audubon Society. Now, with more than half a million members, five hundred chapters, ten regional offices, a twenty-five million dollar budget, and a staff of two hundred and seventy-three, the Audubon Society is a powerful force for conservation research, education, and action.

The Society's headquarters are in New York City; the legislative branch works out of an office on Capitol Hill in Washington, D.C. Ecology camps, environmental education centers, research stations, and eighty sanctuaries are strategically located around the country. The Society publishes a prize-winning magazine, *Audubon*, an ornithological journal, *American Birds*, a newspaper of environmental issues and Society activities, *Audubon Action*, and a newsletter as part of the youth education program, *Audubon Adventures*.

The Society's mission is expressed by the Audubon Cause: to conserve plants and animals and their habitats, to further the wise use of land and water, to promote rational energy strategies, to protect life from pollution, and to seek solutions to global environmental problems.

National Audubon Society   950 Third Avenue   New York, New York   10022

Publication of the various volumes in the Living with the Shore series has been greatly assisted by the following individuals and organizations: the American Conservation Association, an anonymous Texas foundation, the Charleston Natural History Society, the Coastal Zone Management Agency (NOAA), the Geraldine R. Dodge Foundation. the Federal Emergency Management Agency, the George Gund Foundation, the Mobil Oil Corporation, Elizabeth O'Connor, the Sapelo Island Research Foundation, the Sea Grant programs in North Carolina, Florida, Mississippi/Alabama, and New York, The Fund for New Jersey, M. Harvey Weil, and Patrick H. Welder, Jr. The Living with the Shore series is part of the Duke University Program for the Study of Developed Shorelines.

Library of Congress Cataloging in Publication Data
Main entry under title:
Living with the California coast.
(Living with the shore)
Bibliography: p.
Includes index.
1. Shore protection—California. 2. Coasts—California.
3. Coastal zone management—California. I. Griggs,
Garry B. II. Savoy, Lauret E. III. Series.
TC224.C2L58   1985   333.91'7'09794   84-28814
ISBN 0-8223-0632-8
ISBN 0-8223-0633-6 (pbk.)

Living with the California coast

Living with the Shore

Series editors
Orrin H. Pilkey, Jr.
William J. Neal

The beaches are moving: the drowning of America's shoreline,
new edition
Wallace Kaufman and Orrin H. Pilkey, Jr.

From Currituck to Calabash: living with North Carolina's barrier islands,
second edition
Orrin H. Pilkey, Jr., et al.

Living with the Texas shore
Robert A. Morton et al.

Living with Long Island's south shore
Larry McCormick et al.

Living with the Alabama-Mississippi shore
Wayne F. Canis et al.

Living with the West Florida shore
Larry J. Doyle et al.

Living with the Louisiana shore
Joseph T. Kelley et al.

Living with the South Carolina shore
William J. Neal et al.

Living with the East Florida shore
Orrin H. Pilkey, Jr., et al.

Contents

Figures and tables

Figures

Tables

Preface

Standing at the edge—looking west

California's physical grandeur, wealth of resources, and mild climate have drawn countless numbers to the state since the first explorers sailed her coastal waters over 4 centuries ago. We are now the nation's most populous state. Owing its initial boom in growth to manifest destiny and the discovery of gold in the 1800s, California is still one of the fastest growing states, and its shoreline has been the site of much of this rapid development. Today the only pristine coastline exists in the northern portion of the state or along the central coast where the topography is too rugged to allow extensive development.

The southern California shoreline today from San Diego to Malibu is little more than a backdrop of riprap, seawalls, highways, and cliffs and beaches covered with homes, condominiums, and restaurants. Far too often these structures were built with little regard for the dynamic nature of the coast and the physical processes that continue to shape it. The long-term result is significant and repeated damage to coastal property, public and private. The winter storms of 1977–78 and 1982–83 have made it clear that living directly on the shoreline involves a heavy tax and risk. Relative to the southern portion of the state, northern California fares somewhat better. There, development is scattered and comparatively light, and opportunities still exist to learn from earlier mistakes and promote safe, rather than hazardous, unplanned development.

The present book is the ninth in the Living with the Shore series being published by Duke University Press. The books in this series are concerned with shoreline planning and development in each of the nation's coastal and Great Lakes states. Our intent is that this book will help all Californians to understand the processes and problems inherent in the coastal zone so that they may make more educated decisions about building, buying, and living on the shoreline. The attraction of the coast is clear, but the hazards present and costs of living in this dramatic setting are not always so obvious, or are too quickly forgotten.

As an umbrella book to the series, Duke University Press has reprinted with an updated appendix the classic *The Beaches are Moving: The Drowning of America's Shoreline* by Wallace Kaufman and Orrin H. Pilkey, Jr. This book covers the basic issues dealt with specifically in the state-by-state books. *Coastal Design: A Guide for Builders, Planners, and Home Owners*, a general construction volume by the series editors, is intended to apply to all coasts and is available from Van Nostrand Reinhold.

The first 4 chapters discuss the processes operating in the coastal zone and the hazards that are present. The next 2 chapters deal with some alternatives for combating or mitigating shoreline damage or erosion, and there is a discussion of land use planning and regulation along the coast.

The second half of the book consists of 12 chapters that discuss the various regions of the state's coastline from north to south. In order that the book be up to date and accurate we have asked the coastal geologists most familiar with each stretch of coast to evaluate and write those particular sections. Detailed maps accompany each chapter that delineate hazardous or potentially hazardous areas and the existing erosion rates. As writers and editors we attempted to fill in the voids and tie the individual chapters together. We have also attempted to avoid complicated terminology that we feel would limit the book's usefulness.

Three appendixes have also been included to make this book more useful: a glossary of words commonly used in the book, a collection of useful reference articles and books arranged by topic for additional information, and a list of public agencies that can provide information on or have responsibility for some aspect of the coast.

Four generations of coastal geologists have contributed to this effort. Francis Shepard, of the Scripps Institution of Oceanography, now in his eighties and widely recognized as one of the fathers of marine geology, is coauthor of the San Diego chapter. One of his early graduate students at Scripps Institute of Oceanography and for years a professor of geology at the University of California, Santa Barbara, Robert Norris, has written the Santa Barbara to Ventura chapter. The senior author studied under Professor Norris at UC Santa Barbara, while the junior author has just completed graduate work at UC Santa Cruz.

Much of the information presented in the following pages is the result of previous research by the authors as well as other scientists who are not directly referenced. For additional information the reader should consult the original publications listed in the bibliography at the end of the book.

As authors and editors we would like to acknowledge the efforts of a number of people who assisted us with this book. Orrin Pilkey of Duke University deserves a great deal of credit for initiating this entire series of coastal books and encouraging and supporting the entire effort from beginning to end. We are grateful to all of our colleagues throughout the state who contributed to the book by writing particular chapters. A number of other individuals deserve thanks for their assistance in our research and data collection efforts. These include George Armstrong, Gary Jones, Rogers Johnson, John Tinsley, Doug Pirie, Dave Wagner, Dick McCarthy, and Bruce Fodge. We are grateful to Barbara Gruver and Tonya Clayton for their cartography and drafting throughout the book and to Alice Ingerson for her conscientious and careful editing. We would like to give special thanks to Robbie Marshall, who typed and retyped more drafts than we care to mention without ever losing her patience or asking why.

Gary Griggs
Lauret Savoy
Santa Cruz, California
August 1984

Living with the California coast

Part One
1. Introduction

Gary Griggs and Lauret Savoy

The California coast has something for everybody along its nearly 1,100 miles of land and water (fig. 1.1): beaches to escape the heat of the city on a summer day; fertile near-shore waters that support an extensive sport and commercial fishing industry; and private property on the ocean, a place to build your home for recreation or retirement. These are but a few of the shoreline's benefits. This coastal magnet or attraction is clearly seen in a single statistic: 80 percent of our state's population now lives within 30 miles of the shoreline, and this number continues to increase.

The conflict between accelerating coastal development and the inherent geological instability of the shoreline is developing into a dilemma of increasing magnitude. Many diverse forces and processes interact on the coast, making the shoreline the world's most dynamic environment. Waves, tides, wind, storms, rain, and runoff act to build up, wear down, and continually reshape the interface of land and sea. Of nearly 1,100 miles of California coastline, 86 percent or 950 miles is undergoing erosion, and 125 miles of this erosion area is deemed critical, meaning that structures or utilities are threatened!

The ongoing natural processes of rock weathering and cliff retreat were either not recognized, appreciated, or completely understood or were ignored by most coastal builders, developers, and homebuyers in the past. Within the past 10 years this problem has come into clear focus along virtually the entire coastline. Public and private losses during the 1978 storms amounted to over $18 million. More recently, the high tides and storm waves during the winter of 1983 inflicted over $100 million in damage to ocean-front property! Damage in these recent storms was not restricted to broken windows and doors—33 ocean-front homes were totally destroyed.

California is not the only state with shoreline erosion problems. Much of the shoreline of each of our coastal states is eroding, and the low, sandy coastlines common along the East and Gulf coasts often erode more rapidly than California's. These East and Gulf coast residents also live under the added threat of a hurricane season.

The conflict between developing the coastline and hazards such as erosion associated with this development is increasing for several reasons: (1) an increased migration to coastal communities and desirability of ocean-front property; (2) the progressive erosion of shorefront yards and vacant property such that structures and utilities are now being undercut and encroached upon; (3) the human-induced acceleration of seacliff erosion due to cliff-top construction with its associated roof and street runoff and landscape watering; and (4) an apparent recent climatic change, well documented in southern California, to a period of

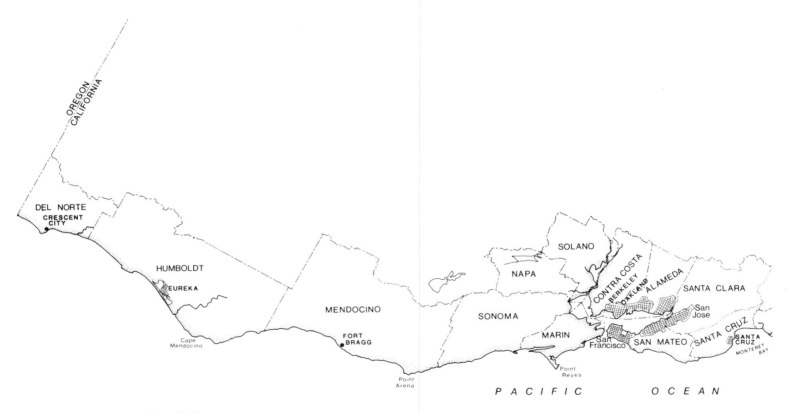

Figure 1.1. Index map of the California coast.

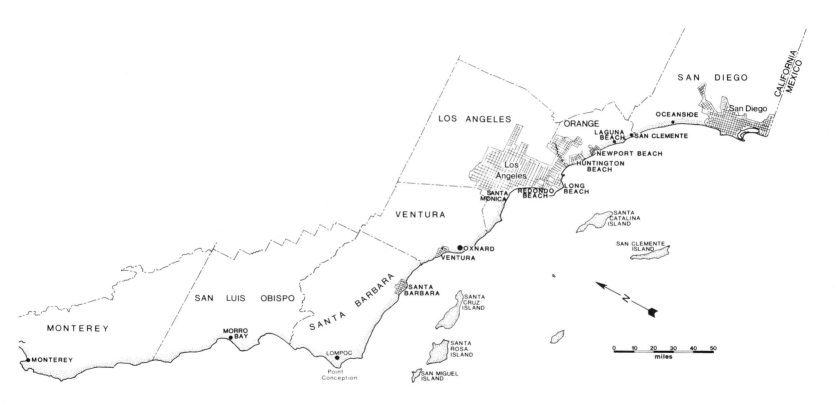

Figure 1.2. Over $100,000 was spent on rock to protect this cliff-top home. Photo by Gary Griggs.

Figure 1.3. A new $1,500,000 timber bulkhead was heavily damaged at Seacliff State Beach in Santa Cruz during the winter of 1983. Note the remnants of 2 earlier walls in the surf zone. Photo by Gary Griggs.

Figure 1.4. The pile of wreckage on the left is all that remains of a $350,000 beach house in Santa Cruz that was built too close to sea level with almost no protection. Photo by Gary Griggs.

more frequent large storms bringing heavy rainfall and larger waves. In addition, coastal engineering projects at a number of locations have directly or indirectly accelerated erosion rates in adjacent areas, principally by trapping sand and starving downcoast beaches.

The climate in southern California has drawn more people to this portion of the coastline than elsewhere in the state. The warm days and wide summer beaches can be deceiving, however, as cliff-top and beach-front dwellers in places like San Diego and Malibu discover during the winter months. Beach sand moves offshore, leaving the deck and plate glass windows behind to absorb the full force of the destructive storm waves.

One cliff-top property owner in the Santa Cruz area recently spent $100,000 for riprap to protect a single home (fig. 1.2). This protection effort was not the first for the property, but was reinforcing and repairing an earlier project. Experience now indicates that some poorly engineered structures are not effective in halting shoreline erosion for any extended period. For example, a timber bulkhead at Seacliff State Beach in Monterey Bay has been damaged or destroyed 10 times in the past 58 years (fig. 1.3). The replacement or repair of riprap, revetments, and seawalls at many locations is an ongoing expense, a natural property tax of sorts, for building in a hazardous location.

For many, living on the ocean is a dream come true. For others it has become a nightmare and an expensive habit to maintain. Take Joe Maschutes, for example. In early January 1983 he moved into a $350 thousand beach-front house in Santa Cruz, and 10 days later, after a storm, his house slid into Monterey Bay. Nothing remained but a small pile of wreckage (fig. 1.4).

Events of this sort are tragic, but we can learn something from them. The Pacific Ocean is a very powerful force to reckon with. Nearly all of our protection efforts are, in the long run, temporary and simply buy a little more time; time bought at great expense. Unless we are wealthy or have some government subsidy or aid, most of us cannot long afford to protect an ocean-front home in an area of *active coastal retreat*. Yet there are literally thousands of homes in such hazardous locations, and many other ocean-front areas where development is being planned or proposed.

Before we plan, develop, construct, or purchase in ocean-front areas along the California coast, we must have a clear sense of the physical processes active in those individual areas. We should recognize the evidence for beach erosion, dune migration, and cliff retreat, and be able to evaluate how rapidly individual areas are eroding or changing. What effect does human activity have on these natural processes and what can we learn from California's past history of coastline development and construction? What options exist for protecting our valuable ocean-front property and how effective and costly are different methods? What "nonstructural" solutions can be utilized in areas where intensive development has not yet taken place, for example, setbacks or buffer zones based on measured long-term retreat rates? And finally, what types of land-use planning regulations do we need to consider and what agencies must/can we work with in our efforts to build, plan, buy, or protect our beach house?

This book discusses the processes, problems, and issues that we feel anyone living or planning to live directly on the coastline must understand. This book should allow anyone to make a well-informed decision about the future of their particular home or homesite and learn something of the past, present, and the future of the individual geographic areas along the state's 1,100-mile coastline.

2. Understanding the shoreline

Gary Griggs, Lauret Savoy, Gerald Kuhn,
Francis P. Shepard, and Donald Disraeli

California is blessed with one of the most spectacular and diverse shorelines in the country—one of high mountains plummeting to sheer rocky cliffs, long stretches of sandy beach, and extensive marshes and wetlands. It is a land defined by the sea. There is little wonder that such an environment continues to draw people to the shore. Yet, the shoreline is also the battleground between land and sea. As more and more people move to the coastal areas, the dynamic processes operating at the shoreline will have an increased effect on development; in addition, our increased use of and demands on coastal areas will in turn affect the stability of the shoreline itself. It is imperative, then, that we live with and carefully use this coastal environment with a full understanding of its processes, rather than demanding too much of it.

The mere existence of a sheer seacliff fronting the ocean is evidence that the coastline is generally retreating by erosion. The presence of loose sand and driftwood on a beach indicates that this landform is regularly washed over by wave and tidal action. Although a particular beach may appear high, wide, and stable during the summer months, it can disappear virtually overnight during a severe winter storm (figs. 2.1 and 2.2). Similarly, a cliff can retreat many feet during a single storm (fig. 2.1). It is vital that we know what happens at the shoreline, why it happens, and how the shoreline may be expected to change during a severe storm, a typical winter, or over a number of years. Only with an understanding of such processes will we be able to identify the limitations that the coastal environment may place on our activity, and thereby prevent costly and repeated attempts to protect or save structures built too near the ever-changing shoreline.

Take a walk along the shore. Whether it is along Imperial Beach, Malibu Beach, Gold Bluffs, or countless other places along California's coast, you are likely to see sandy beaches backed by either low bluffs or abrupt cliffs on one side of you, and the repeated rush of the surf onto the shore on the other. Perhaps this scene evokes, besides a sense of beauty and energy, a feeling of permanence or even timelessness. Such a feeling can be misleading, however. Each feature of the beach—whether breaking waves, steep cliffs, sand dunes, or the beach itself—is dynamic and delicately balanced with the other parts, often changing from day to day, month to month, or year to year.

Waves

Let's first turn our attention offshore, to the driving force that continually shapes and reshapes the shoreline—waves.

Anyone who has seen storm waves batter a beach realizes the tremendous power that the sea expends. A 10-foot-high wave

Sandy beach backed by dunes

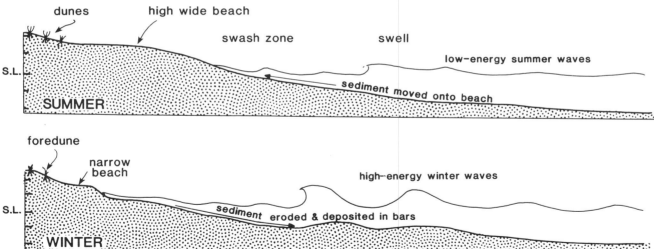

dunes

high wide beach

swash zone

swell

low-energy summer waves

S.L.

sediment moved onto beach

SUMMER

foredune

narrow beach

high-energy winter waves

S.L.

sediment eroded & deposited in bars

WINTER

A

Cliffed coast

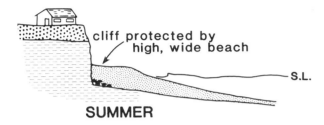

cliff protected by
high, wide beach

S.L.

SUMMER

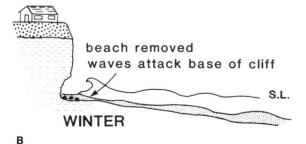

beach removed
waves attack base of cliff

S.L.

WINTER

B

can exert over 1,000 pounds of pressure per square foot; wave pressures as high as 12,700 pounds per square foot have been recorded. Storm waves on the coast of France have been observed to throw 7,000-pound boulders over a 20-foot wall.

Nearly all waves form when wind creates friction as it blows across the ocean's surface. The size of the waves that break on the beach on any particular day depend mainly on the offshore wind characteristics: how long, how fast, and over what distance of the sea's surface the wind blows. The longer, harder, and farther the wind blows, the larger we can expect the waves to be. Waves formed where and while the wind is blowing are typically irregular and choppy and are usually called *sea*. These waves will gradually move out away from a storm area and sort themselves out into a more regular pattern known as *swell*. These latter waves that we see breaking on our beach may have traveled hundreds or thousands of miles across the ocean from their stormy point of origin.

As waves *shoal*, or reach the shallow water near the shoreline, their height increases until they become unstable to the point that they break. Waves usually break where the ratio of wave height to water depth is about 3:4; in other words a 3-foot-high wave will break in about 4 feet of water.

Figure 2.1. Seasonal changes in beach profiles along the California coast. (A) Sandy beach backed by dunes. (B) Cliffed coast.

A   B

Figure 2.2. Seasonal changes in Seabright Beach, Santa Cruz. (A) Summer—wide sandy beach. (B) Winter—beach eroded by storm waves. Photos by Gary Griggs.

Large waves can be particularly damaging when they occur simultaneously with very high tides. This coincidence of severe storm waves and very high tides occurred during January and February of 1983 and was responsible for much of the coastal damage. With the beaches either absent or greatly reduced in size, and the huge waves riding atop a 6.6-foot high tide, many areas formerly above the reach of the waves were heavily hit.

Most waves approach the coast at an angle. Along much of the California coast for example, summer waves often arrive from the northwest. As the water gets shallower near the coast, the portion of the waves closest to the beach "feels" the sea floor first and begins to slow down; meanwhile, the seaward portion of the wave crest continues to travel at almost its original speed. This results in the bending or *refraction* of the wave toward the shoreline (fig. 2.3A). On an irregular coast, refraction causes wave energy to be concentrated at promontories and dispersed in bays (fig. 2.3B). The bending and gradual breaking of the waves around a point or promontory produces some of the best surfing locations in California—Malibu, Rincon, and Santa Cruz's Steamer Lane to name a few.

Waves are rarely completely refracted, and so still break on the beach at a slight angle. As a result, a current flowing parallel to the shoreline, known as a *longshore current*, forms within the surf zone (fig. 2.3). These longshore currents exist along most of California's beaches and are quite important in transporting sand along beaches. This process is also known as *littoral drift* and can be thought of as a river of sand moving along the coast.

Tidal wave? No, tsunami

Wind does not create all waves. Just as you can make ripples by either shaking a glass of water or blowing across the water's surface, any large disturbance or sudden movement in the ocean can also create waves. Enormous oscillations of water caused by particular types of seafloor earthquakes, underwater landslides, or exploding island volcanoes result in waves known as *tsunamis*. Popularly, but incorrectly called tidal waves, for tides have nothing to do with them, tsunamis can come in all sizes. In the open sea a tsunami wave can travel at speeds up to 500 miles per hour and is almost unnoticeable, with a height of only several feet. As the tsunami shoals, or reaches the shallow water near the shore, however, its height increases dramatically. Waves of this type over 100 feet high have at various times hit the coasts of Japan, Alaska, Chile, and the West Indies. Tsunamis that reach California can originate almost anywhere in the Pacific Ocean, from Alaska to Chile to Japan. Because of their great speeds and wave heights, tsunamis possess considerably more power than normal wind-driven waves. So although they might travel thousands of miles across an ocean from their source, tsunamis can cause extensive damage upon reaching the shoreline.

In 1812 a tsunami generated by an earthquake in the Santa Barbara channel created 30-to-50-foot waves that pounded the shore near Refugio and Santa Barbara. The most recent destructive tsunami to batter California's coast accompanied the large 1964 Alaskan earthquake. Crescent City, on the northern Cali-

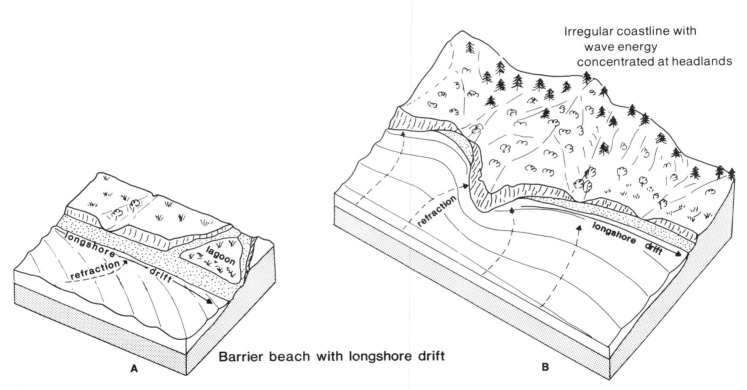

Irregular coastline with
wave energy
concentrated at headlands

Barrier beach with longshore drift

A

B

Figure 2.3. Waves that approach the coastline at an angle are refracted or bent as they enter shallow water.

fornia coast, was the hardest hit, being inundated by several waves with a 20-foot runup. The worst waves struck the waterfront area at 1:45 a.m., drowned 10 people, demolished 150 stores, and littered the streets with huge redwood logs from a nearby sawmill. Most of the city's downtown either was damaged or totally destroyed, and rather than being rebuilt, the blocks nearest the harbor were made into a park. According to the U.S. Army Corps of Engineers, property losses approached $27 million. Southward, damages from the same tsunami ranged from $125,000 at Noyo Harbor in Mendocino County to $250,000 at Los Angeles Harbor.

Although destructive tsunamis do not occur frequently in California, they happen more often than you might think. At least 19 have reached California over the past 199 years, and many more will come in the future. Only the tsunamis of 1812 and 1964, however, have caused major damage during historic time.

Beaches

Beaches are the buffer zones of shock absorbers that protect the coastline or seacliff from direct wave attack. The protection or exposure of the coastline, its orientation with respect to the dominant northwesterly waves, and the presence or absence of sand are major factors affecting whether or not beaches form at any particular location. Where no protection is offered (Mendocino and Big Sur for example) and cliffs receive full wave attack,

beaches may be absent altogether (fig. 2.4). At other locations, rivers or streams have cut through the coastal cliffs, or waves have eroded embayments or bays forming small protected pocket beaches (fig. 2.4).

Figure 2.4. A high-energy cliffed coastline with small pocket beaches formed in the protected coves where streams enter the ocean.

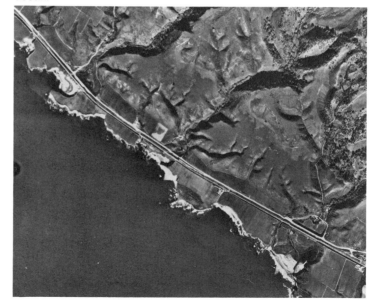

At still other locations, where resistant headlands extend into the ocean, waves approaching from the northwest commonly bend around the headland and erode the weaker downcoast material. In this case a hooked or spiral-shaped bay and long sandy beach form; good examples include Half Moon Bay, Bolinas Bay, and Mission Bay (fig. 2.5).

Last but not least, long, straight, sandy beaches form where the coastline relief is low, the orientation with respect to wave approach is appropriate, and ample sand is available. These beaches are often backed by extensive dunes, such as those at Silver Strand in San Diego, Zuma Beach near Malibu, Seacliff and Sunset beaches near Santa Cruz, the Humboldt Bay beaches, and Pelican Bay Beach north of Crescent City (fig. 2.6).

Where does the beach sand come from?

Along the California coast most beach sand comes either from river and stream runoff or from erosion of coastal cliffs and bluffs. Measurements of sediment transport in rivers indicate that coastal streams, particularly during times of flood flows, are the major suppliers of sand to our beaches (fig. 2.6). Along most of the California coast it has been estimated that 75 percent to 95 percent of the beach sand was originally derived from streams. Sand and gravel erode from rocks and soils high in the coastal watersheds and are transported downstream by rivers. After months and miles of abrasion and sorting these particles ultimately reach the shoreline where they become beach sand. Beaches have often

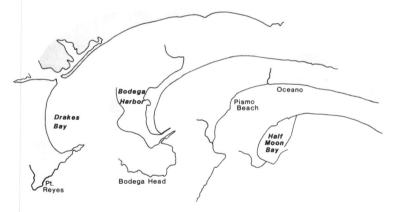

Figure 2.5. Examples of spiral-shaped bays and beaches along California's coastline.

been observed to be much wider in the summers following winters with high rainfall due to the delivery of large amounts of sand to the beaches by high streamflow.

Coastal cliffs can also be important if they consist of or break down into sand-sized material (sandstone and granite, for example). Cliffs and bluffs on the other hand, which are composed of silt or clay-sized material (shales or mudstones, for example), will not contribute significantly to the beach. Although the contributions of beach sand by coastal cliffs can be important locally, particularly where the cliffs are rapidly eroding, this source probably doesn't account for more than 25 percent of our beach sand.

The beach—a river of sand

Once sand arrives at the coast, waves and wave-induced currents provide the energy necessary to form the beach and to move the beach materials along the coast. The direction of this movement or littoral drift of sand along the beach is determined by the dominant angle of wave approach. For example, along much of the coast, summer waves from the northwest drive littoral drift southward along the beaches. During the winter months, waves often arrive from the west-southwest, resulting in a northward littoral drift.

It is essential to know something of the rate and direction of littoral transport prior to any human intervention in the nearshore system. Decisions on sand mining, planning, and design of coastal structures to prevent erosion, and questions of beach

Figure 2.6. Sediment plume from the San Lorenzo River at flood stage as it enters northern Monterey Bay. Note waves breaking offshore from river mouth due to sand deposition. Photo by U.S. Geological Survey.

nourishment and long-term stability are all tied to this littoral drift system. Virtually anything we insert into this river of sand is going to interrupt and disrupt the natural system. Unfortunately, littoral drift rate was either not considered or not well understood prior to construction in a great many cases. The results have quickly become apparent, and dredging, erosion control, or damage costs have amounted to hundreds of millions of dollars. There is no excuse, however, for making these kinds of mistakes in the future.

Average yearly littoral drift rates along California's coast show a considerable range, from about 30,000 cubic yards at Redondo Beach to nearly 1 million cubic yards at Ventura (table 2.1). One million cubic yards, if spread evenly through the year, amounts to 275 dump trucks full of sand moving along the beach daily, or about one every 5 minutes. The effects of halting or disrupting

Table 2.1   Estimated annual littoral drift rates and directions along the California coast

Location	Annual rate (cubic yards)	Direction
Santa Cruz	300,000	East
Santa Barbara	400,000	East
Ventura	700,000–1,000,000	Southeast
Santa Monica	275,000	Southeast
Redondo Beach	30,000	South
Camp Pendleton	100,000	South

this incredible flow of sand may now be clearer. At Ventura harbor, for example, between 1978 and 1983, over 4 million cubic yards of sand was dredged at a cost of about $4.2 million. This is only one of many such sand traps along the coast.

Where does the sand go?

We now know that beaches in most places are being continually supplied with sand (although in decreasing amounts in southern California) and that littoral drift is constantly moving the material somewhere. Where is all this sand going and why aren't the beaches growing wider and wider? If we follow the paths of typical grains of sand, we will observe several possible routes through which sand may leave the beach permanently.

Sand dunes. Dunes are connected to beaches in some locations and act as a sink where beach sand may be permanently lost (fig. 2.7). As a beach widens and the area of dry sand expands, transport by wind may begin to occur. Generally, dunes form wherever ample dry sand is available, wind is persistent, and a low-lying area landward of the beach exists where the sand can accumulate. If the wind direction is steady enough, sand will move, grain by grain, over the dune's surface, resulting in a downwind movement of the dune. Much of this sand is permanently lost from the beach. It has been estimated, for example, that 20,000 cubic yards of sand are blown inland each year along the 35-mile coastline from Pismo Beach to Point Arguello.

Dunes serve as an important coastal buffer because they are

flexible barriers to storm waves and provide protection to the lower backdune areas. In fact, wherever dunes can be created they often work better than expensive seawalls. Dunes also maintain a large stockpile of sand, which feeds the beach during a severe storm. Under storm attack, the beach is first cut back, and if wave erosion continues, then portions of the dune may be eroded. This sand is moved offshore where it is stored in sandbars that tend to reduce the wave energy impinging on the shoreline (fig. 2.1). As a storm subsides, smaller waves begin to transport sand back onto the beach, which is ultimately rebuilt. With time or the absence of large storms, wind will move the excess sand back onto the dunes. The process of natural dune rebuilding will take several years, however. Dune erosion, either during storms or because of a reduction in sand supply, can be destructive to any structures placed on the dunes.

Sand dunes are ephemeral or temporary landforms. Storms will recur, sandbars will shift, dunes will move. In addition, dune instability may result from human construction or recreation. The frontal or foredune is particularly prone to change as has been discovered in recent years by owners of new condominiums and houses perched on active sand dunes. Ian McHarg, in his classic book *Design with Nature*, suggests that no development, recreation, or human activity of any type should occur on the primary or secondary dunes, which are the least stable and contain the most fragile vegetation (fig. 2.8). This principle has been widely violated along the California coast, with costly consequences. Development, if it is to occur at all, should take place

Figure 2.7. Sand dunes migrating inland along the central California coast near Pismo Beach. Photo by California Department of Fish and Game.

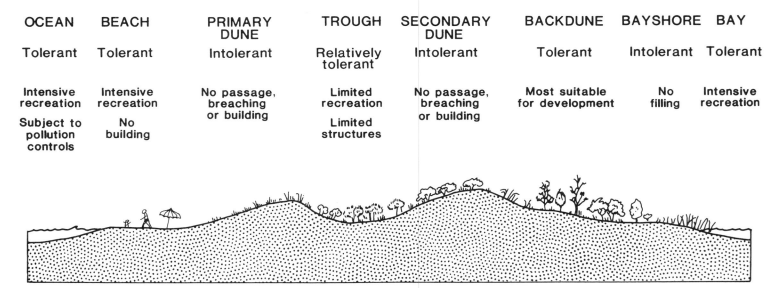

OCEAN	BEACH	PRIMARY DUNE	TROUGH	SECONDARY DUNE	BACKDUNE	BAYSHORE	BAY
Tolerant	Tolerant	Intolerant	Relatively tolerant	Intolerant	Tolerant	Intolerant	Tolerant
Intensive recreation	Intensive recreation	No passage, breaching or building	Limited recreation	No passage, breaching or building	Most suitable for development	No filling	Intensive recreation
Subject to pollution controls	No building		Limited structures				

Figure 2.8.   Tolerance of the dune environment to disturbance. Each dune complex consists of a number of distinct areas that vary considerably in their stability or ability to withstand encroachment or alteration. A typical dune profile shows the backdune area to be the most suitable for human habitation. Adapted from I. McHarg, *Design with Nature*. Garden City, N.Y.: Natural History Press, 1969.

on the backdunes, which have the advantage of protection from winter storms. Limited cluster development might also occur in the trough between dunes, provided groundwater withdrawals would not adversely affect dune vegetation, and the dunes themselves are not breached by roads, utilities, or human trampling.

Recreational impact comes in the form of pedestrian traffic and off-road vehicular (ORV) use. Most of the foot traffic impacts come from uncontrolled crossings of the dune from the backdune to the beach (fig. 2.9). Large numbers of crossings, dissecting the dune field with numerous paths, lead to the development of large barren areas such as *blowouts*, decreased dune growth, and no development of new dunes. Heavy foot use has been shown to decrease the amount of plant material and accelerate dune destabilization.

Sand mining. In addition to these natural processes, sand is also quarried directly from some California beaches. Three major sand mining companies remove sand directly from the beach face in southern Monterey Bay. The high-purity quartz sand is in high demand for a number of industrial uses. There is concern, however, as to whether the approximately 330,000 cubic yards of sand being removed each year is greater than the volume entering the area naturally. If the removal volume is greater than that added each year, then shoreline retreat will occur.

Submarine canyons. A major but invisible loss of sand takes place through submarine canyons that indent much of the southern California's offshore continental shelf (fig. 2.10). Where these canyons extend close enough to shore, they intercept the littoral

Figure 2.9.   Abused dune fenced off for recovery. Photo by Lauret Savoy.

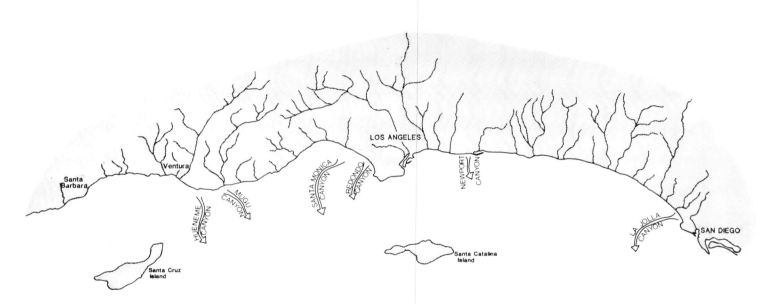

Figure 2.10.   Submarine canyons of the southern California coastline.

drift and funnel it away from the beach into deep basins. Every year, for example, the Scripps Submarine Canyon at La Jolla swallows as much as 115,000 cubic yards of sand, enough to form a beach 3 feet deep, 200 feet wide, and 1 mile long. Monterey Submarine Canyon, which crosses Monterey Bay, is perhaps one of the world's greatest submarine canyons and is large enough to hold the Grand Canyon of the Colorado. Much of northern Monterey Bay's littoral drift is probably carried offshore by this deep canyon system.

Under certain conditions, particularly during large storms, waves and wave-induced currents may be sufficient to carry sand offshore into deep water. This is believed to be most common at points or projections along the coast. If carried far enough offshore the sand may be lost permanently from the beach.

Are there threats to our beaches?

In recent years we have begun to realize that many of our activities are rapidly or gradually affecting beach sand supply and therefore beach stability. Since the 1950s many of southern California's rivers and coastal streams have been dammed for water supply and flood control. The dams effectively impound the water but also trap the sand destined for the coastal beaches. Dams also control the high-velocity flood flows that are responsible for transporting large volumes of sand to the beach. Thus, the benefits of flood control and/or increased water supply have been countered by the gradual reduction of sand input to the coastline.

The nearly complete damming of the few streams of southern California has almost eliminated the natural beach sand supply from this source.

As an example, the total sediment supply of the Santa Clara River southeast of Santa Barbara has been reduced 37 percent by dams. In the past 20 years alone this coastal region has been starved of enough sand to build a beach 300 feet wide, 3 feet deep, and 60 miles long. The Ventura River and Santa Maria River just to the north have had 66 percent of their drainages blocked by dams. Throughout Southern California, beach materials are trapped by 311 water supply lakes and flood control reservoirs and an additional 77 sand and gravel quarries. The construction industry alone annually mines over 20 million tons of sand and gravel from beaches, dunes, and riverbeds.

The northern California sand supply situation is better. Major dams are few, and sand and gravel mining is limited. Along the central coast between San Francisco Bay and Point Conception, a number of dams have partially reduced streamflow and therefore sand input, although not nearly to the degree true of southern California. Portions of the watersheds of the Salinas, Carmel, and San Lorenzo rivers have been dammed.

Douglas Inman, director of Scripps Institution's Center for Coastal Studies, predicted years ago that the beaches of southern California would gradually disappear as this river sand supply was cut off. What he did not anticipate, however, was the boom in construction of small boat harbors. Sand derived from harbor construction and maintenance has usually been pumped directly onto the beaches. Thus the beaches of the Los Angeles and San Diego areas have been partially subsidized by the dredging of an estimated 50 million cubic yards of sand from new harbors and marinas (Oceanside and Marina Del Ray, for example). However, this supply, too, will come to an end.

Protective barriers such as riprap or seawalls that were emplaced to halt or control seacliff erosion have also reduced sand supply. These structures do help to prevent the breakdown of the cliffs, and thus contribute to the beach, but they may also produce wave wash or *reflection* that actually removes sand from the beach in front of a seawall or similar structure.

What you should remember about beaches

The two most important things to remember about beaches are (1) they are temporary or ephemeral features that undergo regular, expected, and sometimes dramatic seasonal changes, and (2) where fronting a bluff or seacliff, they act as effective buffers or shock absorbers against wave attack.

As a dynamic and fragile feature, a beach may grow, shrink, alter its shape, or even disappear in a single storm. During the winter, large, steep, closely spaced waves scour away and remove beach sands to form one or a number of low offshore bars. When the weather calms in the spring, smaller waves that are less steep and more widely spaced push the available sand back onshore and rebuild the wide summer beach. This is a natural seasonal process by which the beach strategically adapts to high winter storm waves and low summer swell. Remember that the beach and the ocean are in a *dynamic equilibrium*, such that when one changes, the other must adjust. So if a house is built on a wide beach during the summer, it should be no surprise to the owner to find the ocean in the living room during a winter storm.

Weather and the California coast

Most serious coastline erosion occurs during major storms combined with high spring tides. For this reason it is important to realize where these storms come from and how they affect the shoreline. At present there appear to be four principal tracks along which storms advance on the California coast.

Waves from Aleutian source area

The most common type of storm in recent years originates in low pressure areas south of the Aleutian Islands and advances from the northwest down the coast of California, often bypassing the southern part of the state before turning eastward. Possessing great energy, particularly along California's north coast, these storms generate 20- to 30-foot offshore waves along with winds above 40 miles per hour. Local wind-generated waves usually approach the coast from the southwest and have a great capacity for producing extensive coastal erosion when coupled with south-west storm swell and spring tides. A severe storm of this type occurred during December 1940 and January 1941 and caused considerable damage along the entire coast, particularly the central and south coast. In January and December 1978 the north and central coast experienced severe storms. Again, during the winter of 1982–83, more intense storms did great damage along almost the entire length of California's coastline.

Waves from Hawaiian Island source area

These storms originate in the open Pacific and often pass through the Hawaiian Islands, sometimes causing more damage than a tsunami. They approach the southern California coast from the west. Such storms were responsible for extensive damage to homes, piers, and roads along the southern California coast during the winter of 1977–78 and again in 1979–80. During these winters, high rainfall initiated coastal landslides and greatly accelerated cliff erosion.

Hurricane-generated storm waves from the west coast of Mexico

In this type of storm, the violent winds come from the south, but occur in the summer and early fall rather than during the winter months. Such storms have rarely come to our coast in recent years, but often devastate the coast of Baja California to the south before moving off eastward or westward and dissipating. If this type of storm does reach southern California, it is generally accompanied by southerly winds and huge southwest waves that can be disastrous to south-facing coasts such as Malibu, Newport, Laguna, and Long Beach. Prior to 1983 the most recent hurricane-generated storm of this type reached the southern California coast in September 1939. Despite the presence of groins, jetties, breakwaters, and other coastal structures, beaches were overrun and numerous homes and structures at Long Beach and Newport were severely damaged or destroyed.

With the warm water event of 1982–83 in the Pacific, storms of this type have become more frequent. The Hawaiian Islands were hit by such a storm in November 1982, causing the most storm damage of the century to the island of Kauai.

Waves coming from southern hemisphere hurricanes

This type of storm swell originates in the Antarctic–New Zealand area and occasionally causes great damage at selected sites along the southern California coast. These waves approach the coast from the south-southwest and occur predominantly during the summer.

We apparently experienced such a swell in August and again in September of 1934. Thirty-foot breakers pounded the Newport-Balboa area, doing great damage to roads, piers, and beach cottages.

A large storm swell of this type also hit the southern California coast more recently. On 6–9 August 1983 damage was reported only from Malibu, Capistrano Shores, Oceanside, and along a one-eighth-mile section of coast in South Carlsbad. The bluff top at the Carlsbad site eroded as much as 26 feet during this time.

The great southerly storms of the early 1800s

During the first half of the last century violent "southeasters" were a common occurrence along southern California's coast. Historic accounts, such as *Two Years Before the Mast* by Richard Henry Dana, note that these storms were the "bane of the coast of California" and describe waves 50 to 60 feet high. It is highly probable that these storms were of the same type that still hit Baja California today, called *chubascos*. There is considerable evidence that the storms, which ceased to happen regularly around 1856, occurred during a period when the water along the southern California coast was unusually warm.

Is the weather changing?

The storm damage inflicted on virtually the entire California coast in recent years has produced widespread concern and discussion regarding whether or not the climate is changing and whether winters of this sort may become typical.

Marine geologists at Scripps Institution of Oceanography in La Jolla are predicting that the southern California climate is changing and that in the future we can expect a more variable or extreme climate, in contrast to the "mild" conditions that prevailed from 1946 to 1976. This 30-year period was also one of rapid development and construction along the southern California coastline. The climatic record for the central and northern California coast does not indicate any clear trends. However, there have been several periods in the past 100 years when these coastal areas experienced winters of unusual storm severity. For example, rainfall in the Monterey Bay area from 1935 to 1944 was above average every year but one; severe storms battered the coastline in 1939, twice each in 1940 and 1941, and again in 1943.

If we have entered a period of more severe climate throughout California, we can expect more frequent and intense storms with their destructive impact on beaches, seacliffs, and any construction in the southern and central coastal areas. A continuation of winter storms like those since 1978 would have a devastating effect on the developed portions of the coastline of California. Although engineering structures of one sort or another have been used for years in an attempt to protect ocean-front property, these are normally very expensive. The winter of 1982–83 has shown many of them to be temporary and ineffective as well. Many coastal residents will be closely watching the winters of the next few years.

Importance of volcanism in producing climatic changes

One important cause of the changing weather conditions suggested above has been volcanism. For a hundred years some weather scientists and geophysicists have suggested that the great volcanic eruptions of past centuries have played an important part in changing weather conditions, largely in the temporary cooling of the atmosphere after the eruptions. The idea has been generally overlooked by most U.S. Weather Bureau scientists. It is largely since the eruption of El Chichon in Mexico in 1982 that geophysicists have been able to make reliable collections of the gases that were discharged with the volcanic clouds from a volcanic eruption. They discovered that sulphuric acid droplets develop in great abundance in clouds high in the atmosphere after a volcano erupts. The clouds are then carried around the earth and are apparently capable of producing weather changes when these acid droplets move down into the atmosphere circulating above the earth.

In an examination of detailed climatic records for the United States, Gerald Kuhn and Francis Shepard of the Scripps Institution of Oceanography discovered that after each of the great explosive volcanic eruptions of the past 100 years there had been drastic changes in the climate that lasted 1 to 3 years. These changes were well documented in the United States, but have not, as yet, been confirmed for much of the rest of the world. It was noted that temperatures were extremely variable, with rapid alternations between very cold periods and shorter episodes of warmer weather. Along with the changing temperatures, storms greatly increased beyond the level normal before the eruptions. Most of the great episodes of hurricanes and tornadoes fall naturally into these periods after the major eruptions. Precipitation also increased tremendously, and floods produced by great rainfall were reported. Many other unusual conditions, such as record snowfalls and severe and destructive hail and thunderstorms also seemed to be concentrated in these posteruption periods. The following are some examples of the surprising weather changes that occurred after each of the four greatest volcanic eruptions of the past 100 years.

In 1883, after the eruption of the Indonesian volcano Krakatoa, great storms and tornadoes were common throughout the United States and most other parts of the world for several years. During this same period major flooding occurred in southern California. Coastal erosion, as shown by old land surveys, was also severe.

A very curious phenomenon was widely reported within a few days following the major eruption of Katmai in Alaska in 1912. A "dry fog" came down from the volcano and encircled much of the earth. It was accompanied by a great drop in temperature and severe storms. One of these storms created waves that extinguished the light in the lighthouse more than 200 feet above sea level at Trinidad Head in northern California.

Following the great eruption of Agung on Bali in 1963, record numbers of tornadoes were also reported in 1964 and 1965. In addition, many great floods were reported worldwide following this eruption.

During its eruption in 1982 El Chichon ejected large quantities of sulphur gases that have probably contributed to great weather changes. Following this eruption in April 1982, the entire United States, as well as other parts of the world, experienced many catastrophic climatic changes. Along the west coast of the United States storm conditions and coastal erosion were some of the worst during the past hundred years. Tornadoes were even greater than those reported following the eruption of Agung. The great floods, which have been almost worldwide, included one of the worst that has ever been experienced in the Mississippi Valley. It may well be that the recent strange warming of the equatorial Pacific ("El Niño") and the devastating hurricanes in the Pacific are somehow related to this most recent volcanic eruption. The climatic changes following the four greatest eruptions of the past 100 years support the ideas put forth almost a century ago, that volcanic eruptions can significantly alter weather patterns.

3. Seacliff erosion

Gary Griggs and Lauret Savoy

If you have ever watched storm waves batter the coastline you quickly appreciated the tremendous power exerted by the ocean. Even during calm weather, small waves have their effect by constantly washing the sand and gravel across the shoreline, wetting and drying the rocks, and gradually carrying off bits and pieces. Although this day-to-day activity has its impact on the cliffs, it is generally the large winter storm waves and high tides that are responsible for most cliff or bluff retreat. In part, this is because the winter waves are larger and simply have more energy to expend. In addition, winter rains often weaken cliffs, making them more susceptible to failure. A further and key factor is the reduction in width or even the total disappearance of the protective beaches during the winter. With this buffer zone of sand or gravel gone, the waves attack the cliffs directly.

The degree to which a seacliff yields to this attack depends primarily on its exposure to wave attack, the height of the waves that typically reach the area, and the physical properties of the materials making up the cliff. The hardness or consolidation of the cliff rock and the presence of internal weaknesses such as fractures, joints, or faults all directly affect the resistance of the material to wave erosion. Coastlines consisting of hard crystalline

Figure 3.1. The irregular granitic coastline of the Point Lobos area south of Carmel. Photo by California Department of Fish and Game.

rock, such as the granite of the Monterey Peninsula, usually erode very slowly. Within these rocky zones, however, erosion rates can still vary considerably (fig. 3.1). Waves attack the weaker zones, the fractures and joints, to form inlets and coves. The more resistant rock is left behind as points, headlands, islands, or sea stacks. In contrast, erosion can be uniformly rapid where the coastline is made up of weaker sedimentary rock or unconsoli-

Figure 3.2. A straight coastline eroded in unconsolidated sedimentary rock south of Half Moon Bay. Photo by U.S. Army Corps of Engineers.

Figure 3.3. Cliff failure in sandstone at Bolinas. Photo by Gary Griggs.

dated material. In these zones the cliffs often retreat in a more linear fashion, producing relatively straight coasts (fig. 3.2).

It has become clear that we are capable of altering natural erosion rates or processes. Chapter 5 deals specifically with coastal protection and the effects of various engineering structures on the shoreline. Often in the process of protecting buildings from a retreating shoreline we have inadvertently accelerated erosion.

The retreat or erosion we observe along much of California's coast is due not only to waves gnawing at the cliff base, but also to landsliding, slumping, and rockfalls originating higher on the cliff (fig. 3.3). For example, the combination of jointing patterns, percolation of surface water from street runoff and yard watering, wedging by tree roots, and undercutting by waves has led to massive rockfalls along the seacliffs flanking northern Monterey Bay (fig. 3.4). These rockfalls commonly occur after intense storms when rainfall and runoff have been high and the surf heavy. During March of 1983 a slab 100 feet long and 6–12 feet wide collapsed at Capitola and fell to the beach below, removing part of a city street. Cliff failure that occurred several years earlier at this same location led to the relocation of a 2-story duplex to a safer site 2 blocks inland. Many coastal communities have lost entire ocean-front streets through continuing erosion over the years.

Landslides can be much larger than the Capitola example, as indicated by the Portuguese Bend landslide in the Palos Verdes Hills along the Los Angeles county coastline (figs. 3.5, 17.17, 17.18). This massive failure, which covered about one half of a

Figure 3.4. Large rockfalls are common in the sandstones and mudstones making up the cliffs in the Capitola area. Photo by Gary Griggs.

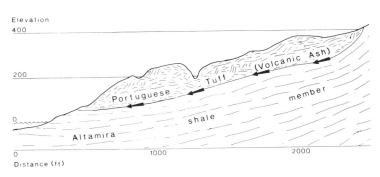

PORTUGUESE BEND LANDSLIDE

Figure 3.5. Geologic cross-section of the massive Portuguese Bend landslide near Los Angeles. Arrows show the plane of sliding.

square mile, began to move in 1956 and continued to move for a number of years. More than 200 homes were destroyed or damaged and property damage in 1956 dollars exceeded $10 million.

Street or storm drain runoff, septic tank leaching or landscape watering, and the alteration of normal drainage away from natural gullies so that more water passes through or across the bluffs or cliffs, all have their impacts. Each of these processes can erode the rock directly or lead to weakening of the cliff material followed by slumping or sliding (fig. 3.6). Although native vegetation usually acts to stabilize the bluffs or cliffs, at some locations nonnative vegetation has been planted in an attempt to stabilize the seacliff material and has had the opposite effects. Excessive watering may weaken the cliff-top materials. The roots of some trees may penetrate into the weak zones in the rocks and act as wedges to pry away large blocks (fig. 3.7). In addition, ground cover, such as ice plant, will often hang down over the cliff with such weight that it will uproot and carry down the soil and loose cliff-top material with it.

How fast is the coastline eroding?

Throughout California's 1,100 miles of coastline, extreme variations in wave exposure and rock resistance have produced wide-ranging erosion rates. At some locations erosion has been negligible for the 60–75 years of good historic record; at other locations, average rates may be as high as 5 to 10 feet per year.

Figure 3.7. Tree roots may act as wedges along joints in the bedrock (arrow) to promote cliff failure. Photo by Rogers Johnson.

Figure 3.6. Failure of the loose sediments at the top of this low cliff has been accelerated by storm drain runoff (arrow). Photo by Gary Griggs.

In general, the sedimentary rocks that form much of California's seacliffs are retreating at average rates of about 6 inches to 1 foot per year.

Erosion rates within local areas have been recognized as having changed drastically during particularly stormy periods. For example, erosion in the resistant sandstones and siltstones at Sunset Cliffs, San Diego, had averaged about a half an inch per year for a 75-year period prior to 1973, and three-fourths of the area studied had undergone no significant erosion during this period. However, the U.S. Corps of Engineers in 1976 indicated that at the foot of Del Mar Street, Sunset Cliffs, the toe of the bluff had retreated landward 38 feet and the bluff top 40 feet between 1962 and 1976, for an average rate of 2.7 feet per year.

At Point Año Nuevo, in southern San Mateo County, the erosion rate has averaged about 9 feet per year for about the last 300 years—one of the highest natural rates along the state's entire coastline (fig. 3.8). The seacliffs at this point are only several feet above sea level, so that storm waves can easily and quickly erode the material. It is extremely important to know the long-term erosion rate prior to buying ocean-front property. Chapter 5 discusses this issue in more detail.

In this book we have compiled the existing data on coastal erosion from earlier investigations and reports and have also made an effort to evaluate areas for which no data were previously available. Shoreline erosion rates are determined by using historic data, such as old maps, or from successive photographs

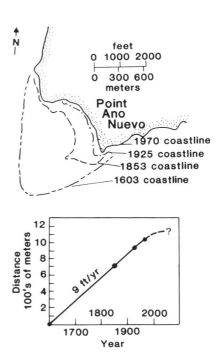

Figure 3.8. Coastal erosion at Point Año Nuevo, 1603–1970. Old maps and charts have been used to determine the progressive retreat of the coastline.

beginning in the 1920s. The longer the available record, the more accurate the results.

City and country assessors, surveyors, public works and planning departments usually have very accurate historical surveys of developed or improved beachfront areas. If the depth of an ocean-front lot or the location of the seacliff has been recorded on early maps of known dates, then an accurate present-day remeasurement and comparison of the two positions may be all that is necessary to determine an erosion rate. Dividing the amount of land lost by the number of elapsed years between surveys will produce an average annual erosion rate for the particular site.

At the town of Bolinas, for example, north of San Francisco in Marin County, comparing the location of the present cliff edge with that depicted in the 1927 subdivision maps indicates that at one particular location near Duxbury Point the coastline retreated about 120 feet during this 57-year period.

Aerial photographs are often one of the most accessible sources of data for determining changes in the shape of the coastline, and therefore rates of coastline retreat. Offices of the U.S. Army Corps of Engineers, the U.S. Geological Survey, the Soil Conservation Service, and other federal, state, and local agencies normally have aerial photographic collections available. Aerial photographs are difficult to work with, and it is a good idea to hire a geologist with experience in aerial photo interpretation to carry out an investigation of this sort. Although "average" erosion rates for some of the more populated southern and central California coastline areas are known, this is not the case for most of California. The later sections of this book contain the most comprehensive data available on erosion rates. Most of this has never before been published, and much of the research was done specifically for this book.

The erosion rates presented in the later chapters are all based on readily available historic maps and photographs. The values presented are based on actual measurements and observations and should be considered as reasonably representative of the adjacent areas. Nonetheless, local variations in the characteristics of the rock constituting the cliffs can produce wide ranges in erosion rates in adjacent areas. If the rates given in the vicinity of your parcel are high and/or if a hazard designation has been given to that particular stretch of coast, you would be well advised to look further into any potential problem.

It is important to keep in mind exactly what we mean by "average" erosion rates. As stated earlier, geologists normally will measure the amount of coastline retreat or erosion over the time interval covered by the available maps and/or photographs, and simply divide this by the number of years to get an average yearly rate. However, years of observation, particularly during severe winters such as 1983, have shown that erosion is usually episodic and irregular. Although the average rate of retreat along the cliffs above Capitola in northern Monterey Bay is about 1 foot per year, large slabs may collapse overnight, suddenly moving the cliff edge back 5 to 10 feet, followed by little change for a number of years. Short-term rates will often be much different than long-term (30–50 year) averages. This realization is important in planning a safe setback from an eroding cliff for any coastal construction.

Many of the old piers and other structures that were damaged during the storms of early 1983 had survived earlier severe weather, making it clear that this particular winter's damage was extreme by any measure. The 1983 damage resulted from the repeated coupling of high tides, storms, and large waves. The rapid erosion that occurs during major storms of this sort can greatly alter shorelines that survived many years of more moderate winters. As an example, one section of seacliff at Santa Cruz had eroded about 25 feet in the interval from 1931 to 1982, or at average rate of about 6 inches per year. During the January 1983 storms, waves removed about 46 feet of bluff top. This single storm increased the "long-term" erosion rate from 6 inches per year to over 16 inches per year! These kinds of observations argue for using average erosion rates with great caution and making every effort to research the historic changes as far back in time as possible.

4. Building or buying on the coast

Gary Griggs and Lauret Savoy

As we have seen in earlier chapters, the coastline is a dynamic and ever-changing environment. These changes occur both over short time intervals, (for example, the changes from a single storm) and over longer intervals as well (for example, the progressive erosion of a particular unstable bluff area over a number of years). Both types of changes can affect an individual property or building and both should be seriously evaluated prior to investing your life savings. Although you probably fall in love with that special ocean-front house or new lot during those long, sunny summer days, you should wait until the winter storms come before making your final decision, or at least investigate what happens at that time. That wide, protective beach can disappear quickly during a major storm, and before long your concrete patio and redwood deck have been undercut by waves (fig. 4.1). Many ocean-front residents have discovered too late that sliding glass doors and half-inch-thick plywood siding are no match for the large driftwood logs thrown about by the surf crashing through their front yards.

Two areas of major concern that are addressed in this book are the particular site itself and the structure, either existing or proposed. Considering both the hazards that affect coastal areas and the very high cost of ocean-front property, anyone contemplating such an investment is strongly advised to hire a professional with experience in the coastal zone to evaluate the stability of the property and its structures.

Along the California coast there are three particular physical environments where widespread development has taken place but that are potentially hazardous to any structure. These include (1) the beach front itself, (2) dunes, and (3) eroding cliff or bluff tops. The processes, stability, and concerns present in each of these settings will be treated individually.

Beach-front construction

Throughout many areas of coastal California, houses have been built too close to the beach. These homes have been built directly on slab or grade constructions, or on pilings above beach sand and driftwood. These beach-front areas are no different than the flat floodplains adjacent to our creeks and rivers. Just as floodwaters during major rainstorms overflow the stream banks and wash over the floodplains, large waves combined with high tides will periodically sweep completely across the beach to the flanking sea cliff. This may not happen on the beach every winter, but the sand beneath your house and the driftwood you see tell you that it has happened before and will happen again.

Much of the $18 million of damage in 1978 and the $100 million in the 1983 storms occurred when waves attacked beach-front homes in places like Stinson Beach, Rio Del Mar, Aptos, Solimar, Faria, Malibu, Del Mar, Oceanside, and Imperial Beach, to name a few. A look at older newspapers indicates that the same developments are regularly damaged or threatened. In both 1978 and 1983 the entire state coastline experienced extremely high tides (7–8.5 feet), which included a storm surge of a foot or more. Four or five days of storm waves 12 to 16 feet high occurred simultaneously. Much of the beach sand is usually removed by waves during the first day or two of such a storm. With this cushion or shock absorber gone, the wave energy is expended directly against the homes and other buildings, or against any erosion control structures.

Damage in 1983 included undermining of shallow pilings or piers so that homes totally collapsed onto the beach. Homes on low pilings were also uplifted by waves at high tide and smashed through pilings as they fell. In addition, waves overtopped low, protective seawall barriers and either damaged or destroyed the home fronts facing the sea (fig. 4.2). In some cases structural damage of this sort also led to house collapse. Nonstructural damage such as loss of decks, patios, yards, and landscaping was very widespread.

There are logical questions to ask now. Can we recognize such inundation-prone areas in advance and avoid them altogether, and is there anything that can be done to protect existing structures?

The answer to the first question is a clear yes. As discussed earlier, those beach and backbeach areas covered with sand and driftwood are easily recognized. Some may even be partially vegetated. All, however, are at the mercy of the waves, some more

Figure 4.1. The beach was removed in front of and beneath this home built directly on the backbeach of northern Monterey Bay during the winter of 1978. Photo by Gary Griggs.

Figure 4.2. High waves combined with storm tides battered the front of these homes at Aptos-Seascape despite their elevation and protective riprap. Photo by Gary Griggs.

often than others. The existence of a wide beach between a house or potential house site and the ocean should not be accepted as a permanent feature or buffer (figs. 4.3A and 4.3B). It won't always be there. A brief investigation into the past storm history of a particular area can provide more specific information that will help you assess risk. A simple measurement of the site or house elevation relative to sea level, when compared to information on expected tidal ranges, storm surges, and storm-wave heights will also give you a clearer sense of the hazard or risk of a particular site. The recent damage to beach-front homes throughout the coastline zone is clear testimony that either these risks were not adequately evaluated or that the hazards were disregarded in the design, planning, or repair processes.

There are some structural solutions as well for protecting homes that are planned or already exist on beaches. These include both foundations and support structures as well as erosion control structures such as seawalls. Coastal protection will be covered in detail in chapter 5 so our discussion here will be somewhat brief.

If one is to build at all on the beach, it only makes sense to design a foundation and structure that minimize exposure to wave action. A piling foundation that extends well below any potential depth of wave scour is the most sensible starting point. These pilings should be adequately cross-braced and then the structure itself elevated safely above any maximum inundation or wave impact level. Shallow piers can collapse from scouring. Conventional concrete slab or footing foundations can easily be

A

B

Figure 4.3.   Ocean-front homes at Pot Belly Beach along northern Monterey Bay. (A) Summer. (B) Winter from same location. Note loss of beach sand. Photos by Gary Griggs.

undermined, placing the house at beach level, directly in the path of storm waves. The value of elevating homes on deep pilings or concrete piers above any potential wave impact was clearly revealed at Stinson Beach in Marin County. Houses that faced the sea but were elevated in this manner were virtually undamaged (fig. 4.4), whereas homes built on conventional foundations further landward received major damage.

There are no specific coastal building codes at present. The common procedure in any location where a nonconventional foundation is proposed (such as pilings or piers) is to require an engineered design. There are no doubt countless situations where homes were built without benefit of such engineering, and other locations where engineers used inadequate design parameters. A good example of this is underestimating the depth of wave scour so that pilings lose support and collapse, or underestimating wave runup so that homes are built too close to sea level (fig. 4.5).

The construction of some protective wall or barrier is another common approach to dealing with the problems inherent in living on the beach. It should be mentioned here that all protective works provide only a certain degree of protection, that this limited protection is provided at great cost, and that all such structures have finite life spans. The questions usually come down to (1) what will it take to protect my house or property? (2) how much is it going to cost now and in the future, and can I afford it? (3) will it damage my beach, and what will happen to adjacent beaches? Other major questions that are not usually considered but are of equal concern are (1) will the protective structure have

Figure 4.4. Variable wave impact on homes at Stinson Beach due to foundation differences. House in foreground is elevated on piers and received almost no damage. Home in background is built directly on beach and suffered heavy damage. Photo by Gary Griggs.

Figure 4.5. Large waves combined with high tides removed the bluff beneath this house and portions of the foundation, floor, and deck at Bolinas in 1983. Photo by Gary Griggs.

Figure 4.6. Emergency riprap was brought in to save homes built directly on the frontal dunes at Pajaro Dunes in central Monterey Bay in 1983. Photo by Rogers Johnson.

an impact on neighboring houses or property? and (2) how long will the structure last, and is this structure going to protect me from all storms? The longevity of particular engineering structures will be discussed in the next chapter. The second set of questions just listed was answered during the winter of 1983 when many protective structures were simply overtopped or completely destroyed by the severe tide and wave conditions. Other structures, however, served their purpose reasonably well. The point is that the presence of some protective structure in itself does not guarantee complete protection. If you live behind such a structure, do not do so with a false sense of security. Evacuate the structure in the face of a storm that may overtop or destroy the seawall and flood your house.

Construction on dunes

Because the coastline of California is a geologically active area, much of the coast is characterized by steep cliffs and coastal mountains, occasionally broken by valleys and river mouths. As a result, sand dunes, which need both a dominant wind direction and a source of sand, as well as a flat, low-lying area to form on, are somewhat restricted along California's coast. The more geologically stable, low-relief east coast of the United States, by contrast, has far more extensive sand dune development.

The dunes that are closest to the ocean and tied to the beach for their sand supply must be viewed as constantly changing landforms. Further inland we may observe older dunes, which have been stabilized (held in place) by vegetation and no longer depend on the beach for fresh sand. The ocean-front dunes will regularly change. Storm waves cut away at the front of the dunes during the winter, and summer wind and swell bring the sand back onto the beach to rebuild the dunes. During severe winters the most seaward dunes may be severely eroded or breached altogether (fig. 4.6).

Dune stability is easily affected by human intervention, as has been common along stretches of the New Jersey coast. Houses were built on conventional foundations atop beachfront dunes, destroying the grasses and other stabilizing vegetation. These dunes were breached for beach access, groundwater was withdrawn, and large areas were paved. In March 1962 a violent storm attacked the entire east coast with 60-mile-per-hour winds combined with high tides and waves 30 to 40 feet high. In 3 days, 2,400 homes were destroyed or damaged beyond repair, and 8,300 homes were partially damaged. Foundation exposure was the most persistent problem. Those houses that sat high on the dune, with the best view, found the sand swept out from under them until their foundations or pilings collapsed or were damaged beyond repair.

The most extensive area of ocean-front dune development in California is probably central Monterey Bay. Although the developed dune area is relatively small (a total of about 2 miles of ocean front) compared to east coast development, the problems and hazards are identical. Where individual houses and condominiums have been built on conventional foundations and directly on the frontal or primary dune, periodic dune erosion will threaten or undermine the structures. A careful look at the old aerial photographs and historical records at Pajaro Dunes, for example, indicated that one large area of low dunes now occupied by 87 condominiums had been completely washed over about 1930! Since the Pajaro Dunes development was initiated in 1969, severe storms have threatened major portions of the ocean-front development during 3 different years. Up to 40 feet of dune erosion occurred during the winter of 1983 at this location. Dozens of structures were threatened, foundations were undermined, and about $2 million worth of emergency rock was brought in quickly and laid on the beach to protect the development (fig. 4.6). The dynamic nature of the dune environment was given little if any consideration in the planning of this development. The dunes fronting the beach at any location should be seen as active, moving landforms that do not provide guaranteed or long-term stability.

Dune building and stabilization

For centuries shoreline inhabitants have sought methods to protect themselves using the sand reservoir stored in coastal dunes. Queen Elizabeth I established a law prohibiting the removal of marram grass (often called European beach grass) every spring. The Dutch, however, are the best-known users of this natural dune defense from the sea. The best-known Dutch use of dunes is the development of dikes and dunes that protect the Wadden Zee.

The quickest way to build a dune, besides using a bulldozer, is to erect picket sand fences (similar to snow fences) perpendicular to the dune-building winds. Sand comes quickly to rest at the base of the fence, which significantly alters the wind flow. There are two common fence styles, the old-fashioned slat fence and the newer fabric fences. The most efficient spacing of slats occurs when the open area is slightly less than the width of each slat. In 1978 dollars the slat fence cost between $1.25 and $1.50 per yard. Although fabric fences are easier to install, they may cost up to twice as much, and there are at least two other disadvantages: (1) under heavy sand loads, the fence tends to sag, reducing the effective height, and (2) in populated areas fabric fences appear to be more susceptible to vandalism.

Sand fences have been used extensively and are usually quite successful. Yet they become useless as soon as sand accumulation reaches the top of the fence. At this point sand simply shifts from one side of the fence to the other with changes in wind direction. To trap more blowing sand, vegetation should be planted, or another fence must be built on top of the old fence. An alternate measure is just to establish plantings of beach grass. For optimal protection, uniform beach grass cover should be maintained.

Methods that use both fence and plants in sequence have many advantages. First, the fence can be put in place at any time during the year and is fully effective as soon as it is erected. Plants, on the other hand, may only be planted in late winter or early spring in many locations and do not become effective sand traps until they develop later on in the season. Second, wind velocities may be too great to allow establishment of vegetation. Fences significantly slow the wind, preventing scour and providing time for full below- and above-ground plant development. Plants are important in stabilization because they are the most practical long-term measures. They can be less expensive and more durable, as well as more aesthetically appealing and self-repairing.

Eroding cliffs or bluffs

The number of existing homes and "buildable" lots atop eroding seacliffs along the California coast is far greater than in either the dune or beach-front environments. Bluff or cliff tops throughout much of southern California have been intensively developed with single-family homes, condominiums, apartments, hotels, and motels. Although the extent of development is far less along the central and northern coast, these ocean-front cliffs have still been heavily utilized. Major portions of the bluff or cliff tops of Santa Cruz and San Mateo counties, for example, are covered with homes.

Again, to put the eroding seacliff problem into a statewide perspective, of California's 1,100 miles of coastline, 86 percent or 950 miles is undergoing erosion. Of this, about 125 miles is deemed "critical" by the U.S. Army Corps of Engineers, meaning structures or utilities are threatened (fig. 4.7). As discussed earlier, the erosion rates vary considerably and may pose a severe threat when they average 1–2 feet per year, but no immediate threat when the average rate is only several inches each year.

Figure 4.7.   Severe erosion has taken place at the end of an unprotected county roadway that has affected adjacent property owners despite their own protective rock. Photo by Gary Griggs.

Again, however, as was discussed earlier, average rates must be used with caution. Years of observation indicate that erosion is an episodic and irregular process and that short-term rates will often be much different than long-term averages.

A professional evaluation of long-term coastal erosion rates is strongly recommended before seriously considering ocean-front property in either case. Local planning departments or the Coastal Commission will require this type of information or investigation prior to any major land use changes or construction proposals. The maps and references included in the later sections of this book will enable you to make some preliminary assessment of potential problems, or to find the results of more detailed studies in other works. In general, you will observe that we know most about erosion in those areas where intense development has taken place, for example, much of the south coast and portions of the central coast. Knowledge of much of the area north of San Francisco and other areas of the central coast is more limited. These latter areas, however, are in the initial stages of development and need to be looked at carefully.

Cliffs consisting of weak rock or loose material may fall grain by grain or chip by chip, and also in the form of large blocks, which can be far more destructive. Evidence for recent cliff failure may be visible after the winter months either as fresh or unvegetated scars on the bluffs or cliffs themselves, or as blocks or debris at the base of the cliff (figs. 3.3 and 3.4). This type of evidence again should alert you to potential erosion problems. Other observations or evidence can be useful as well. Locations where the cliffs consist of layered sedimentary rock, such as shale, tilted down toward the beach (also known as daylighting rock layers) are notoriously unstable (see fig. 4.8). Subsiding or cracked roads, sidewalks, or patios near the cliff edge more than likely reflect slow failure in the underlying material. The age or permanence of the vegetation on the cliffs may also provide evidence of their stability or recent history. Unvegetated cliffs or those with only very young or immature vegetation probably have suffered recent failure or erosion. Older trees on the other hand suggest at least some extended period of stability. It is also important to look not only at the immediate cliff in front of your prospective house or property, but also at the area to either side, which is probably experiencing the same processes and may provide additional evidence.

Prior to purchasing a home or homesite you should evaluate: (1) how close the house and property are to the cliff edge relative to the average erosion rates, (2) if those rates continue, how close will the cliff edge to your home in 5 years, 10 years, or 50 years? (3) what if a major storm occurs and 10–20 feet of failure occurs overnight? (4) is there clear evidence of instability? These questions should be evaluated and your fears laid to rest or resolved. Some cliff-top areas cut in resistant rock, for example, and protected by a wide, sandy beach have experienced little measurable erosion in 50 years of photographic record. Others, however, are retreating all too quickly and homes have collapsed, been relocated, or are now threatened.

In the Yorkshire area of England, for an extreme example, the seacliffs are cut into easily eroded glacial deposits. Typical erosion rates average 3 to 6 feet per year. Yet in one area, 35 to 100 feet of cliff was lost overnight during the great storm of 1953. Accurate surveys and records have been kept of this area since the time of Roman occupation about 2,000 years ago. The erosion history indicates that 82 square miles of land and 28 towns have been lost to the sea during this period. The immediate seacliff should not be the only point of concern, because at some locations large slides, slumps, or earthflows may extend for hundreds of thousands of feet inland (fig. 3.5) and may be as hazardous to property and structures as failure of the cliff face itself.

The same erosion rate questions you would ask regarding the purchase of a house on the seacliff should also be asked prior to buying a lot. A typical, and we feel sensible, approach to new construction in areas of this sort is to build with a large enough setback behind the cliff edge so that the structure will survive for at least 100 years, based on long-term erosion rates. Along the Atlantic coast some areas have utilized a rolling setback, establishing a 100-year setback line each time a new project or building is proposed. Since the shoreline is continually eroding, the setback line is continually moving inland.

In addition to reducing erosion rates through various engineering structures, we also have the ability to increase erosion rates through excess water or loading at the cliff edge as well as through the secondary effects of coastal engineering structures immediately upcoast or downcoast. This is discussed in detail in chapter 5.

Figure 4.8. Shale that is tilted seaward (arrow) and therefore prone to failure forms the cliffs immediately upcoast from Bolinas. Photo by Gary Griggs.

5. Shoreline protection and engineering

Gary Griggs and Lauret Savoy

Coastal erosion or retreat is a natural ongoing process that has only become a problem because we have built permanent structures in areas that are prone to erosion or wave impact. Beaches, dunes, low bluffs, or high cliffs are all temporary features that will continually be shaped or altered by wave and tidal forces. Although cliff retreat or beach erosion does not necessarily occur with regularity, all of our knowledge and experience with the past indicate that much of the coastline is constantly changing, some areas slowly and others more rapidly. The more rapid or frequent the changes, the greater the potential impact on anything we build in this active environment. Unfortunately many homes and other structures were built literally within a stone's throw of the waves, and herein lies the problem. The Pacific Ocean is 10,000 miles wide and not too concerned about 100 or so yards of shoreline at the edges. In California and elsewhere around the country, however, we have built right at the edge. Where we have made that decision, there are going to be some inevitable and expensive consequences.

What if I do nothing?

Once a home or structure has been built in a location that is prone to wave damage, several options exist. One is simply to take your chances and do nothing. Depending upon the particular location, the setback from the sea, and the past erosion or inundation problems, this may work for some period of time. The cost is nothing until a major storm finally does occur, and then either a rapid emergency response is necessary or everything may be lost overnight.

Pajaro Dunes, an expensive development of vacation homes, townhouses, and condominiums built along a mile of ocean-front sand dunes in central Monterey Bay, is a case in point. Some dune erosion occurred during the first year after development was initiated in 1969. A few old automobile bodies were brought in to halt the localized erosion. During the winter of 1978, storm waves attacked the dunes at 3 locations and came to within 10 or 15 feet of several groups of homes. Consultants were hired to evaluate the problem and recommend alternatives. By the time the report was completed, however, summer had come, and most of the property owners had forgotten about the threats to the development from the previous winter. The homeowners decided to do nothing. Finally in January 1983, the combination of very large waves and high tides took its toll, as up to 40 feet of dune was removed. This left a near-vertical scarp 15 to 18 feet high that came right to the foundations of a number of homes; at least one house began to collapse and had to be shored up. At the end of the storm, riprap was emplaced along the entire seaward frontage of the development at a cost of more than $1 million.

As mentioned earlier, however, had the project planners and their consultants looked at the historical record, including the early aerial photos of the area, it would have become clear that frontal dunes are periodically eroded and then are slowly rebuilt. The erosion problem and the need for costly coastal protection could have been avoided altogether had the site's history been thoroughly studied. A setback of several hundred feet from the front edge of the dunes, for example, would have provided a reasonable buffer zone.

From the standpoint of all of the people of California, the do-nothing approach has some distinct advantages. As will be pointed out in the remainder of this chapter, attempting to halt shoreline erosion through structural means, such as the riprap just mentioned for Pajaro Dunes, may have some secondary impacts or side effects. Two immediate effects are those of making beach access temporarily more difficult and changing the beach's aesthetic appearance (fig. 5.1). There are also potential side effects on adjacent unprotected properties, as well as the effect of a protective structure on the beach itself. The narrowing of the beach and the scouring of the sand by wave reflection may occur, for example.

It is legitimate to ask whether the people of California should allow every beach-front property owner to build an engineering structure to protect his or her individual site; particularly when much of the work is subsidized through low-interest loans, dis-

Figure 5.1.   Some coastal homeowners have tried a variety of protection techniques; none of these have survived. Photo by Gary Griggs.

aster relief, or joint public-private projects. The property being saved is private and belongs to a very few, whereas the beach being degraded belongs to all California citizens. At least 2 east coast states, Maine and North Carolina, are actively trying to prevent people from erecting additional engineering structures along the coastline. In other words, these states are enforcing the do-nothing alternative in order to save public beaches.

Relocating or elevating a structure

A second alternative response to an eroding shoreline or bluff is to move or in some cases elevate the structure. In a number of beach-front, backbeach, or even sand dune developments, homes have been built directly on the sand on conventional slab or perimeter foundations. Homes along the sand spit at Stinson Beach in Marin County and in the Beach Drive area of Rio Del Mar in northern Monterey Bay are 2 examples. When the storm waves and high tides of the January 1983 storms struck these beach areas, many low-lying homes took the complete brunt of the attack. A number were total losses or were torn completely off their foundations. Houses that were elevated high on pilings or piers, however, survived with little or no damage in most instances (fig. 4.4). Two homes in Rio Del Mar, although set on concrete piers, collapsed because the piers were not deep enough and consequently were undermined (fig. 5.2). Certainly, any reconstruction or new construction in beach-front settings, if

Figure 5.2. The failure of a protective timber bulkhead followed by scour beneath shallow piers led to collapse of this Rio Del Mar house during January 1983. Photo by Gary Griggs.

allowed at all, should be elevated well above any potential area of wave impact.

If your threatened property is large enough for you to move your house far enough back to extend significantly the structure's life span, this should definitely be considered. Certainly the size, condition, and physical setting of the dwelling are critical considerations that need to be assessed by a professional house mover and possibly a structural engineer. During the winter of 1983 as a result of 30 to 50 feet of rapid bluff retreat at Pacifica, a 3-story building was moved away from its ocean-front site, as were 23 mobile homes (figs. 5.3 and 5.4). The cost of moving a house may be far less than a major coastal protection structure or the cost of damage if nothing is done. If your parcel is not large enough to relocate the structure, another parcel will obviously have to be purchased. Relocation may still be more economical in the long run, however, depending upon the magnitude of the erosion problem and costs of providing long-term erosion protection. Typical house moving costs for a moderate-sized structure may be in the range of $10,000–$20,000.

We suspect that this option really has not been seriously considered by most threatened ocean-front property owners simply because of a desire to preserve their home and setting at any cost. Much of California's shoreline construction has also taken place within the past 20 to 30 years, a period now believed to have been one of a calm and mild climate, at least in southern California. The storms and coastal damage since 1978 may indicate a return to more severe winters with greater storm intensity

Figure 5.3. A 3-story structure in Pacifica was relocated as winter waves cut back weak bluffs. Photo by Gary Griggs.

and wave impact, and consequently more erosion, inundation, and damage in the coastal zone. The initial and recurring costs of providing protection under these conditions may eventually exceed the value of a structure or the cost of its relocation. It is important, therefore, to look at the real costs, and at the advantages and disadvantages of various coastal protection measures.

Controlling coastal erosion

The most common approach to coastal erosion problems where public or private structures or utilities are at stake has historically been some type of erosion control structure. Efforts of this sort vary considerably in cost, size, effectiveness, and life span. At one extreme, large pieces of broken concrete, asphalt, car bodies, or other debris have simply been dumped at the base of a cliff in an attempt to reduce wave impact. At the other end of the spectrum are massive, carefully engineered concrete seawalls such as those along Ocean Beach in San Francisco (fig. 5.5). The purpose of any of these structures is essentially the same—to reduce, minimize, or halt shoreline retreat and thereby to protect structures and property from wave attack. A permit is required for virtually any coastal protection structure, although under emergency conditions these may be obtained very quickly. Either the city or county within which the property is located would normally be the primary permit-granting agency; coastal commission approval is required as well. A set of engineered plans for the proposed structure is required for any such permit applica-

Figure 5.4. Up to 50 feet of cliff retreat took place at this trailer park in Pacifica during the winter of 1983. Twenty-three mobile homes had to be removed from their ocean-front sites. Photo by Gary Griggs.

tion, as is an evaluation of the physical conditions to which the structure will be exposed and therefore must resist.

The emphasis in the following pages is on reducing or controlling marine erosion due to wave impact on beaches, sand dunes, or seacliffs. Structures built on bluffs or cliffs, however, must also contend with erosion or cliff failure due to terrestrial processes. The major concern is usually with erosion or gullying from surface runoff, or landsliding. Countless examples from one end of the state's coastline to the other indicate that excess water, whether surface or subsurface (from landscape watering or roof or street runoff, for example), is nearly always the catalyst.

Any and all efforts to control on-site or off-site runoff at cliff-top locations will act to reduce the potential for erosion or failure of the slope and is the single most important step you as a homeowner can take. In particular, the runoff from rain gutters, patios, walkways, and driveways should be collected and routed either well away from the cliff edge and house foundation, or if necessary and feasible, carried to the base of the cliff or bluff in an enclosed pipe. Uncontrolled runoff and drain pipes are often the cause of slumps or landslides at cliff-edge locations (fig. 3.6). Few landslides occur during the dry summer months, simply because the excess water is not normally present. It is during the winter months, particularly during major storms, that the bluffs or cliffs become saturated to the point of failure (fig. 5.6).

Although large-scale landslides are difficult to control, soil slips and gullying can be minimized in some cases through a number of widely used erosion control measures. Jute netting, for exam-

Figure 5.5. This massive, concrete, curved-face seawall was built along San Francisco's great highway in 1929. It is still functioning today as a result of regular repair and maintenance. Photo by U.S. Army Corps of Engineers.

Figure 5.6.   A landslide during the intense rainstorm of 3–4 January 1982 destroyed the house at the base of the bluff and removed most of the backyard of the home at the bluff top. Photo by Gary Griggs.

ple, combined with planting of deep-rooted native vegetation will help to stabilize soil and reduce erosion. The vegetation serves 2 functions: (1) it utilizes soil moisture, thereby drying out the ground surface, and (2) the roots anchor and stabilize the soil. The planting of exotic or ornamental plants or lawns that require large amounts of water is not recommended and should be avoided.

Although plastic sheeting is sometimes seen draped over a crumbling hillside, this should be seen as only a very temporary measure. This method simply transmits all the water to the base of the bluff or cliff where additional problems such as erosion or inundation may be initiated.

Riprap

Where protective beaches are either narrow, seasonal or nonexistent, and property on the beach or bluffs is disappearing or endangered, the emplacement of large blocks of rock called riprap has been a common solution (fig. 5.7). In fact, more of California's coastal property has probably been protected with riprap than any other method. Riprap has been used on a widespread basis both at the base of retreating bluffs or cliffs and also in front of structures built on the beach or sand dunes. The rock is usually stacked at a 26- to 34-degree slope (2:1 to 1.5:1) against the bluff, dune, or beach with the intent of absorbing most of the wave impact. Although expensive, as are all coastal protection

Figure 5.7.   Protective riprap below a roadway in the city of Capitola. Photo by Gary Griggs.

measures, this is one of the most common methods used by individual property owners.

For optimal results, the individual rock must be large enough (usually 3 to 5 tons) to remain stable under the most extreme wave conditions to be experienced at the particular location. The rock must also be extremely durable (for example, granite or marble) so that it does not decompose readily. The availability of durable rock in the sizes needed places severe constraints on usable rock and at some locations may add considerably to the costs of such an approach. A typical 1983 price along the central California coast was $65 per ton for riprap in place. Normally at least 3 to 10 tons of rock per front foot of property are required, which amounts to from $200 to perhaps $650 per foot. Riprap for a property 60 feet wide would cost on the order of $12,000 to $39,000. These figures will vary somewhat from place to place as a function of rock availability and required rock volume.

Some of the most important design considerations are that (1) the riprap be large and dense enough to remain stable under storm wave conditions, (2) the riprap be high enough that waves do not overtop it and damage the structures being protected, and (3) that the rock be placed in such a way (either on bedrock or deep enough in the sand) that it is stable and will not be susceptible to undermining or collapse.

Riprap has limited effectiveness or a short life span at many locations because the criteria mentioned above have not been met. For one reason, riprap is often emplaced under emergency conditions in the midst of a storm. The rock is simply dumped

Figure 5.8. Riprap is often brought in during a storm and simply dumped in front of the threatened structures. Timber bulkheads in front of these Rio Del Mar homes were destroyed in January 1983. Photo by Gary Griggs.

quickly on the sand in front of threatened structures at low tide and no real design is followed (fig. 5.8). These "emergency" walls may never be rebuilt. Placing rock during the winter months, when the beach has been cut back and underlying bedrock may be exposed as a solid base or foundation, will produce a more stable barrier. If riprap is placed directly on the high part of the beach or berm, wave scour will eventually remove the underlying sand, causing the rock either to settle or to collapse and tumble onto the beach. Protection is then either lost or greatly reduced. The removal of the supporting beach sand followed by collapse is no doubt the most common mode of riprap failure. As a result, poorly planned and installed riprap will often need to be rebuilt or supplemented with additional rock after 5 to 10 years of wave attack. This appears to be a particular problem where no bedrock exists beneath the beach sand for support. The initial construction costs may, therefore, be only the beginning.

At Aptos-Seascape in Monterey Bay a group of expensive homes was built on an elevated fill terrace directly on the beach in 1968. Riprap about 15 feet high was placed on sand along the entire ocean front, and 21 homes were built on the terrace with virtually no setback from the top of the riprap. There were no problems for 15 years. During the 1983 storms, however, 12- to 15-foot waves combined with several days of 6.6-foot high tides overtopped the riprap and washed through most of these homes, breaking out windows, doors, and walls (fig. 4.2). Continued wave action scoured the sand off the beach and from beneath the riprap so that approximately 350 feet of the rock settled and collapsed onto the beach.

Another potential problem with riprap and virtually all other structural solutions is discontinuous placement or construction. In other words, one property owner has a barrier constructed, whereas his or her neighbor does not. With time, wave energy may be focused or reflected onto adjacent "unprotected" property or erosion may proceed around the structure and attack the protected property from the sides (fig. 5.9). A coordinated and integrated approach to erosion problems is normally the best solution, although individual property owners and even small municipalities do not normally have either the technical knowledge or financial capability to deal with the problem in this manner. As a result, we can observe today a combination of approaches, some effective and some ineffective, side by side. Despite your own best efforts, a poor design or inadequate protection by a neighbor may greatly reduce the effectiveness of your solution. In addition, collapsed structures and riprap strewn across the sand also reduce the aesthetic value of the beach.

Figure 5.9.   Although riprap is protecting the front of this Half Moon Bay apartment building, lateral erosion continues to occur. Photo by Gary Griggs.

Revetments

A revetment is really nothing more than neatly placed and layered riprap. Instead of simply stacking large rocks on the beach or at the base of a cliff, a layered structure is built up that is designed to minimize scour beneath the structure, minimize wave reflection, minimize overtopping, and absorb the wave energy. A

Figure 5.10. Installation of a revetment with successive layers of filter cloth, small core stone, and large cap rock. Photo by Gary Griggs.

revetment usually consists of a basal filter cloth, which is porous, covered by a graded layer with smaller rock at the base and large (2–5 ton per rock) armoring the surface (fig. 5.10). Large rocks should also be placed at the toe or seaward edge of any structure, whether a riprap wall or revetment, in order to reduce scour. The size, shape, and objectives of a revetment are identical to those of riprap. It simply has the added advantage of the graded rock and filter cloth, which reduce scour and make the structure less subject to undermining and subsequent settlement or collapse. The other design considerations for riprap hold for revetments as well. The rock must be durable and in large enough angular blocks to remain stable under extreme wave conditions, and the structure needs to be high enough to prevent overtopping.

Seawalls and bulkheads

Seawall and bulkhead are terms that are often used interchangeably. These are structures that may be constructed, parallel to the shoreline, of concrete, timber, and pilings, or even steel, and have a wide variety of shapes or configurations. Their purpose is primarily to protect seacliffs, bluffs, buildings, roads, or other improvements from direct wave attack. In many cases they also serve as retaining walls to keep weak cliffs or fill from slumping or failing.

Concrete seawalls have been used widely in many developed areas along the California coast with varying degrees of success. The exposed face of such a seawall may be vertical, concave,

Figure 5.11. January 1983 waves at high tide are overtopping and destroying a timber bulkhead at Seacliff State Beach. All of the logs in the photograph have been washed over the bulkhead by storm waves and deposited in a camping area for recreational vehicles. Photo by Gary Griggs.

stepped, or some combination of these. Each design has its own advantages and disadvantages.

The principal difference between riprap and a seawall or bulkhead is that the loose riprap actually absorbs some of the wave energy whereas the solid, impermeable face of a seawall reflects the wave energy. During storms, however, this absorption by riprap can be greatly reduced as indicated by the damage to oceanfront structures behind riprap walls during the 1983 winter.

There are some predictable side effects associated with seawall construction. Although these structures certainly are of value in protecting the cliff, bluff, beach, or structures behind them, they do not protect the beach in front of them. As the waves or surf break against the vertical structure, the wave energy is deflected and reflected both downward and seaward, which actually accelerates beach erosion. This side effect is a particular problem in the winter months when any protective beach seaward of the seawall has been removed. The sand removal or scour by the reflected waves produces deeper water in front of the seawall, so that larger waves now may break closer to the structure. Overtopping of low seawalls or bulkheads is thus a common occurrence during severe winter storms (fig. 5.11).

The possibility of overtopping must be taken into account in designing drainage from behind the wall. Seawalls and bulkheads must be capable of withstanding the pressure exerted on the face of the structure by waves, logs, and other debris, and they must also be protected against the action of water behind them. Rain water, runoff, or water from wave overtopping percolating into

Figure 5.12. Seacliff Beach State Park after storm waves in the winter of 1940–41. Pilings on the right were part of bulkhead built in 1940 and destroyed that winter. Structure on left is the remains of a seawall built in 1927 and destroyed the following winter. Photo by Gary Griggs.

the fill material will build up high pressure behind an impermeable wall and lead to failure. Drain or weep holes at the base of a seawall should be installed to relieve this water pressure.

Concrete seawalls that extend deep enough to prevent undermining or scour, and high enough to prevent overtopping, have survived for years at many places along the California coast. Two concrete seawalls at Bolinas, for example, north of San Francisco, have survived almost 50 years with little if any maintenance.

On the other hand, seawalls or bulkheads constructed of pilings connected by heavy timbers have had a poor record at many locations. A piling and timber bulkhead protecting Seacliff Beach State Park south of Santa Cruz has been destroyed or seriously damaged 10 times in the past 60 years. The typical design used at Seacliff consists of 12-inch diameter pilings set about 8 feet apart and tied together with 3×12 timbers that are simply nailed in from the back (landward) side (see fig. 5.13 and a similar structure in fig. 5.14). A timber bulkhead of this type was built at Seacliff in 1927 and destroyed the following winter; it was rebuilt in 1940 and destroyed the next winter (fig. 5.12). Again in 1978 and in 1980, similar structures were destroyed. An identical structure was rebuilt and completed in late 1982 at a cost of over $1.5 million and projected to last 20 years. Six weeks after completion, the storm waves of late January 1983 destroyed 700 feet of the 2,600-foot-long structure and inflicted about $740,000 in damage (fig. 5.13).

The failure pattern of the piling and timber bulkhead at Seacliff State Beach has been identical in recent years. Large logs

Figure 5.13. Partial destruction of new timber bulkhead at Seacliff State Beach in January 1983, less than 2 months after construction. Photo by Gary Griggs.

Figure 5.14. Destruction of a timber bulkhead and parking lot at Rio Del Mar in January 1983. Timbers have all been removed and filter cloth was ineffective in preventing scour of fill. Photo by Gary Griggs.

Figure 5.15. A seawall constructed of vertical steel H beams and heavy timber logging with a concrete cap was overtopped in Rio Del Mar during 1983 storms but suffered only minor damage. Photo by Gary Griggs.

Figure 5.16. Where pilings have not been driven deep enough, storm-wave scour can lead to collapse. Photo by Gary Griggs.

and other debris batter the wall from the front. Simultaneously, waves and logs overtop the structure and begin to break through the asphalt parking lot behind the bulkhead. As the timbers are loosened and the asphalt is torn up, the underlying fill begins to wash out, even with filter cloth behind the bulkhead. Once the integrity of the structure is lost and the back fill is removed, the waves easily batter and remove the timbers, leaving the pilings behind (fig. 5.14). In high-energy environments, timber and piling bulkheads constructed in this fashion have not proven effective over the long term.

A more resistant seawall can be constructed utilizing steel H beams as pilings or vertical supports to hold the connecting timbers. With the timbers or logging contained within the slots of the H beams (see fig. 5.15), they cannot be battered easily from either side as is the case with wood pilings.

A common shortcoming with many piling-supported bulkheads, whether using steel, wood, or concrete pilings, is the failure to set them deeply enough. Six to 8 vertical feet of beach sand can be removed during a major storm, removing lateral support and toppling the structure (fig. 5.16).

Another common seawall or bulkhead problem occurs where a variety of different techniques are used side by side. They probably will not be adequately tied together and will present different degrees of resistance to wave action. Once a single wall falls and the waves can break through the structure, the attack on other structures and removal of sand from their rear will produce additional failure. Where protective structures such as seawalls or bulkheads are being planned in front of a number of properties, homeowners are strongly advised to develop a uniform and permanent approach rather than a haphazard and irregular one. If a solution of this sort is well planned for the particular conditions, it will be more economic and will also survive longer. Keep in mind, however, that you may lose some of your beach to protect your property, and that, with a few very expensive exceptions, even the best of walls is not permanent.

Combined approaches and other structures

Erosion control measures may also be used in combination, either to complement one another or to accomplish together what a single measure might not do alone. A common beach-front approach is to utilize riprap, for example, in front of a bulkhead. The riprap will absorb some of the wave energy while the bulkhead serves as a retaining wall in front of beach-front fill and buildings. Again, although a combined structure that is high enough and adequately constructed is certainly more costly, it can provide far more protection and support than either structure individually.

Over the years, coastal homeowners, an occasional maverick engineer, and the Army Corps of Engineers demonstration projects have experimented with a number of other, often low-budget devices in an attempt to reduce erosion and provide protection for their particular problem area. While some of these may provide short-term protection at some locations, we can probably safely assume that were any of these approaches highly success-ful, they would have been utilized quickly on a widespread basis. This is not to say that the existing, most commonly used methods are fail-safe, or that we should not try to develop better ways to cope with coastal erosion, but rather that many novel approaches have been tried and all appear to have their limitations.

For example, at some locations where steep cliffs are under-going rapid erosion, the entire cliff has been sprayed with gunnite or concrete in order to protect it completely (fig. 5.17). This has proven locally effective, although it obviously is not visually at-tractive or popular with neighbors. If adequate drainage is not provided for water seeping through the cliff, excess water pres-sure can develop behind this coating, creating potential prob-lems. In addition, erosion will still occur at the margins of the concrete, eventually undermining it or removing its support (fig. 14.18). Again, an isolated approach of this sort on a single parcel has its limitations.

Large (2×2×4 feet) concrete blocks have been used at other sites as protective barriers. Unless these interlock in some fashion and are set upon a solid base, however, ultimately they will be undermined or battered by large waves to the point of collapse (fig. 5.18).

At several locations in northern Monterey Bay steel oil drums were brought in to provide temporary protection for beach-front homes during a storm. Even though these drums were cabled together they were simply too light and unstable to present any

Figure 5.17. Gunnite has been used by this San Mateo County homeowner to reduce erosion. Adjacent sandy cliffs continue to erode. Photo by Gary Griggs.

real obstruction. Large waves simply scattered the drums across the beach (fig. 5.19).

Before embarking on a coastal protection project on any scale or at any cost, it is critical to question the effectiveness and probable life span of the structure being recommended or proposed. The owner should search out a coastal engineer with a successful track record because coastal engineering is a highly specialized profession. It is clear from the failure of countless shoreline structures, however, that most or many of the structures simply are inadequate to withstand the forces in the surf zone. It also should be emphasized that there is really no such thing as low-cost shore protection in a high-energy environment. All evidence indicates that every structure has a finite life span, with the most important influence probably being the uncertain occurrence of the next major storm, with simultaneous large waves and high tides. What is the cost of rebuilding or repairing the structure you are considering? How often will this need to be done? You need to ask yourself these questions prior to investing in any protection project.

Figure 5.18.   These large concrete blocks were originally stacked in front of a beach residence for protection. Winter waves scoured the underlying sand, and partial collapse followed. Photo by Gary Griggs.

Figure 5.19.   Storm damage in 1980 at Rio Del Mar. A temporary barrier of oil drums failed almost instantaneously. Photo by Gary Griggs.

Groins

Groins are shore protection structures designed to trap the littoral, longshore drift of sand and thereby either build or stabilize a protective beach. Groins are usually built perpendicular to the shoreline and may be up to 150 feet or more in length. Because a groin forms a trap or barrier to littoral drift, sand tends to accumulate on the updrift side to form a beach. The amount of sand trapped and the size of the beach formed depends upon the height, length, and permeability of the groin. Short, low, permeable groins, for example, will allow considerable sand to move past the structure. These structures can be built of concrete, pilings, steel sheeting, or even rock.

Although single groins can be constructed, they are often built in groups (called groin fields) in order to form or stabilize a beach along a long stretch of coastline. If the downcoast beaches are narrow and the littoral drift or sand transport rates are low, then it may be necessary to place sand between the groins or fill the beach artificially. Because of the barriers they form and possible downcoast effects, groins should not be built unless properly planned for the particular site and until aftereffects on adjacent beaches have been thoroughly considered.

The predictable impact of a single groin or of a groin field when the individual groins are placed too far apart is not only deposition upcoast from each groin, but erosion downcoast as well (fig. 5.20). Depending upon the nature and use of the beach or coastline in these downcoast areas, a certain amount of liability may be expected by those constructing groins. Erosion adjacent

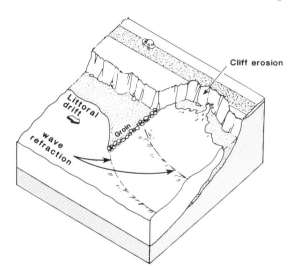

Figure 5.20. Effect of a single groin on the coastline. A groin will trap littoral drift and is therefore used to (1) stabilize a beach or (2) form a beach, and thereby protect the coastline. Downcoast erosion may follow groin emplacement, however.

to the downcoast side of the groin, for example, is common and can be expected.

Although groins can be very effective in forming or stabilizing a beach, they do not provide total protection from wave action. For example, when large storm waves approach the coast directly or head-on, rather than from an upcoast direction, there may be no littoral drift component and the beach may be removed altogether. This has happened repeatedly at the small resort town of Capitola near Santa Cruz. A single groin was constructed in 1969 to trap sand in order to form a permanent beach. As long as the waves came from the northwest, the protective beach remained. During at least 3 winters since that time, however, major storms have attacked the coast from the southwest. The 200-foot-wide sand beach was quickly removed, and the waves overtopped a small seawall, battered the beach-front buildings, and flooded the city streets with debris (see fig. 5.21). In the spring and summer as waves from the northwest begin to dominate, littoral drift again is trapped to form a wide beach. The important things to remember are that groins trap sand but do not provide complete protection in themselves, and that there are predictable downcoast effects. Very long groins may force sand seaward so that it is effectively lost from the beach.

Figure 5.21. A protective beach at Capitola was removed during the storms of January 1983, and waves washed over a low seawall into the business district. Photo by Gary Griggs.

Jetties and breakwaters

Jetties and breakwaters have totally different purposes from any of the structures or measures previously discussed. Jetties are built in pairs perpendicular to the shoreline to protect or stabilize an entrance channel against wave action or sedimentation, and thereby provide safe navigation (fig. 5.22). The jetties at Bodega Bay and Moss Landing harbors are good examples. An additional goal is to prevent or reduce shoaling of a channel by littoral drift. Jetties, because of their length, initially impose a total littoral barrier to sand transport. Accretion or sand buildup will occur updrift or upcoast from the jetties where there is a dominant littoral drift direction. Erosion will occur downcoast as the beaches are starved unless sand-bypassing is initiated. Planning for jetty construction where littoral drift rates are high and the coastline is erodible should include some method of bypassing the sand to minimize both channel shoaling and downcoast erosion.

The effects of jetty construction at a small craft harbor at Santa Cruz provide clear evidence of the direct and indirect effects of coastline alterations of this sort (fig. 5.22). The combined effects of this project include (1) formation of a wide, permanent, upcoast beach that now protects a previously eroding stretch of urbanized cliffs; (2) shoaling of the harbor entrance channel with annual dredging costs now in excess of $500,000; (3) increased erosion rates of unprotected but developed downcoast cliffs; and (4) the immediate loss of a public beach at the town of Capitola farther downcoast. The harbor jetties produced a major impact on private and public property for several miles downcoast.

Breakwaters are structures built to protect a harbor or anchorage from wave action. Crescent City, Half Moon Bay, Santa

Figure 5.22.   Jetties at the entrance to the small craft harbor at Santa Cruz have trapped considerable sand coming down coast (at arrow). Photo by U.S. Army Corps of Engineers.

Barbara, and Long Beach harbors are all examples of major harbors protected and formed by breakwaters. Regardless of the placement and orientation of a breakwater it will affect the adjacent coastline in several ways. The primary and intended effect is that of greatly reducing or nearly eliminating wave action within the protected area behind the breakwater. This reduction of wave energy, however, often produces the secondary or indirect effect of interrupting littoral drift because the energy is no longer available to move the sand through the harbor area.

At Santa Barbara an initially detached breakwater built over 50 years ago proved quickly to be an effective sand trap (fig. 5.23; also see chapter 14). Sand moving eastward first filled a large embayment west of the breakwater. A college stadium and large parking area were subsequently built on this sand fill, which extended the beach seaward over 300 yards from the original shoreline. As the sand continued its path downcoast it swung around the tip of the breakwater and began to settle out in the protected waters now cut off from wave energy. The harbor has been dredged continuously since 1935 and is presently costing about $750,000 annually. In the early years of sand accumulation prior to the initiation of dredging, erosion of the protective beach downcoast damaged many homes. The artificial filling of the area upcoast from the breakwater and the bypassing of littoral drift, although costly, could have mitigated the downcoast erosion damage.

The situations at Santa Barbara and Santa Cruz are not unique but provide well-documented examples of the direct and indirect

Figure 5.23. The breakwater at Santa Barbara created a major sand trap and has been dredged on a continuous basis for over 50 years. Dashed line delineates preharbor shoreline. Photo by the U.S. Department of Agriculture.

impacts of major coastal engineering structures that often have not been anticipated in advance. While jetties and breakwaters are designed to provide protected harbors, in many locations they have had significant impacts on littoral drift and coastal erosion in adjacent areas. Thus, beach-front homes or structures on eroding seacliffs have often been the indirect victims of this coastal roulette.

The other direct cost associated with these harbors, which have intercepted large quantities of littoral drift, is the annual dredging of sand. Present federal dredge expenditures along California's coast now total some $7–8 million annually and are increasing regularly (see table 5.1). Federal funding, however, may be eliminated in the near future and this would return these costs to the harbors themselves.

Table 5.1 California harbor dredging data

Harbor	Average annual yardage (cubic yards)	Average annual cost
Channel Islands	1,135,000	$1,635,000
Ventura	700,000	$ 915,000
Santa Barbara	400,000	$1,000,000
Humboldt Bar and Entrance	315,000	$ 450,000
Santa Cruz	167,000	$ 500,000

Some final thoughts on shoreline protection

It has become abundantly clear in recent years that shoreline protection is a very expensive business. The damage or failure suffered by many structures indicates that these projects are often temporary as well. Because most of California's shoreline will continue to erode, the costs of protecting ocean-front property in the long run will exceed the value of the property. In fact, without government or insurance assistance (through disaster relief, low-interest loans, joint projects or direct subsidies) it is doubtful that most coastal homeowners could protect their property for long. If the more severe winter weather of recent years continues, many ocean-front residents are going to be faced with some difficult decisions: Can we afford to live here?

There are also the indirect effects that accompany many protective structures: reducing beach access and severely altering the natural aesthetics; potential impacts on adjacent unprotected properties; wave reflection from impermeable structures leading to beach scour and larger waves breaking closer to shore; and possible sand impoundment with its associated downcoast effects and removal costs. All of these concerns, potential impacts, and added costs must be evaluated before embarking on a project. To do otherwise is foolhardy and may end in expensive disaster.

6. Coastal land-use planning and regulation: reducing the risks of environmental hazards

James Pepper

An overview of coastal planning in California

Although the problems associated with human settlement in the coastal zone have been present for a long time, the perceived need for coastal planning and regulation grew slowly. The first clear public statement of the need for coastal protection in California was reported to the state legislature in 1931 by a special Joint Legislative Committee on Seacoast Conservation. Sustained citizen and governmental attention did not, however, gain momentum until the 1960s. Ironically, the state population tripled during this 30-year period, and much of the growth took place in coastal communities.

During the 1960s various state and federal governmental advisory commissions were established to determine the most effective planning approach for utilizing coastal resources while maintaining a high-quality coastal environment. When the state legislature failed to pass any important coastal bills during the 1960s, environmental groups developed a strategy to pressure the legislature into instituting coastal planning and management. Following 3 years of failure, these groups mounted a campaign to place a citizen initiative measure mandating coastal planning and management on the ballot. An all-out petition-circulating effort was successful and the initiative known as Proposition 20 was placed on the ballot. In November 1972, less than 2 weeks after final approval of the Federal Coastal Zone Management Act, the California voters, by a respectable majority of 55.1 percent, established a Coastal Zone Conservation Commission. This commission was responsible for making coastal protection policy and preparing a coastal plan; it was also empowered to exercise interim controls through a permit system.

A wide range of concerns and issues, including geologic hazards, were addressed in the policies and plans produced by the commission during the 3-year planning period mandated by the initiative. The 1975 Coastal Plan recommended some 162 policies to guide future coastal resource use and development. Those portions of the plan concerned with geologic hazards and protection of resources contained 9 policies ranging from sand movement and shoreline structures to landslides and bluff-top erosion.

Although the California Coastal Zone Conservation Act mandated the preparation of a statewide comprehensive and enforceable plan for the orderly, long-range conservation and management of the coastal zone, the plan was required to be adopted by the state legislature, the very governmental body that had previously failed to pass coastal conservation legislation. Furthermore, the initiative contained a provision that called for its own repeal on 1 January 1977, when the entire coastal planning enterprise would go out of existence unless the legislature and governor were to extend its life or create a successor agency. Thus when the commission completed its planning document in 1975, it recommended the plan to the state legislature.

In 1976 the legislature enacted the California Coastal Act, based on the recommendations in the coastal plan prepared under Proposition 20. Although the Coastal Act was less sweeping and environmentally responsive than the coastal commission's plan, it remained relatively strong in many areas, including coastal hazards.

Although the Coastal Act declared that the California coast "is a distinct and valuable natural resource of vital and enduring interests to all people," the difficult task was translating broad policy statements into land-use controls appropriate to the characteristics of a particular coastal property. In order to deal effectively with the specific coastal planning issues of each local community, the state returned the planning function to local governments via Local Coastal Programs (LCP). This approach was a pioneering means of bringing *local* plans and regulations into conformity with *statewide* policies as set forth in the Coastal Act. A permanent State Coastal Commission was also established to coordinate and review coastal development decisions and plans to ensure that the Coastal Act policies were met.

Coastal policies related to geologic hazards

Since this book aims to provide a comprehensive survey of the physical processes of beach erosion, dune migration, and cliff re-treat, it will be useful to summarize those Coastal Act policies that are directly relevant to such concerns:

Policies concerning *shoreline structures*:

(Section 30235): Revetments, breakwaters, groins, harbor channels, seawalls, cliff-retaining walls, and other such construction that alters natural shoreline processes shall be permitted when required to serve coastal-dependent uses or to protect existing structures or public beaches in danger from erosion and when designed to eliminate or mitigate adverse impacts on local shoreline sand supply. . . .

(Section 30251): Permitted development shall be sited and designed . . . to minimize the alteration of natural land forms. . . .

Policies concerning *hazard areas*:

(Section 30253): New development shall: (1) Minimize risks to life and property in areas of high geologic, flood, and fire hazard.

(2) Assure stability and structural integrity, and neither create nor contribute significantly to erosion, geologic instability, or destruction of the site or surrounding area or in any way require the construction of protective devices that would substantially alter natural landforms along bluffs and cliffs. . . .

In developing these policies the coastal commission acknowledged that minimizing the impacts of coastal hazards simply made economic sense. The 1975 plan recognized, for example, that beach erosion alone costs property owners and governmental bodies several million dollars each year for building seawalls, bulkheads, and other erosion-combating structures, and for importing sand.

The 1975 plan cited specific costs related to beach sand loss as well as general costs associated with natural hazards. As noted in an earlier chapter, the state of California suffered over $100 million worth of coastal damage from the winter 1983 storms. Approximately 3,000 homes and 900 businesses were damaged, and 11 of the 15 coastal counties were declared state and federal disaster areas.

As discussed in earlier chapters, the scale of the damage is staggering. For example, waves caused 20–40 feet of bluff recession in sand dunes at Pajaro Dunes, a condominium and second home development in central Monterey Bay. Storms reduced the beach profile by 10–15 vertical feet at the city of Del Mar where waves also damaged the 3-year-old, $94,000 Longard Tube, a structure specifically designed to prevent erosion. In addition, the seawall at Seacliff State Beach in Santa Cruz was again severely damaged at a location where a seawall or bulkhead has been rebuilt 8 times in the last 58 years.

Two major planning questions emerge from these coastal development problems: (1) What form of land use control can best prevent or minimize impacts on future development in known hazardous areas? (2) Should public liability and disaster relief funds be waived in areas of known hazard in order to provide economic disincentives to development and redevelopment? The next pages address these questions through a brief discussion of techniques for implementing coastal plans.

Land use control techniques for reducing risks and impacts of geologic hazards in the coastal zone

No plan dealing with controversial matters, particularly when property rights are involved, is likely to be self-enforcing. Thus, the successful implementation of coastal plans is the final test in the planning process. Three major types of land-use controls are relevant to the issue of reducing or avoiding coastal hazards: (1) regulation; (2) acquisition; and (3) economic incentives or disincentives. Various combinations of these types of controls constitute a fourth approach.

Regulation

The state of California, like all other states, has the constitutional power to regulate land use. This power, commonly referred to as the police power, provides states with the power to regulate public and private activity to protect the health, safety, and welfare of the general public. Although the exercise of the police power is often contested as denying property rights to those regulated, both the U.S. and state constitutions protect property owners against the taking of property without due process and just compensation. The fact that property rights are not absolute, but rather undergo constant redefinition, complicates the understanding of this issue. An examination of the rapid urbanization of the United States reveals a corresponding increase in restrictions on the use of private property—restrictions held to be constitutional by courts at all levels of government. Frequently the issue is not the violation of rights but rather the reduced expectations of persons who cannot use their property as fully or intensively as they might like.

This power to regulate is manifest in several forms, including, but not limited to, zoning, control over land division and subdivision, building codes, performance standards, design review, and various special programs such as growth management. Each form can be useful in addressing the problems of coastal geologic hazards. For example, zoning can be used simply to deny development in hazardous areas, or to set forth specific provisions that limit land uses to those that are either unaffected by the hazard or do not increase the risk of hazard; subdivision regulations could guide land division through requiring site investigations of potential geologic hazards, and requiring setbacks and other provisions necessary to avoid hazardous areas; building codes, performance standards, and design review processes can assist in assuring that when development is permitted in hazardous areas, the greatest care is taken to minimize risks; and special programs could be used to direct development from hazardous areas to more environmentally sound areas.

One solution for minimizing the hazard of coastal erosion is to require an adequate setback for any new development in eroding areas. In this way natural processes can continue, and any undesired impacts from protective devices can be avoided. A setback should be based on the local geologic conditions, specifically long-term erosion rates.

The "stringline" is one method used to determine an appropriate setback. The statewide interpretive guidelines provide the following definition of this method:

> In a developed area where new construction is generally infilling and is otherwise consistent with Coastal Act policies, no part of a proposed new structure, including decks, shall be built farther onto a beachfront than a line drawn between the most seaward portions of the adjoining structures. Enclosed living space in the new unit should not extend farther seaward than a second line drawn between the most seaward portions of the enclosed living space of the adjoining structures.

This method has been used in the past when development was proposed for an area that was already substantially developed, since it was generally considered inappropriate to deny a permit for a project if it would be subject to the same hazards as existing development.

However, a "rolling setback" may be a more reasonable approach. Under this approach, the setback would be established with respect to the expected lifetime of the proposed structures. For example, using a 100-year structure lifetime as a guideline, the required setback would be located behind a line delineating the future coastline resulting from 100 years of erosion, using the local long-term erosion rate to make this calculation. Thus, the setback line would move farther landward each year as erosion proceeded, rather than allowing new construction to occur at existing setbacks in hazardous areas simply because adjacent structures were already present.

Relocating endangered structures constitutes a second nonstructural method. Relocating structures threatened by erosion is usually possible when the cost of another means of protection would exceed the cost of relocating or rebuilding damaged structures, or if protection is not practical or feasible.

A third method for minimizing the hazards from coastal erosion is to prohibit land divisions in areas subject to rapid erosion. In such areas, some sections are generally more stable than others. If large parcel sizes are maintained, a structure can be sited with greater safety by avoiding the more hazardous section. It is more difficult to do this with smaller parcels, since there is less land area to work with. Thus, if land divisions are prohibited in these areas, there is a greater degree of flexibility in siting a structure, and the hazard can be minimized.

A final nonstructural solution requires the owner to assume liability if development occurs in known areas of erosion. In the past, many individuals who have experienced loss because of damage caused by hazards have sought government aid to recover their loss. This is generally not unreasonable if the risks of a certain area were not known. However, the public should not be expected to subsidize development or rebuilding in areas that are known to be hazardous. A memorandum on this subject prepared by the state commission staff elaborates on this point:

> In those situations where people are aware of the hazards of a site, and elect to develop regardless, staff believes that it is not reasonable for those people to petition the government for relief in the event their investment is damaged or destroyed. Rather, the staff believes that when individuals persist in building despite the hazards and the discouragement of the government, that they should assume full liability for their actions.

In some cases, the state commission has required owners to assume liability as a condition of development. This has most often occurred in areas affected by severe erosion and seacliff retreat. In order to warn subsequent purchasers regarding the risks inherent in the site, the waiver has been incorporated into a deed restriction.

Although zoning and land division regulations are the most common forms of land-use regulation, performance standards, design review, and conservation regulations are equally important tools for protecting or minimizing the alteration of sensitive coastal environments when development is permitted. For example, the environmental complexity of coastal dune systems warrants a regulatory approach that reflects their ecological requirements. The dune profile is generally in dynamic equilibrium with both the coastline and the native site vegetation, in addition to being highly susceptible to disturbance. Outright prohibition of development through zoning may be the only reliable means for maintaining the fragile foredune region in a natural state and thus maintaining the storm protection provided by the natural dune system. But if development is permitted in backdune areas, revegetation and dune stabilization programs may be necessary. These programs could require the use of native plant species for landscaping, encouraging natural plant distribution and succession endemic to the particular dune system. High volume pedestrian paths and off-road recreational vehicle trails frequently produce major environmental impacts in dune systems, requiring structural solutions such as fences and raised wood walkways, or nonstructural solutions such as trail rotation programs to minimize adverse environmental changes.

Acquisition

Public acquisition of private property constitutes a second type of land-use control. This technique involves the transfer of all or partial rights to use through (1) gifts, (2) purchase, or (3) condemnation.

Although gifts are a useful and usually a low-cost form of land-use control, they are voluntary and are thus too random and unpredictable for implementing a coastal plan. It is conceivable that once high-hazard lands were identified and corresponding regulations were adopted, gifts of affected properties might increase, but such an approach is simply too problematic for assuring the level of hazard protection needed in the coastal zone.

Negotiated state purchase is among the most enduring forms of control. However, considering the fiscal condition of state and local government, this approach is also problematic. Mandated purchase or condemnation with or without compensation could also be used either to acquire property or to prohibit its use in hazardous areas. This technique poses additional difficulties since the power of condemnation is strictly limited by statutory provisions, and its exercise constitutes land-use control on the most inclusive and dramatic scale. However, given the magnitude to which existing development is subject to coastal hazards, condemnation may well become an increasingly common control technique.

Economic incentives and disincentives

This type of control technique involves the creation of economic incentives or disincentives to encourage property owners to make particular land-use decisions. The most dramatic but effective form (relevant to coastal hazard issues) would require development applicants to waive public liability and disaster relief funds for developments in hazardous areas as described earlier. The primary issue in disaster relief funding is that of the concomitant potential use of public funds for the repeated reconstruction of damaged or destroyed structures on the same hazardous sites. In effect, disaster relief funds may encourage increased shoreline development since the risks are shifted from the private sector to the public sector. If public subsidies for rebuilding were reduced, the private sector would probably respond rationally through the establishment of insurance mechanisms that reflect the true costs and risks of developing in hazardous areas.

Combination techniques

Various combinations of regulation, acquisition, and economic incentives or disincentives can lead to the design of control techniques that are specific to the particular characteristics of a given hazardous area and its existing or proposed uses. The transfer of development rights (TDR) is one such "hybrid" technique. Under a TDR program, development rights previously allocated in hazardous areas could be transferred via the economic market to lands identified as capable of bearing more intensive development.

Summary

All 4 types of land-use control have been used to minimize or avoid hazardous construction in coastal areas. However, specific environmental, economic, social, and political conditions determine which control technique will be used for each particular situation.

When do I need a permit?

Under the 1976 Coastal Act, each of the state's 67 coastal cities and 15 coastal counties was mandated to prepare its own Local Coastal Program (LCP) through a land-use plan and zoning ordinances. The state coastal commission is required to determine that these plans and ordinances conform with policies set forth in the Coastal Act. A city or county can issue its own coastal permits only after its LCP is approved by the state commission.

A development permit is required for most structures, or for activities that modify land or water use in the coastal zone. Permits are required, for example, for the installation of any coastal protection measure. Most repairs and improvements to single-family residences or other structures and to public works facilities are exempt from such permits. When permits are required, their granting or denial is based on specific Coastal Act policies or the policies in an approved Coastal Land Use Plan for the subject property. Only specific types of permits may be appealed to the state commission, including appeals related to projects on the immediate shore front and along coastal bluffs. The state commission has also adopted *interpretive guidelines* to assist permit applicants and local governments in understanding previous commission decisions.

Persons wanting to build in the coastal zone can determine whether or not a coastal permit is required by checking with the closest commission district office. District commission staff provide applicants with detailed explanations of the permit process (outlined in the next subsection), and assist in preparing necessary application forms.

What is involved in getting a permit?

The procedure for obtaining a Coastal Development Permit (hereinafter referred to as "permit") from the coastal commission is set forth in Title 14, Division 5.5, Chapter 5 of the California Administrative Code, the official publication containing the regulations of all state agencies. This procedure includes provisions for permit applications, permit processing, public hearings, staff recommendations, commission review and voting, appeals, violations, and enforcement.

The balance of this chapter outlines the steps involved in obtaining a Coastal Development Permit as required under California Coastal Commission procedures. It is important to note, however, that this permit authority is transferred to the applicable local government after final certification of its respective Local Coastal Program. To determine which public agency has coastal permit jurisdiction over a proposed project, potential applicants should determine the status of the Local Coastal Program by contacting the local government within which the proposed project is located.

In order to obtain a Coastal Development Permit, an applicant must complete the following procedure:

1. Determine if the proposed project is a coastal "development" as defined in the code. If the project qualifies as a "development,"
2. determine if the development is exempt from a permit, excluded from a permit (either categorical or geographical),

authorized for a permit waiver, or constitutes an emergency. If a permit is required,
3. determine the type of permit required (regular, consent, or administrative). Following this determination, the applicant must
4. complete the permit process of the applicable local government (city or county), and
5. submit a completed application form to the Coastal Commission (appendix A) including payment of a filing fee (as per an established fee schedule).
6. Following a 5-day review period to determine if the application is complete, a public hearing date is set for the commission to consider the application. This date can be no later than the forty-ninth day following the date on which the application is filed.
7. A commission staff analyst is then assigned to the project to conduct a field inspection.
8. A report is prepared by the commission staff, including a recommendation for approval, conditional approval, or denial.
9. During steps (7) and (8) above, adjacent landowners and residents located within 100 feet of the perimeter of the parcel on which the development is proposed are notified of the public hearing date. The applicant is responsible for providing the commission with a list of land owners and residents, and posting a public hearing notice as close as possible to the site of the proposed development.
10. The public hearing is then held, with a view toward securing

all relevant information and material necessary for the commission to reach a decision without unnecessary delay. This hearing includes a staff presentation, a presentation by or on behalf of the applicant, other speakers for the application, and speakers against the application. Following the hearing the commission votes for approval or conditional approval, denial, or continuance of the application. If a continuance is made, a final action must be made within 180 days or a maximum of 270 days from the initial filing date.

11. After final commission approval, the Coastal Development Permit is issued within 5 working days.

Federal programs

The coastal problems of shoreline erosion, flooding, and losses due to the destructive forces of storm winds and waves are not unique to California. Imprudent coastal development with respect to natural hazards and the destruction of valuable natural habitat and resources have led to several national regulatory programs. Coastal residents, property owners, officials, and developers should be aware of their responsibilities under these regulations.

The previously discussed California Coastal Management Program complies to the federal *Coastal Zone Management Act of 1972* (P.L. 92-583) that encouraged all coastal states to manage their shoreline. The program is administered through the National Oceanic and Atmospheric Administration, Office of Ocean and Coastal Resource Management, and requires land-use planning to conserve coastal resources.

The *National Flood Insurance Act of 1968* (P.L. 90-448) as amended by the *Flood Disaster Protection Act of 1973* (P.L. 92-234) was passed to encourage prudent land-use planning and to minimize property damage in flood-prone areas, including the coastal zone. Local communities must adopt ordinances to reduce future flood risks to qualify for the National Flood Insurance Program. The NFIP provides an opportunity for property owners to purchase flood insurance that generally is not available from private insurance companies.

The initiative for qualifying for the program rests with the community, which must get in touch with the Federal Emergency Management Agency (FEMA). Any community may join the National Flood Insurance Program provided that it requires development permits for all proposed construction and other development within the flood zone and ensures that construction materials and techniques are used to minimize potential flood damage. At this point the community is in the "Emergency Phase" of the NFIP. The federal government makes a limited amount of flood insurance coverage available, charging subsidized premium rates for all existing structures and/or their contents, regardless of the flood risk.

FEMA may provide a more detailed Flood Insurance Rate Map (FIRM) indicating flood elevations and flood hazard zones, including velocity zones (V-zones) for coastal areas where wave action is an additional hazard during flooding. The FIRM identifies Base Flood Elevations (BFEs), establishes special flood hazard zones, and provides a basis for floodplain management and establishing insurance rates.

To enter the Regular Program phase of the NFIP, the community must adopt and enforce floodplain management ordinances that meet at least the minimum requirements for flood hazard reduction as set by FEMA. The advantage of entering the Regular Program is that increased insurance coverage is made available, and new development will be more hazard-resistant. All new structures will be rated on an actual risk (actuarial) basis, which may mean higher insurance rates in coastal high-hazard areas, but generally results in a savings for development within numbered A-zones (areas flooded in a 100-year coastal flood but less subject to turbulent wave action).

FEMA maps commonly use the "100-year flood" as the base flood elevation to establish regulatory requirements. Persons unfamiliar with hydrologic data sometimes mistakenly take the "100-year flood" to mean a flood that occurs once every 100 years. In fact, a flood of this magnitude could occur in successive years, or twice in one year, and so on. The repeated flooding along the California coast as a result of recent winter storms illustrates this point. If we think of a 100-year flood as a level of flooding having a statistical probability of occurrence of 1 percent in any given year, then during the life of a house within this zone that has a 30-year mortgage, there is a 26 percent probability that the property will be flooded. The chances of losing your property becomes 1 in 4, rather than 1 in 100. Having

flood insurance makes good sense.

In V-zones new structures will be evaluated on their potential to withstand the impact of wave action, a risk factor over and above the flood elevation. Elevation requirements are adjusted, usually 3 to 6 feet above still-water flood levels, for structures in V-zones to minimize wave damage, and the insurance rates also are higher. When your insurance agent submits an application for a building within a V-zone, an elevation certificate that verifies the elevation of the first floor of the building must accompany the application.

The insurance rate structure provides incentives of lower rates if buildings are elevated above the minimum federal requirements. General eligibility requirements vary among pole houses, mobile homes, and condominiums. Flood insurance coverage is provided for structural damage as well as contents. Most coastal communities are now covered under the Regular Program. To determine if your community is in the NFIP and for additional information on the insurance, get in touch with your local property agent or call the NFIP's servicing contractor (phone: [800] 638-6620), or the NFIP Region IX Office at (415) 556-9840. For more information, request a copy of "Questions and Answers on the National Flood Insurance Program" from FEMA.

Before buying or building a structure on the coast, an individual should ask certain basic questions:

Is the community I'm located in covered by the Emergency or Regular Phase of the National Flood Insurance Program?

Is my building site above the 100-year flood level? Is the site located in a V-zone? V-zones are high-hazard areas and pose serious problems.

What are the minimum elevation and structural requirements for my building?

What are the limits of coverage?

Make sure your community is enforcing the ordinance requiring minimum construction standards and elevations.

Nature has a way of exceeding human expectations. Storm waves coming from just the right direction at just the right time on a high spring tide may cause flood levels that exceed the predicted levels of the flood maps. In 1983 a moderate storm at Stinson Beach destroyed buildings beyond the V- and A-zones. Similarly, the same year in Hawaii, Hurricane Eva's flooding damaged and destroyed buildings beyond the A-zone. California is revising the flood maps and the new designations should be available by 1985. Homeowners should regard the requirements to obtain flood insurance as *minimal*, and go beyond those requirements when elevating and flood-proofing their structures.

Most lending institutions and community planning, zoning, and building departments will be aware of the flood insurance regulations and can provide assistance. It would be wise to confirm such information with appropriate insurance representatives. Any authorized insurance agent can write and submit a National Flood Insurance policy application. All insurance companies charge the same rates for national flood insurance policies.

The National Flood Insurance Program states its goal as "to . . . encourage state and local governments to make appropriate land use adjustments and to constrict the development of land which is exposed to flood damage and minimize damage caused by flood losses" and "to . . . guide the development of proposed future construction, where practical, away from locations which are threatened by flood hazard." To date, development in the flood-hazard areas continues at a rapid rate.

Revision of minimum flood elevations in the V-zones of coastal counties takes into account the additional hazard of storm waves atop still-water flood levels. Existing FEMA regulations stipulate protection of "dunes and vegetation" in the V-zones, but implementation of this requirement by the local communities has not always been strong. The existing requirements of the NFIP do not address other hazards of "migrating" shorelines, for example, shoreline erosion or shifting of inlets. Thus, buildings may meet the minimum FEMA elevation requirements, but at the same time can be located near highly exposed and eroding shorelines. In addition to recognizing the flood hazard, the need exists to incorporate location and structural codes that reflect migrating shorelines, hurricane-force winds, wave uplift, horizontal pressures, and scouring to minimize the loss of structures as well as the dollars that have supported the insurance program. This is not to say that state and local codes and ordinances have overlooked the latter.

In the past the National Flood Insurance Program has been subsidized and has grown to become a large federal liability. As of August 31, 1981, more than 1.918 *million* flood insurance policies valued at $97.972 *billion* had been sold nationwide.

Coastal counties had 1.165 million of these policies valued at $64.667 billion.

The FEMA Disaster Assistance Program Division serves as an advisory agency for the reduction of impacts due to natural hazards (for example, flooding, landslides, earthquakes), as well as exerting some regulatory control to reduce future property damage. Under the authority of the federal *Disaster Relief Act of 1974* (P.L. 93-288) the agency evaluates potential hazards and determines plans to mitigate the effects of such hazards. Reduction of loss due to flooding is specifically addressed under the authority of the *Inter-Agency Agreement for Nonstructural Flood Damage Reduction* and *Executive Order 11-988 Flood Plain Management* that designates FEMA as the lead agency to determine actions that will reduce the impact of flooding. For more information call FEMA, Region IX, Disaster Assistance Program Division (415) 556-9830.

The federal *Water Pollution Control Act Amendments of 1972* (P.L. 92-500) control any type of land use that generates, or may generate, water pollution. They also regulate dredging and filling of wetlands and water bodies. The U.S. Army Corps of Engineers administers the program through a permit process. Before dredging, filling, or placing any structure in navigable water (almost any body of water), contact the Army Corps of Engineers.

In addition to the laws noted above, other federal regulations may be important locally. Coastal residents should check with the California Coastal Commission.

Summary

During the past 10 years, land-use planning and regulation in the California coastal zone have been increasingly effective in reducing the development risks of environmental hazards. Although it is impossible to calculate the long-term benefits of this coastal planning, it is increasingly clear that new coastal development is generating fewer environmental costs, reducing maintenance and mitigation costs, and providing greater protection of unique and fragile natural landforms. By reducing both the adverse environmental impacts and the risks of environmental hazards of coastal development, land-use planning and regulation have achieved a significant degree of success in protecting the valuable natural resources of the coast that are of such a vital and enduring interest to the people of California.

Coastal residents and property owners should expect additional regulations in the future. Current revision of flood maps may expand V- and/or A-zones or increase the minimum elevation requirement for flood insurance. Building codes may be strengthened or more strictly enforced. A return of very large storm waves, such as those that occurred in the 1940s, or earthquake-induced tsunamis would cause extensive property loss and might result in new setback requirements. Local problems of how to respond to shoreline erosion, provide public access, or protect changing habitat will necessitate continued changes in coastal regulation.

Part Two
7. The northern California coast: Oregon border to Shelter Cove

Lauret Savoy and Derek Rust

Oregon border to Redwood National Park

The northwesternmost corner of California (fig. 7.1) is a land of coastal fog and heavy winter rains, steep coastal cliffs and unspoiled beaches, meandering rivers and lush forests of redwood, Douglas fir, and spruce. Before the arrival of the first European explorers, several Indian groups inhabited the forested coastal lands of northernmost California. Thriving on the abundant food from the sea, these included the Tolowa, Yurok, Chilula, Wiyot, and Mattole Indians. The earliest European explorers sailed the north coast waters in search of new land and navigable harbors during the sixteenth and seventeenth centuries. Spain showed the greatest interest in exploring the northwest coast of the New World, and the Cabrillo-Ferrelo voyage of 1542–43 is generally considered the first such attempt to reach this coastline of Alta California (High California). Many historians believe members of this voyage sited and named Cape Mendocino, and perhaps ventured as far as southern Oregon.

Although few records remain of the voyages, some researchers believe that Sir Francis Drake in 1579, and later Sebastian Rodriguez Cermeno in 1595, may have landed near Trinidad Head in Humboldt County. The last of the early explorers, Sebastian Vizcaino, probably reached Cape Mendocino in 1603. Nearly 200 years passed, however, before renewed interest in northern California exploration brought European vessels once again to this area. The Spaniards Juan Francisco de la Bodega y Cuadra and Bruno Heceta "discovered" the protected anchorage of Trinidad Bay on Trinity Sunday in 1775, and named it the Puerto de la Trinidad. Over 25 years later, trade ships for the Russian-American Company discovered the entrance to Humboldt Bay. The Russians did not disclose their knowledge of the bay's existence until 1852—2 years after it was rediscovered from the land by members of the Josiah Gregg party.

It was not until the 1850s, however, after the discovery of gold on the Trinity River and the Josiah Gregg expedition into the Humboldt Bay area, that the north coast area was settled permanently. It was easier to reach the rugged north coast by ship than by traveling overland. Thus, shiploads of miners and settlers descended on the coast, establishing port settlements like Crescent City, Trinidad, Eureka, and Klamath City. The prosperous redwood logging industry also began to flourish, and with the increasing demand for lumber created by California's booming population, eventually surpassed mining as a major industry.

Physical setting and development. Much of the rugged northwesternmost corner of coastal California is still sparsely popu-

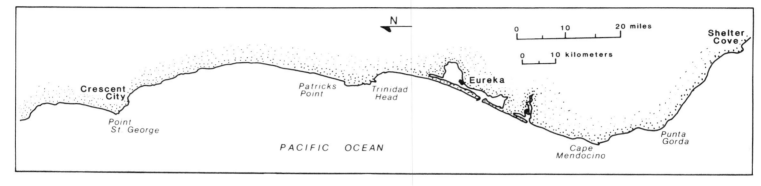

Figure 7.1. Location map for the coastline between the Oregon border and Shelter Cove.

lated. Coastal development in this region is centered on Crescent City and the Smith River mouth in Del Norte County. The southern portion of the county south of Crescent City and much of northern Humboldt County were recently protected within the 50-mile-long Redwood National Park. These areas are now unavailable for further development.

Crescent City lies on the southwest corner of a broad coastal lowland nestled between wooded ridges approximately 20 miles south of the Oregon border. The Smith River drains the lowland at its northern end, flowing across a wide floodplain. To the north of the river mouth, a low, flat coastal terrace is cut at the base of the mountains. The southern terminus of the lowland is marked by steep, rugged ridges rising abruptly from the sea at Enderts Beach in Redwood National Park.

Fur trapping brought the first white explorers, led by Jedediah Smith, to the area in 1828. No permanent settlement existed, other than Tolowa and Yurok villages, until 1850 when the discovery of gold in the nearby Trinity River brought countless prospectors and fortune-seekers. Crescent City was laid out in 1853, and within a year it became a major trade port for shipping gold and supplying the nearby mines. The town grew so rapidly that it was even promoted as a site for the state capital. Today the principal industries are logging and commercial fishing. Coastal development before the last 30 years was limited to scattered homes near the cliffs along Crescent City's western shoreline. By the 1960s houses and a fronting roadway (Pebble Beach Drive) lined most of this ocean front.

Geologic setting. The coastal area is underlain by a rock that is known as the Franciscan Melange, which is exposed in the seacliffs at the Smith River mouth, along Point St. George and part of the Crescent City shoreline, and in innumerable offshore rocks and islands. The melange, by definition, is a disrupted mixture of boulders "floating" in a mud matrix. These boulders or blocks form the many prominent points and seastacks, as waves easily erode away the surrounding weaker matrix. Along the mile-and-a-half length of Pebble Beach to the south of Point St. George, layered mudstones and siltstones occur at the shoreline. Generally, along this entire coast, loosely consolidated sands and clays overlie the melange, the siltstone, and the sandstone.

Oregon border to Smith River (fig. 7.2)

Between the Oregon border and the Smith River mouth, the coast is characterized by a sand and cobble beach backed by bluffs of up to 40 feet. Scattered seastacks lie offshore along the entire reach but are abundant at the mouth of the river. Highway 1 parallels the shore, and although most of the coastline has been subdivided (except for 2 parks) approximately 50 percent of the area has been developed.

Along most of this segment, the nonresistant sands and clays that cover the melange form the bluffs and are exposed to wave attack. The occurrence of large driftwood logs in the backbeach indicates that waves can reach the bluff base and do possess the potential to extensively erode the bluffs during major storms (fig. 7.3).

Just north of the mouth of the Smith River the coast assumes an irregular outline where the Franciscan Melange is exposed at the beach level. Massive boulders or "knockers" within the melange form prominent points and seastacks like Pyramid Point and Prince Island. With limited development, few erosion problems have been noted as of yet. Large, cliff-battering waves could easily cut away the weak sand and clay bluffs, however. So although the scenery is quite beautiful, any bluff-top development should give careful consideration to the erodibility of these cliffs.

Pelican Bay beaches (figs. 7.2 and 7.4)

Unlike the beaches to the north of the Smith River, which are backed by low, wave-cut bluffs, the 12-mile-long beach between the Smith River and Point St. George is wide and backed by extensive sand dunes, wetlands, and 2 large fresh-to-brackish water lagoons. Much of this scenic area was subdivided in the 1960s.

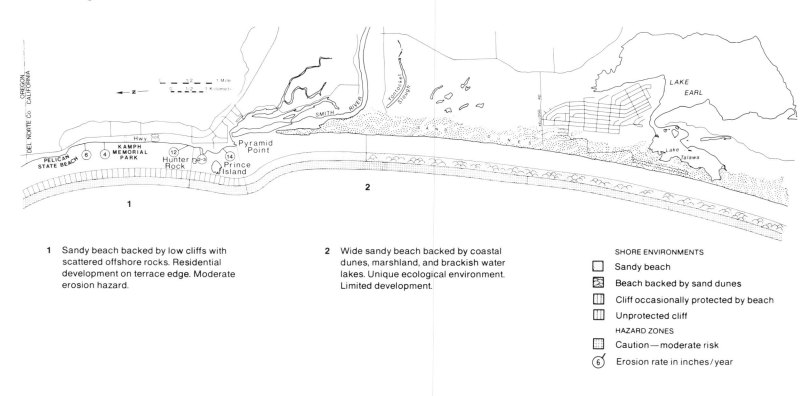

1 Sandy beach backed by low cliffs with scattered offshore rocks. Residential development on terrace edge. Moderate erosion hazard.

2 Wide sandy beach backed by coastal dunes, marshland, and brackish water lakes. Unique ecological environment. Limited development.

SHORE ENVIRONMENTS

☐ Sandy beach

▨ Beach backed by sand dunes

▥ Cliff occasionally protected by beach

▥ Unprotected cliff

HAZARD ZONES

▦ Caution—moderate risk

⑥ Erosion rate in inches/year

Figure 7.2. Site analysis: Oregon border through Lake Talawa.

Figure 7.3. Northward view of bluffs near the Smith River mouth. The presence of driftwood logs on the backbeach indicates that large waves reach the base of the bluffs. Photo by Lauret Savoy.

Point St. George to Crescent City (fig. 7.4)

The shoreline resumes an irregular, cliffed configuration at Point St. George. This prominent headland and the cliffs between Pebble Beach and Battery Point are exposures of the knobby melange. Along Pebble Beach sandstone forms a level rocky beach backed by a 20–45-foot-high seacliff. Pebble Beach Drive closely parallels the bluff edge along this stretch. This sandstone exhibits extensive cracks or "joints." Wave erosion along these joints has cut embayments, tunnels, and channels into the rock along the northern half of Pebble Beach where the seacliff is low in elevation. Erosion along these channels can, in turn, eventually undermine the cliff. Farther south, residential development is quite extensive along Pebble Beach Drive. While wave attack accomplishes significant erosion of the melange and sandstone, human activities also account for a substantial percentage of the bluff recession, particularly of the weaker unconsolidated sands and clays. Construction activities and urbanization at the cliff edge have promoted erosion by (1) removing extensive vegetation from the slope, (2) overloading the edge with the road and homes (fig. 7.5), and (3) increasing rainfall runoff on impervious surfaces such as roads, which causes severe, localized gullying on the slopes and, along with septic leachate, adds more water than can erode or saturate the slope.

During the winter of 1978 high waves endangered 1 home situated at the bluff edge. Erosion has threatened the roadway at a number of locations during the storms of the past several years.

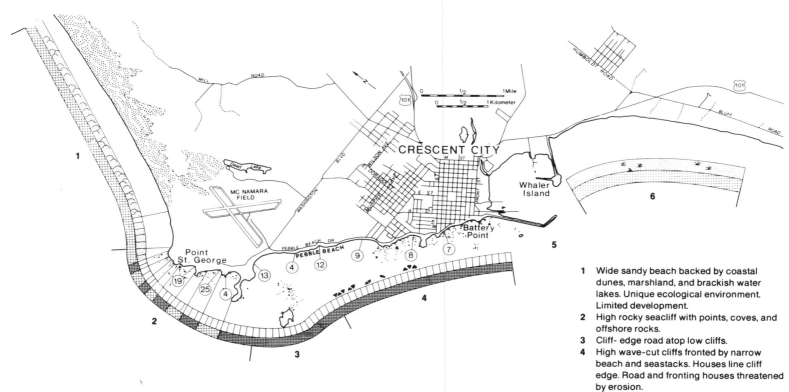

Figure 7.4. Site analysis: Point St. George through Del Norte Coast Redwoods State Park.

1 Wide sandy beach backed by coastal dunes, marshland, and brackish water lakes. Unique ecological environment. Limited development.
2 High rocky seacliff with points, coves, and offshore rocks.
3 Cliff- edge road atop low cliffs.
4 High wave-cut cliffs fronted by narrow beach and seastacks. Houses line cliff edge. Road and fronting houses threatened by erosion.
5 Breakwater and harbor
6 Sandy beach backed by wetlands and cliffs to the south.

Figure 7.5. Numerous houses line the shoreline along Pebble Beach Drive. Photo by Lauret Savoy.

Limited timber cribbing and riprap have been emplaced in attempts to arrest erosion along the stretch and are now being repaired and rebuilt (fig. 7.6). In 1973–74, the Harbor District dredged 600,000 cubic yards of sediment from the harbor and dumped it at the bluff base in an attempt to check recession of the beach and bluffs fronting the Seaside Hospital north of Battery Point. Much of this material consisted of mud and silt, which, because of the small grain size, is easily removed by wave action. Within 2 years approximately 80 percent of the sediment had been removed, once again exposing the bluffs to wave attack.

Between Crescent City and the boundary of Redwood National Park, the shoreline is defined by an undeveloped beach and marsh area known as South Beach.

Figure 7.6. Rock rubble or riprap has been placed at the foot of the bluffs at scattered locations along Pebble Beach Drive as a protective measure. Photo by Lauret Savoy.

Big Lagoon to Shelter Cove (figs. 7.7, 7.12, 7.13, 7.14, 7.15, 7.16)

Stretching 40 miles south and 65 miles north of Cape Mendocino, this coastline includes a variety of erosion conditions with rapidly eroding cliffs side by side with wide sandy beaches. As the maps show, building on this coast is often risky. Fortunately there is little development at present.

The resistance of the rocks and relief along the coast, and the sand brought to the coast by rivers, are the chief factors that produce the shape of the coastline. These factors vary widely from place to place, dividing the coastline into several distinct segments. These will be considered from north to south using the coastal maps.

Big Lagoon area coastline from Dry Lagoon to Agate Beach (fig 7.7)

The headland between Dry Lagoon and Big Lagoon (figs. 7.7A and 7.8) is made up of a massive, slow-moving earthflow most active during periods of heavy rainfall. Movement has exceeded a rate of 8 feet per year during winters with high rainfall such as 1964, when over 30 inches of rain fell in the area over a 6-day period. In 1973 a portion of Highway 101 in this area was abandoned and rerouted further inland at a cost of $2.5 million.

The sandbar impounding Big Lagoon (fig. 7.8) is periodically overwashed by very large waves during extreme high tides. During the 9 November 1980 magnitude 7 earthquake (centered about 30 miles offshore), a number of slumps occurred along the lagoon side of the sandbar, producing scarps up to 3 feet in height. The sandbar is breached at its north end by large waves about 3 to 5 times each winter.

The coastal cliffs from the community of Big Lagoon to Agate Beach (fig. 7.7B) are made up of loose sands. The base of these cliffs is typically protected by a steep beach. However, erosion can occur at catastrophic rates during particularly stormy winters when large waves and high tides combine to remove the beach sand quickly and attack the cliffs (fig. 7.9). This severe episodic hazard is masked by the average erosion rates shown on the map because they include many intervening years of no significant erosion.

Case study of the Big Lagoon/Ocean View community (fig. 7.7B)

The problems here were readily identifiable and demonstrate the importance of conducting a geologic analysis prior to initiating coastal construction.

The protective beach has given a deceptive impression of security, and in 1929 about 60 beach cottages were built, 15 of these within 30–40 feet of the cliff edge. The left aerial photograph in fig. 7.10 shows the community in August 1931 when the cliffs were protected by a wide beach.

However, in the 1935–36 winter and particularly in the 1939–40 winter, unfavorable combinations of storm waves and high tides removed the beach. All of the most vulnerable cottages were gone in November 1941, probably having been relocated farther inland. Up to 60 feet of cliff was lost between 1931 and 1941. During the 1941–42 winter the lack of a protective beach encouraged further cliff retreat—reaching up to 30 feet by February 1942.

These disastrous winters were followed by a long period of generally benign conditions. In 1962 lots for the Big Lagoon/Ocean View subdivision were laid out along the top of the bluffs to the south of the original community (fig. 7.8).

Over the past several years conditions have once again become malevolent. The 1982–83 winter waves cut back the cliffs up to 33 feet in another catastrophic erosion event. Two more cottages had to be moved and sand was stripped away below the cliffs bordering the new subdivision, threatening a number of large, new homes on the cliff top.

Erosion continued during 1983–84, with up to 30 feet more erosion at the north end of the community (compare the 1984 photograph in fig. 7.10 with the earlier photograph). One cottage now within 10 feet of the cliff edge was over 88 feet away in 1931. This represents an average retreat rate of 18 inches per year for over 50 years.

A return to benign conditions may allow the beach to reaccumulate in the future, but the long-term trend is clear: the cliffs will continue to retreat, and additional building on cliff-top lots in this area should not be approved.

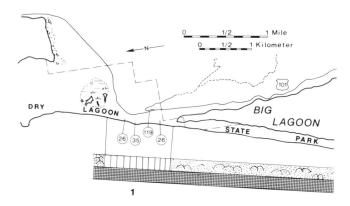

1 The headland between Dry Lagoon and Big
   Lagoon is a massive slump. Downslope
   movement has exceeded 9 feet/year during
   wet winters. In 1973 part of Hwy. 101 was
   abandoned and re-routed inland at a cost of
   $2.5 million.

Figure 7.7A. Site analysis: Dry Lagoon–Big Lagoon area.

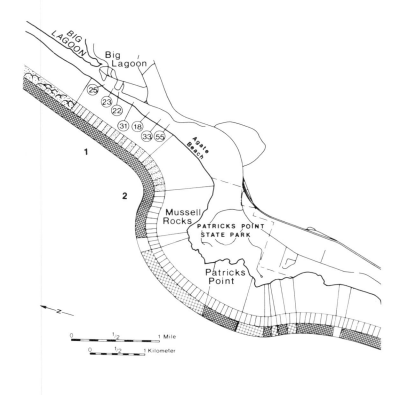

Figure 7.7B. Site analysis: Big Lagoon through Little River State
Beach.

1 Coastal bluffs fronted by sandy beach. High erosion hazard. Storms of recent years have caused extensive bluff erosion, particularly near the Big Lagoon community. Retreat typically occurs at catastrophic rates during particularly stormy winters. Average retreat rates are misleading because they include intervening years of no retreat. During the 1982–83 winter, for example, over 30 feet of cliff retreat occurred.

2 Active slumping from the south end of Agate Beach to Mussel Rocks.

3 The shoreline south of Mussel Rocks is rugged and rocky with numerous points, coves, and offshore rocks. Active landslides are common within some of the coves. Wide variations in erosion rates.

4 Between Trinidad and Moonstone slumping has repeatedly closed the coast road, although shoreline retreat rates are relatively low, even showing periods of advance in some cases.

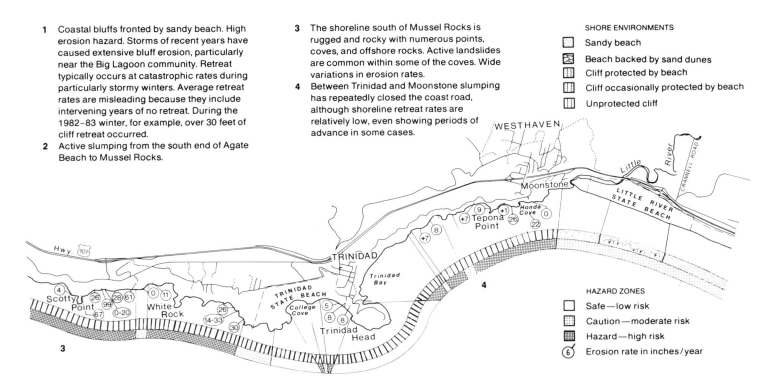

SHORE ENVIRONMENTS

- Sandy beach
- Beach backed by sand dunes
- Cliff protected by beach
- Cliff occasionally protected by beach
- Unprotected cliff

HAZARD ZONES

- Safe—low risk
- Caution—moderate risk
- Hazard—high risk
- ⑥ Erosion rate in inches/year

Figure 7.8.   View looking north over Big Lagoon. The weak, sandy cliffs at the community of Big Lagoon and the Big Lagoon (Ocean View) subdivision depend on the beach for protection. However, the beach is not stable, and during winter storm conditions with high waves and high tides, the beach can be quickly reduced so that waves reach the cliffs. This occurred in 1982–83 and 1983–84 when up to 60 feet of cliff was lost at the community of Big Lagoon (foreground). Similar periods of erosion have taken place in the past. Photo by Don Tuttle, before the 1982 erosion.

The rocky cliffs from Agate Beach to Moonstone (fig. 7.7B)

The character of this 12-mile-long coastline is again defined by the chaotic Franciscan Melange and by sands and gravels. Melange forms the rugged seacliff exposures between Patricks Point and Moonstone Beach, with numerous resistant blocks, ranging in size from boulders to Trinidad Head itself, floating in a muddy matrix. Attacking waves easily erode the relatively weak matrix, forming numerous embayments such as College Cove, while leaving the isolated knockers and the more resistant rocks as headlands and seastacks. The relatively straight cliffs between Agate Beach and Big Lagoon are composed of the weak sands and clays.

The coastal cliffs, with few exceptions, retreat by slumping during the wet winter months. Infiltration of water into the ground substantially weakens the slopes. Segments of the cliff edge break away in slumps and then move downslope to break up into a slowly moving glacierlike mass. At the beach winter waves quickly remove the material, thereby keeping the slopes unstable and encouraging continued landsliding.

Wave heights in a typical winter storm season are likely to exceed 23 feet and can be much higher. Seawater from waves breaking against Trinidad head in December 1914 rose to the lamp housing on the lighthouse at an elevation of 196 feet!

Erosion by coastal streams also encourages slumping, as do the frequent large earthquakes in the region. Twenty-one earth-

A B

Figure 7.9. Downcoast view of Big Lagoon cabins at the terrace edge in December 1976 (A) and February 1983 (B). Since 1976 over 50 feet of the weak bluffs have been eroded here. Note that one house is missing in the 1983 photo. 1976 photo by George Armstrong, California Department of Boating and Waterways; 1983 photo by Lauret Savoy.

Figure 7.10.   Cliff retreat at the community of Big Lagoon from 1931 to 1984. Roads bordering the houses near the cliffs have not changed during this time period, and serve as reference points for cliff erosion. The borders of both photographs are in the same locations. The cliff top beneath 17 houses and parts of the roads visible on the 1931 photograph were eroded away by 1984. Most of the houses were probably relocated further inland, and cliff retreat has continued after April 1984 photograph. 1931 photo by Fairchild Aerial Surveys; 1984 photo by Richard B. Davis.

quakes of magnitude 6 or greater have occurred in north coastal California since 1871.

Erosion hazards on this rocky coast

Because of the variability of the rocks exposed along the coastline, erosion rates change dramatically from point to point. Much of this detail cannot be shown on the map. The variations in erosion rates between Scotty Point and White Rock, for example, reflect the following factors:

(1) Shortness of the period used to determine the average erosion rates, in comparison to the frequency of major coastal slope failures. For example, the cliff top just north of White Rock retreated at an average rate of 42 inches per year between 1870 and 1942, and was then stable from 1942 to 1974 during a relatively calm climatic period. Many cliff localities that appear to be stable now are simply waiting for heavy rainfall (and perhaps earthquake shaking conditions) extreme enough to cause failure.

(2) Effect of localized protection of slopes from waves by large seastacks close offshore.

(3) Formation of deeply indented bays that are relatively protected from the dominant northwesterly waves, e.g., College Cove.

(4) Difference in retreat rate between the top and bottom of coastal cliffs.

(5) Temporary seaward displacement of the shoreline produced by slumped material.

Case Study of the coastal road between Trinidad and Moonstone (fig. 7.7B)

This road exemplifies the problems encountered when development ventures too close to the actively eroding coastline. It was first constructed in 1885 along the coastal slopes as a county wagon road. By 1908 the road was abandoned due to landslides, some of which were probably aggravated by the road construction, and a new road was opened farther inland. This road followed a longer route through Westhaven, and can be seen on the map.

However, road engineers turned their attention again to the coastal route in 1922–23. The California Highway Department built the present coastal road, which also appears on the map. This construction undercut the coastal slope in several places, resulting in renewed landsliding. For example, in 1939 the slopes just south of Honda Cove were reconstructed in order to rebuild a section of the road that had been lost (fig. 7.11).

In 1956–58 the state constructed the present route of Highway 101 farther inland parallel to the coast and turned the troublesome coastal road over to the county. Since then the road has required constant repair, and parts of it have been maintained with a single lane only. Just north of Tepona Point large gates were set up to permit 1 section that failed repeatedly there to be closed off easily for repairs. Several houses bordering the road have also been damaged.

During the wet 1982–83 winter, landslide movement accelerated and the road was disrupted in several places. These sites

Figure 7.11.   Construction of the coastal road across the cliffs south of Luffenholtz Beach increased their instability, and in 1908 a longer inland bypass was opened. The more direct coastal road was rebuilt in 1922–23, although landsliding continued and threatens the cliff-top houses. Photo by Don Tuttle.

have been temporarily repaired with gravel at a cost of almost $22,000, but there has been no attempt to restore the pavement because of an estimated cost well in excess of $4.8 million. Instead the county has proposed that Caltrans (California Department of Transportation) resume responsibility for the road as a scenic alternative to Highway 101; so far, as might be expected, Caltrans has not responded favorably to the proposal.

Moonstone to Centerville Beach: the Humboldt Bay lowlands (figs. 7.7B, 7.12, 7.13, 7.14)

At Moonstone the rocky coast changes abruptly to long, linear beaches backed by dunes and low, sandy cliffs. These beaches extend 35 miles to Centerville Beach and are fed by the great volume of sand brought to the coast by rivers in the area such as the Eel, Mad, and Little. The Eel River transports more sediment for each square mile of its drainage basin than any other river in the conterminous United States.

Case study of the McKinleyville cliffs (fig. 7.12)

Between the Little and Mad rivers the beach and sand dunes are backed by a line of low cliffs cut in relatively weak, loose sands. These cliffs give the appearance of stability because of the width of the beach and dunes, as shown on the map. A number of houses have been built along the cliff top.

The cliffs were cut by a combination of wave action and the

meandering of the Mad River's mouth up and down the coast. Between 1959 and 1982 the mouth of the Mad River moved northward over 3 miles to Vista Point west of the Arcata County Airport, removing the dunes and allowing large waves to attack the base of the cliffs. Two homes built on the cliff top in 1979 overlooked a protective beach and sand dunes almost 1,200 feet wide, yet by 1982 waves were able to reach the cliffs below the houses. Near Vista Point the dunes were cut back up to 575 feet during the 1981–82 and 1982–83 winters.

Since then the river mouth has moved about 3,000 feet to the south, and the beach and dunes are likely to become reestablished. However, these cliffs should be regarded with caution, including the area north of Vista Point where there has been landsliding and gully erosion in the weak sands.

Humboldt Bay coast (figs. 7.12 and 7.13)

Sand brought to the shore rivers in the area feeds the beaches and sand dunes of the North Spit and South Spit that enclose Humboldt Bay. This does not, however, prevent seasonally cyclic changes in shoreline position with changes in storm wave action and sand supply. A buffer zone should be set aside along this shoreline to allow for temporary coastal retreat during periods of intense storm wave activity. South of the Humboldt Bay entrance, large winter waves may wash over portions of the spit north of the Eel River mouth.

Within Humboldt Bay itself the coastline between the Elk River and King Salmon has eroded dramatically in historic time, with almost 1,400 feet lost at Buhne Point from 1854 to 1955. This erosion, which was largely caused by jetty construction and dredging at the entrance to Humboldt Bay, has now been stabilized by riprap.

The mountainous Cape Mendocino coastline from Centerville Beach to Shelter Cove (figs. 7.14, 7.15, and 7.16)

This is a coastline with rugged cliffs, steep coastal streams, and numerous slumps. In some places terraces occur (fig. 7.17) and rocky reefs in the surf zone protect the coastal cliffs. Beach and dune sands often accumulate at the base of these cliffs, adding further protection, particularly near the mouth of the Mattole River. However, these areas are not shown as safe on the map unless the protected cliffs themselves are made up of resistant rocks not subject to slumping during rainy winter conditions. Areas of maximum risk correspond to weak rocks that are too easily eroded by wave action to produce protective reefs.

Most of the coast is undeveloped, and the southern half is preserved in the King Range National Conservation Area. The only significant confrontation between development and erosion occurs on the coastal road 2–4 miles south of Cape Mendocino, and at the community of Shelter Cove.

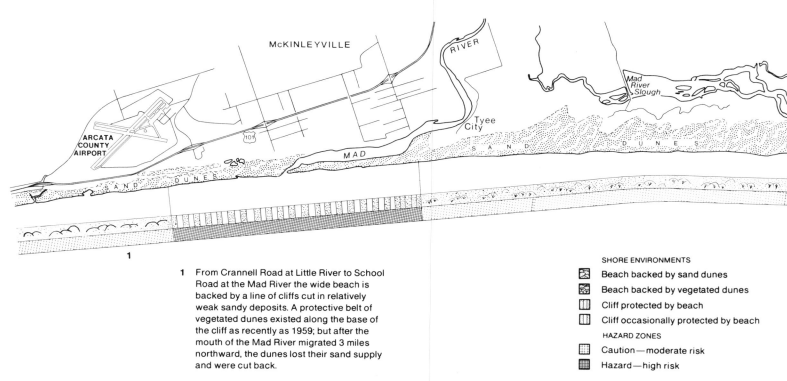

1   From Crannell Road at Little River to School Road at the Mad River the wide beach is backed by a line of cliffs cut in relatively weak sandy deposits. A protective belt of vegetated dunes existed along the base of the cliff as recently as 1959; but after the mouth of the Mad River migrated 3 miles northward, the dunes lost their sand supply and were cut back.

SHORE ENVIRONMENTS

▨ Beach backed by sand dunes

▨ Beach backed by vegetated dunes

▥ Cliff protected by beach

▥ Cliff occasionally protected by beach

HAZARD ZONES

▦ Caution—moderate risk

▦ Hazard—high risk

Figure 7.12.   Site analysis: Arcata County Airport through Humboldt Bay.

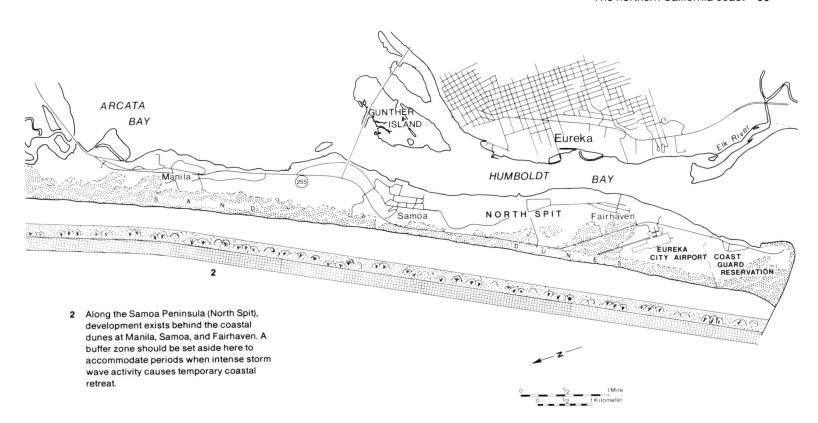

2 Along the Samoa Peninsula (North Spit), development exists behind the coastal dunes at Manila, Samoa, and Fairhaven. A buffer zone should be set aside here to accommodate periods when intense storm wave activity causes temporary coastal retreat.

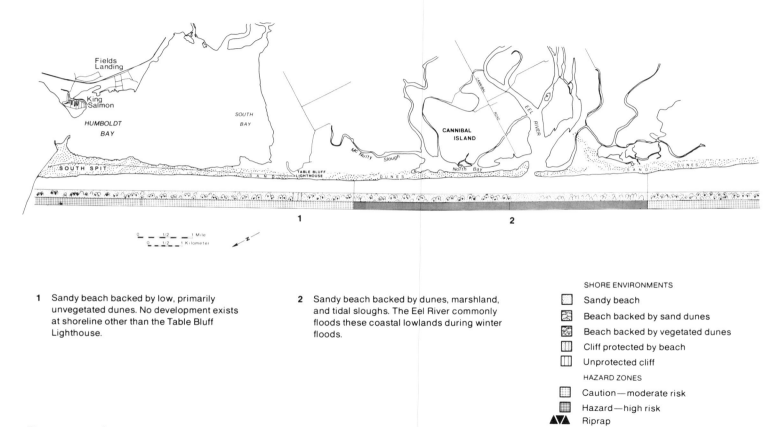

1. Sandy beach backed by low, primarily unvegetated dunes. No development exists at shoreline other than the Table Bluff Lighthouse.

2. Sandy beach backed by dunes, marshland, and tidal sloughs. The Eel River commonly floods these coastal lowlands during winter floods.

SHORE ENVIRONMENTS

- ⬚ Sandy beach
- ▨ Beach backed by sand dunes
- ▨ Beach backed by vegetated dunes
- ⦀ Cliff protected by beach
- ⦀ Unprotected cliff

HAZARD ZONES

- ▦ Caution—moderate risk
- ▦ Hazard—high risk
- ▲▼▲ Riprap
- ▬ Seawall

Figure 7.13. Site analysis: Humboldt Bay through Eel River.

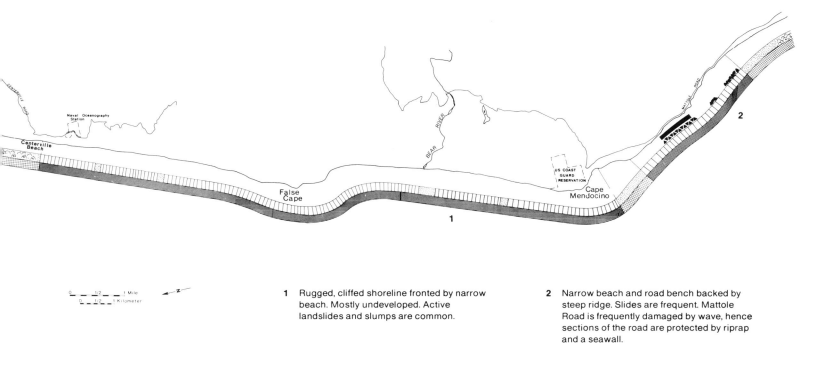

1 Rugged, cliffed shoreline fronted by narrow beach. Mostly undeveloped. Active landslides and slumps are common.

2 Narrow beach and road bench backed by steep ridge. Slides are frequent. Mattole Road is frequently damaged by wave, hence sections of the road are protected by riprap and a seawall.

Figure 7.14. Site analysis: Naval Oceanography Station through Cape Mendocino.

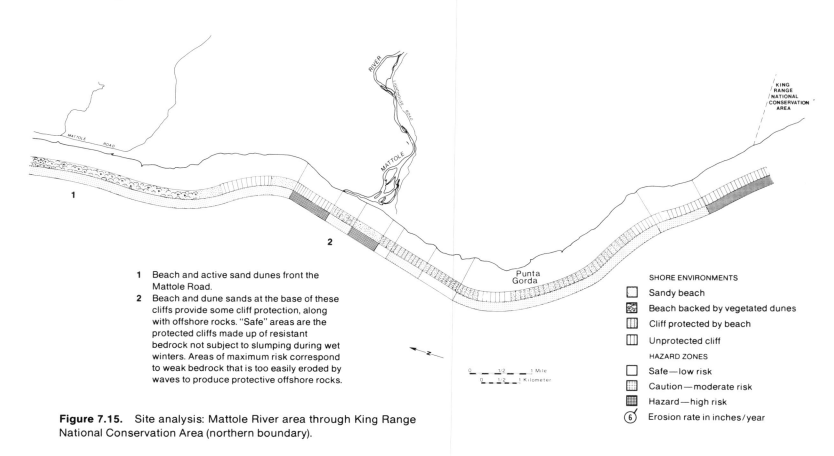

1 Beach and active sand dunes front the Mattole Road.
2 Beach and dune sands at the base of these cliffs provide some cliff protection, along with offshore rocks. "Safe" areas are the protected cliffs made up of resistant bedrock not subject to slumping during wet winters. Areas of maximum risk correspond to weak bedrock that is too easily eroded by waves to produce protective offshore rocks.

SHORE ENVIRONMENTS

☐ Sandy beach
▨ Beach backed by vegetated dunes
⊞ Cliff protected by beach
⊞ Unprotected cliff

HAZARD ZONES

☐ Safe—low risk
▨ Caution—moderate risk
▨ Hazard—high risk
⑥ Erosion rate in inches/year

Figure 7.15. Site analysis: Mattole River area through King Range National Conservation Area (northern boundary).

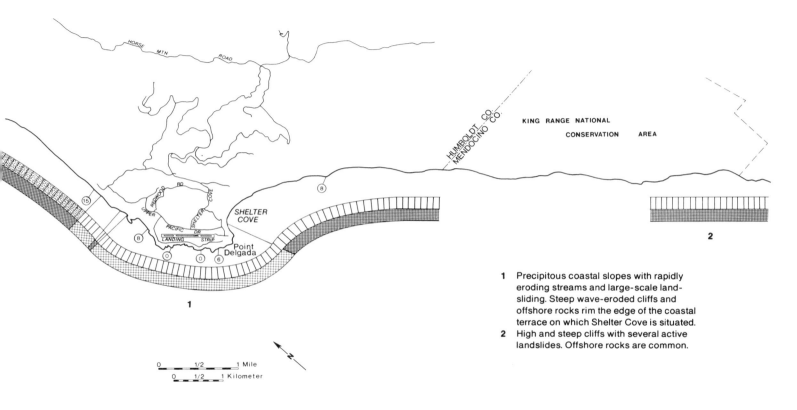

1 Precipitous coastal slopes with rapidly
  eroding streams and large-scale landsliding. Steep wave-eroded cliffs and
  offshore rocks rim the edge of the coastal
  terrace on which Shelter Cove is situated.
2 High and steep cliffs with several active
  landslides. Offshore rocks are common.

Figure 7.16. Site analysis: Shelter Cove area.

Mattole Road south of Cape Mendocino (figs. 7.14 and 7.15)

A short stretch of coastline south of Cape Mendocino serves as an excellent example of how some protection efforts are not a match for attacking waves. For 5.5 miles to the south of Cape Mendocino, the Mattole Road runs along the shoreline near beach level before turning inland north of the Mattole River. It was originally built as a wagon road in the 1860s and eventually connected the town of Petrolia—which holds the distinction of being the site of the first oil strike in California—with towns in the Humboldt Bay area. By the early 1930s the road had eroded so severely that the Humboldt County Civilian Conservation Corps built log cribs, filled them, and replaced the road on top. Since 1935 storm-induced erosion has caused maintenance problems during several winters by exposing the log cribbing to direct wave attack. Once the timber cribs were damaged, the unconsolidated fill underlying the road was easily eroded. For many years the crib wall structures were repaired and replaced, often with riprap dumped at the base of the road. The destructive storms in 1978 undermined and removed up to a 20-foot width of the road at 1 area and severely damaged the cribbing.

In 1980 the preexisting wood cribbing was replaced by concrete cribbing at a cost of $4.5 million. In only 12 days, however, waves damaged the structure to the extent that another $1.5 million was spent on repairs and additional riprap. Further repair work has been required several times since then, and this area re-

Figure 7.17. The Mattole Road follows the coastline just south of Cape Mendocino for several miles before turning inland. The stream in the foreground is an important source of sand for the beach. Photo by Derek Rust.

mains a maintenance headache for the Humboldt County Roads Division (fig. 7.18).

Shelter Cove development (fig. 7.16)

Large-scale subdivision and development began here in 1965, although as yet most of the lots are unoccupied. The slopes in the area are precipitous, with rapidly eroding streams and large-scale landsliding. In 1967 heavy rainfall caused $1 million in damage to roads in the subdivision. Many of the larger slumps probably arose from earthquake shaking along the nearby San Andreas fault zone.

The low headland area that includes Point Delgada is made up of resistant rocks. These rocks also form a protective fringe of emergent reefs in the surf zone so that coastal erosion is limited here. Even in normal conditions the waves around these reefs are hazardous; between 1972 and 1978, 9 people were swept away by waves and drowned.

Figure 7.18.   Riprap and seawall front the Mattole Road near Devils Gate. Large waves commonly batter this coastline, often damaging the road. Photo by Lauret Savoy.

8. Shelter Cove to Point Arena

Lauret Savoy

From False Cape to Cape Vizcaino (fig. 8.1), a distance of approximately 71 miles, the shoreline is typically rugged and mountainous, with dense forests and numerous offshore rocks and islands. Much of this segment lies within the King Range National Conservation Area, the longest roadless stretch of coast in the state with peaks towering 4,000 feet above the shoreline, and in the Sinkyone wilderness. Beaches are scattered and remote, and access is limited to a few roads: the Mattole Road runs along the shore south of Cape Mendocino for 5.4 miles, a small country road leads to Shelter Cove from Highway 1, and a few, mostly unpaved roads perilously skirt the high coastal ridges between Shelter Cove and Cape Vizcaino (fig. 8.1). The entire stretch is relatively uninhabited, with the exceptions of a small development at Shelter Cove, remnants of old mill towns such as Bear Harbor and Usal, scattered cattle and sheep ranches, and active logging operations (fig. 8.2). Many of the steep, narrow roads following the coast north of Cape Vizcaino were originally built over 100 years ago to haul redwood and fir timber to the scattered mills and landings, and some have changed little since then.

South of Cape Vizcaino (fig. 8.1), the shoreline becomes less mountainous and a 70-to-200-foot-high flat terrace is cut into coastal cliffs. Situated on this terrace are the first sizable coastal communities south of the Humboldt Bay area, such as Westport, Fort Bragg, and Mendocino. Highway 1 emerges at this point from its inland route to follow the terraced coastline south to San Francisco.

Geologic setting. Sandstone and shale form the shoreline between Cape Mendocino and Point Arena. Unlike the Franciscan terrain north of Cape Mendocino, this coastline does not exhibit the prominent knobby blocks or knockers that formed the innumerable seastacks farther north.

Erosion. Erosion of the extensively sheared and weak rocks in the seacliffs is the most active form of retreat along the coast. The many northwest trending points and coves reflect the variations in the resistance of the cliff rock. Little erosion can be measured on many of the resistant headlands (fig. 8.3). With few beaches to act as buffers, the waves attacking the weaker zones in the cliff have formed innumerable coves, sea caves, and natural bridges. Rockfalls commonly occur where wave action has undercut the base of the cliffs. Landslides have also been induced by construction activities and increased groundwater flow during the winter rains.

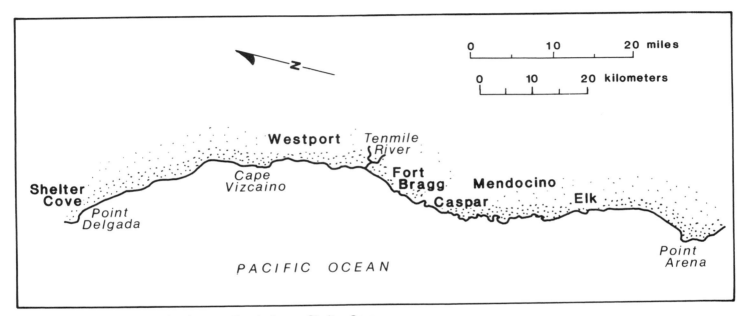

Figure 8.1. Location map for the coastline between Shelter Cove and Point Arena.

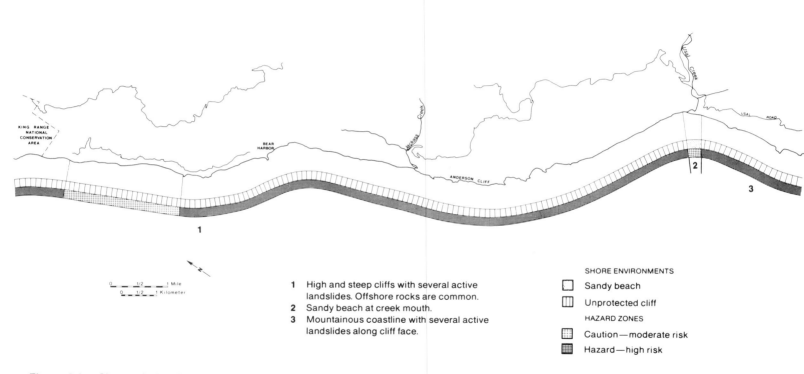

1   High and steep cliffs with several active
    landslides. Offshore rocks are common.
2   Sandy beach at creek mouth.
3   Mountainous coastline with several active
    landslides along cliff face.

SHORE ENVIRONMENTS

☐ Sandy beach

▥ Unprotected cliff

HAZARD ZONES

▦ Caution—moderate risk

▦ Hazard—high risk

Figure 8.2.   Site analysis: King Range National Conservation Area
(southern boundary) through Usal Creek area.

Cape Vizcaino to Tenmile River (fig. 8.4)

Between Cape Vizcaino and the Tenmile River, the coastal terrace contains one sizable community—Westport—and a few rural hamlets that were once productive mill towns and shipping points in the late 1800s. Recent development is limited to small subdivisions along the bluffs just north of the river mouth (fig. 8.5). Highway 1 follows the shoreline along much of this reach, extending inland no more than a quarter mile, and along certain sections precariously skirts the cliff edge (fig. 8.6). Storm-related slumping of portions of the roadway has been common for several years.

Along this shoreline, beaches are few and exist only in small pockets between headlands, with the exception of 1 wide beach north of the Tenmile River. The highest erosion rates occur in the weak rock that exists in the interiors of coves and where small streams and gullies channel through the seacliffs. Prominent points such as Bell Point, and the vegetated bluffs fronted by a large beach near the river mouth, have exhibited little change over the past 100 years.

A

B

Figure 8.3. The rocky headland at Albion Harbor in Mendocino County has changed little between 1882 (A) and 1983 (B). Albion Harbor was once an active shipping point. 1882 photo courtesy of the Mendocino Historical Society; 1983 photo by Lauret Savoy.

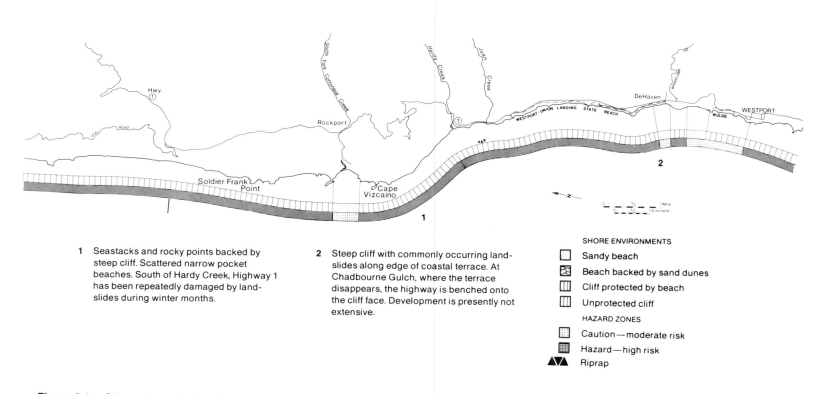

1   Seastacks and rocky points backed by steep cliff. Scattered narrow pocket beaches. South of Hardy Creek, Highway 1 has been repeatedly damaged by landslides during winter months.

2   Steep cliff with commonly occurring landslides along edge of coastal terrace. At Chadbourne Gulch, where the terrace disappears, the highway is benched onto the cliff face. Development is presently not extensive.

SHORE ENVIRONMENTS

Sandy beach

Beach backed by sand dunes

Cliff protected by beach

Unprotected cliff

HAZARD ZONES

Caution—moderate risk

Hazard—high risk

Riprap

Figure 8.4.   Site analysis: Soldier Frank Point through Soldier Point.

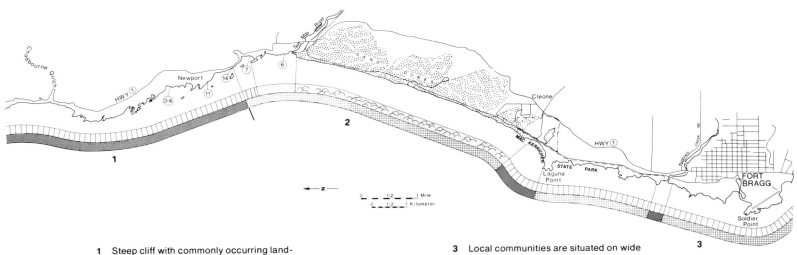

1 Steep cliff with commonly occurring landslides along edge of coastal terrace. At
   Chadbourne Gulch, where the terrace
   disappears, the highway is benched onto
   the cliff face. Development is presently not
   extensive.
2 Wide beach backed by large, active dunes
   and marshland.

3 Local communities are situated on wide
   terrace dissected by several deeply en-
   trenched streams. The cliffed coastline is
   marked by northwest-trending coves,
   points, and offshore rocks. Sandy beaches
   are rare and occur primarily at stream
   mouths. Winter storm waves of the past
   several years have caused damage in Noyo
   Harbor and along Fort Bragg's cliffs. Seacliff
   erosion is a moderate hazard along this
   stretch, particularly where weak rocks form
   cliffs.

Figure 8.5. Houses perched near the edge of steep bluffs to the north of the Tenmile River. Photo by Lauret Savoy.

Figure 8.6. Highway 1 emerges from its inland route at the mouth of Hardy Creek and follows the coastline southward toward San Francisco. At the mouth of Hardy Creek the slope above the highway has been benched for stability. Note the landslide on this slope (arrow). Photo by Lauret Savoy.

Fort Bragg and vicinity (figs. 8.4 and 8.7)

In 1856 the U.S. government attempted to consolidate the Pomo, Sinkyone, and Yuki Indians on a reservation near Noyo. Named after Colonel Braxton Bragg, then of Mexican War fame and later of Civil War fame, the military fort and the reservation that it oversaw were abandoned in 1867 due to interest in logging the area. The Indians were relocated to Round Valley, and Fort Bragg became a prosperous mill town and port. It is now the largest of the few remaining mill towns that dot the north coast. The Georgia-Pacific lumber mill and a few municipal facilities occupy most of the city's shoreline. New coastal residences are being built around Todd Point at the southern edge of the city. Much of the shoreline north of Fort Bragg is contained in Mac-Kerricher State Park.

Winter damage is common here. In 1978, winter waves removed a portion of a seawall below the lumber mill site and damaged part of the Noyo Harbor breakwater. In January 1983, waves caused minor damage to Noyo Harbor and slight bluff erosion at Todd's Point and the mouth of Pudding Creek.

Caspar to Point Arena (figs. 8.7 and 9.2)

The coastline is rugged along most of this reach with rocky cliffs 40–60 feet high dropping abruptly to the ocean at the terrace edge. Along the cliffed coastline, only a few small pocket beaches are present in the large coves. South of Fort Bragg the many

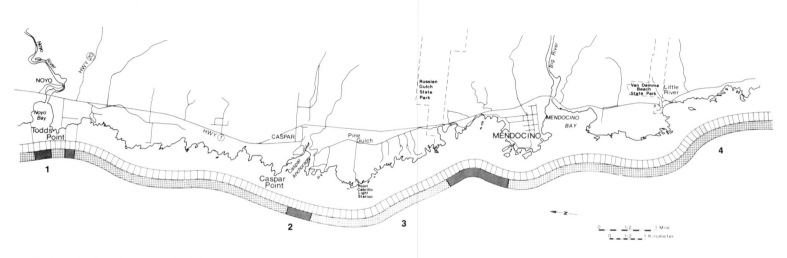

1   Minor bluff erosion occurred at Todds Point during the 1982–83 winter.
2   Steep rocky points and deep coves characterize the coastline. Small communities, new subdivisions, and state parks lie along the coast.

3   Cliff erosion, primarily in the form of landslides, most commonly occurs along the cliffs within the coves. The hazardous areas have been designated where good information on cliffs exists. Other areas may be hazardous, but because erosion information is limited, they are given a moderate hazard rating.

4   Information on shoreline erosion problems is limited south of Mendocino. Prior to development, studies of the short- and long-term cliff stability should be made.

Figure 8.7.   Site analysis: Noyo Bay through Elk area.

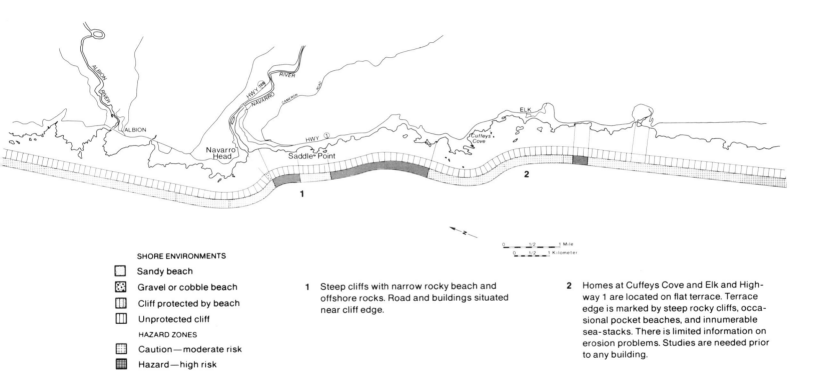

SHORE ENVIRONMENTS

- Sandy beach
- Gravel or cobble beach
- Cliff protected by beach
- Unprotected cliff

HAZARD ZONES

- Caution—moderate risk
- Hazard—high risk

1 Steep cliffs with narrow rocky beach and offshore rocks. Road and buildings situated near cliff edge.

2 Homes at Cuffeys Cove and Elk and Highway 1 are located on flat terrace. Terrace edge is marked by steep rocky cliffs, occasional pocket beaches, and innumerable sea-stacks. There is limited information on erosion problems. Studies are needed prior to any building.

rivers and streams debouch from steep-walled canyons. Most of the coastal communities along this reach were built by the lumber business in the mid 1800s; the town of Mendocino is by far the largest, and a number are now little more than hamlets. Much of the land is in agricultural use or in various state parks, beaches, and reserves (Mendocino Headlands, Russian Gulch, and Van Damme state parks; Caspar, Elk, and Manchester state beaches; and Jug Handle State Reserve). Although the area is sparsely populated, many new subdivisions such as Caspar Headlands, Sea Fair, Coast Highlands, and Irish Beach are being developed at the terrace edge. Seacliff erosion poses a threat to structures that are situated too near the cliff edge, particularly on weak rocks that occur in the interiors of coves (figs. 8.8 and 8.9).

The shoreline between Alder Creek and Point Arena contains over 5 miles of low, sandy beach backed by extensive sand dunes and wetlands. Much of this is protected within Manchester State Beach.

Figure 8.8.   Houses of a Caspar subdivision are clustered around the edge of eroding bluffs in the interior of a cove. Note the massive landslide (center). Photo by Lauret Savoy.

Figure 8.9.   Massive landslide of weak Franciscan rock in cove at Agate Beach near Mendocino. Photo by Lauret Savoy.

9. Point Arena to Point Bonita

Lauret Savoy

When the Spanish explorers landed in northern California, the coastline between Fort Bragg in Mendocino County and the Russian River to the south was inhabited primarily by the Pomo Indians. The Miwoks were widespread along the coastline and in stream valleys to the south of the Russian River (then called Shabaikai or Misallaaka, meaning long snake) as far as the San Francisco Bay (fig. 9.1). These people thrived on the abundant marine life and salmon-filled creeks until Spanish settlement in the early 1800s.

In 1579, almost 40 years after the Cabrillo-Ferrelo voyage along the then uncharted coast, the English pirate Francis Drake sailed along coastal northern California, his ship laden with booty seized in raids on Spanish ports and vessels along the west coast of the southern Americas. Drake stopped to make repairs in a convenient bay, which many historians believe is the bay named for him that lies in the lee of Point Reyes. In 1595 Sebastian Cermeno reached the coast of northern California on a return voyage from trading in the Orient. His Manila galleon, laden with cargo for Spain, supposedly anchored in Drakes Bay for needed repairs. A violent winter storm drove the ship ashore, wrecking her completely. With the entire cargo and several crew

members lost, the survivors sailed south in an open long boat and, after several months, reached Acapulco. Seven years later, Sebastian Vizcaino sailed to northern or Alta California in search of a sheltered harbor in which Spanish trade vessels could safely anchor. For a century and a half after this voyage Spain's interest in the exploration of Alta California waned, until other countries such as Russia and England began extending their holdings to the western portion of the New World. In the mid 1760s, in the hope of maintaining its hold on California, Spain began establishing permanent settlements in much of the region.

In October 1775 Don Juan Francisco de la Bodega y Cuadra apparently found a sheltered anchorage in the lee of Bodega Head. In 1809 Russian traders moved south from Alaska in search of otter and seal furs. They established a village near the present town of Bodega Bay, and in 1812 a trading post/fort, which they named Ross (for Russia), 13 miles north of the Russian River. At that time the Russians referred to the river as the Slavianka, meaning Slav woman. The Russian American Fur Company, as it was called, prospered for nearly 30 years until the near-depletion of the sea otter. After the Russians departed, the Mexican government monopolized the coastal lands with land grants up to the Gualala River, until California obtained statehood in 1849.

Development of the region, particularly north of the Russian River, was then shaped in the mid-to-late 1800s primarily by the several lumber operations scattered along the coast. Most of the small coastal communities began as mill and/or schooner ship-

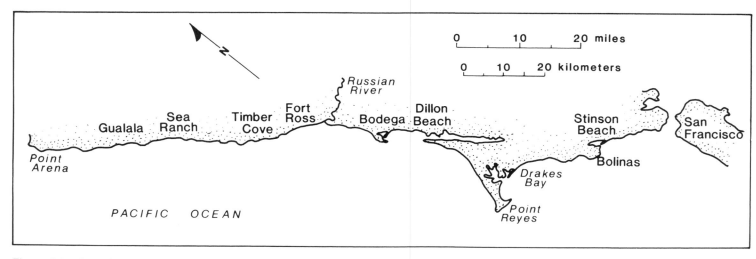

Figure 9.1. Location map for the coastline between Point Arena and San Francisco.

ping points in coves or "dogholes," such as Fisk Mill, Stewarts Point, Timber Cove, Iverson Landing, and Anchor Bay.

Point Arena to Fort Ross (figs. 9.2 and 9.3)

Physical setting and land use. The 57-mile stretch of coastline from Point Arena to Fort Ross is quite rugged and scenic, with innumerable points, coves, and offshore rocks. Nearly vertical seacliffs, between 20 to 100 feet high, are backed by a relatively flat plateau or marine terrace. Beaches are scattered and few. Much of this coastal area remains undeveloped, and it is used chiefly for grazing and agricultural purposes. Many communities, such as Stewarts Point, were once productive lumber mill and shipping points. Highway 1 follows the coastline along the entire reach, and housing developments range from individual lots flanking the highway to large-scale rural residential developments such as Sea Ranch. There are 3 state holdings along the coast —Kruse Rhododendron State Reserve, Salt Point State Park, and Fort Ross State Historic Park, all of which are in Sonoma County.

Geology. Sandstones, mudstones, and minor occurrences of volcanic rocks form the seacliffs along the coast. These cliffs rise vertically from the ocean and few beaches exist, other than within coves, to protect the cliffs from attacking waves. These rocks have been folded and faulted, but usually are not as extensively deformed as the rocks of the Franciscan Formation to the north of Point Arena and to the south of Fort Ross. These cliffs tend to erode actively where the rocks have been weakened by cracks or faults. The many coves and embayments occur where such structurally weakened rocks form the seacliffs, whereas the more competent or resistant rocks form the prominent points.

The entire coastal strip lies in close proximity to the San Andreas Fault, and any development should take the possibility of damage from an earthquake into careful consideration. *Peace of Mind in Earthquake Country*, by Peter Yanev, is an excellent source of data for planning for such an event.

Erosion and protection. Although information is limited, apparently few erosion problems exist along this rugged, undeveloped coast, and many of the rocky points have shown insignificant change over the past century. Cliff retreat, primarily in the form of landslides and rockfalls, has been observed along the cliff-line around the lighthouse at Point Arena, and where sections of Highway 1 and a few residential structures are situated at the bluff edge in coastal embayments (figs. 9.4 and 9.5). Although the winter storms of 1983 did not cause significant erosion, waves were large enough to destroy one-third of the Arena Cove pier and flood several buildings there, temporarily trapping patrons in a wharf cafe. Although waves destroyed all of the beach access stairways at Sea Ranch and locally eroded the bluffs, no homes there were threatened.

Fort Ross to Dillon Beach (figs. 9.3 and 9.6)

Physical setting and land use. The coastline between Fort Ross and the Russian River is more rugged than the preceding stretch

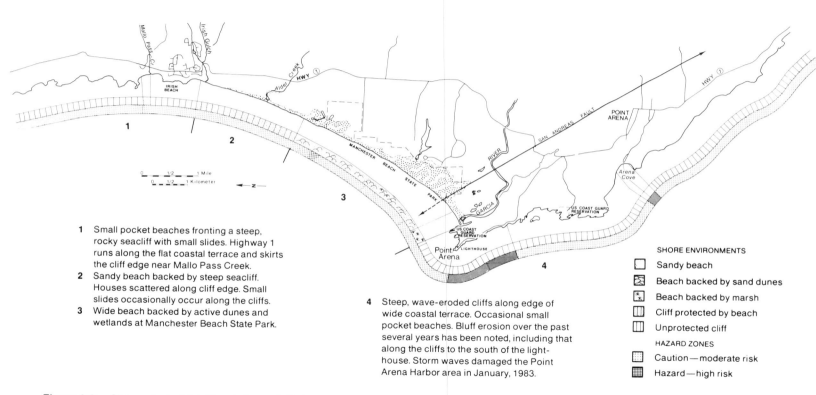

1 Small pocket beaches fronting a steep, rocky seacliff with small slides. Highway 1 runs along the flat coastal terrace and skirts the cliff edge near Mallo Pass Creek.

2 Sandy beach backed by steep seacliff. Houses scattered along cliff edge. Small slides occasionally occur along the cliffs.

3 Wide beach backed by active dunes and wetlands at Manchester Beach State Park.

4 Steep, wave-eroded cliffs along edge of wide coastal terrace. Occasional small pocket beaches. Bluff erosion over the past several years has been noted, including that along the cliffs to the south of the lighthouse. Storm waves damaged the Point Arena Harbor area in January, 1983.

SHORE ENVIRONMENTS

☐ Sandy beach

⊠ Beach backed by sand dunes

✳ Beach backed by march

⊞ Cliff protected by beach

⊞ Unprotected cliff

HAZARD ZONES

☐ Caution—moderate risk

▩ Hazard—high risk

Figure 9.2. Site analysis: Irish Beach through Del Mar Landing.

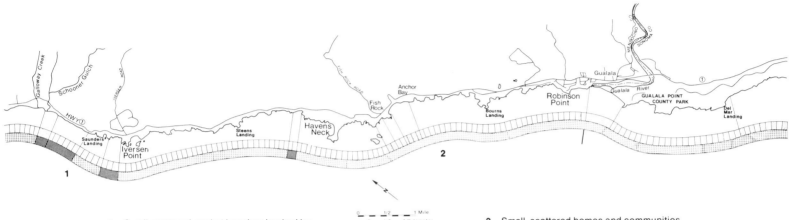

1 Small scattered pocket beaches backed by steep, rocky cliffs. Several streams cut through the cliffs. Scattered houses and subdivisions occur along the cliff edge, particularly near Steens Landing. Although information on erosion problems is limited, slides are particularly prevalent along the cliffs between Galloway Creek and Iverson Point. An evaluation of the instability of the cliffs should be made prior to cliff-edge development.

2 Small, scattered homes and communities and Highway 1 line the steep, rocky shoreline. Information on historical cliff erosion is limited. No structure should be built near the cliff edge without consideration of cliff stability. In low-lying areas such as Anchor Bay beach, storm waves can damage beachfront structures. In January, 1983, a beach house at Anchor Bay was destroyed by waves.

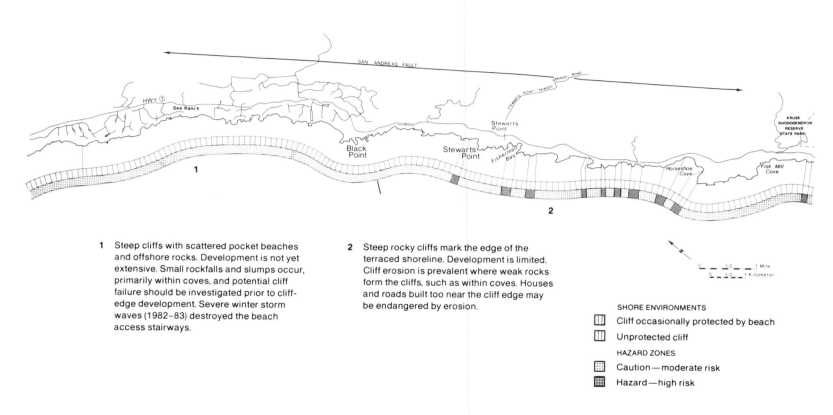

1   Steep cliffs with scattered pocket beaches and offshore rocks. Development is not yet extensive. Small rockfalls and slumps occur, primarily within coves, and potential cliff failure should be investigated prior to cliff-edge development. Severe winter storm waves (1982–83) destroyed the beach access stairways.

2   Steep rocky cliffs mark the edge of the terraced shoreline. Development is limited. Cliff erosion is prevalent where weak rocks form the cliffs, such as within coves. Houses and roads built too near the cliff edge may be endangered by erosion.

SHORE ENVIRONMENTS

Cliff occasionally protected by beach

Unprotected cliff

HAZARD ZONES

Caution—moderate risk

Hazard—high risk

Figure 9.3.   Site analysis: Sea Ranch through Russian Gulch.

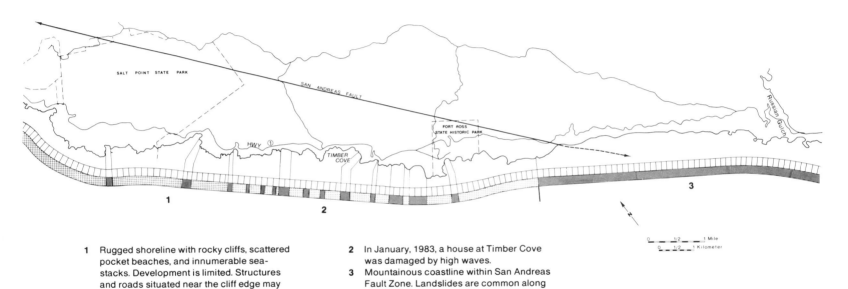

1 Rugged shoreline with rocky cliffs, scattered pocket beaches, and innumerable sea-stacks. Development is limited. Structures and roads situated near the cliff edge may be endangered by cliff erosion, particularly where weak, fractured rocks form the cliffs. These weak areas may erode by creep, landslides, or rockfalls.

2 In January, 1983, a house at Timber Cove was damaged by high waves.

3 Mountainous coastline within San Andreas Fault Zone. Landslides are common along cliff face, and the highway is endangered in several locations by sliding.

Figure 9.4. Houses are being constructed near the edge of eroding bluffs at Timber Cove. Note the slumps in the foreground. Photo by Lauret Savoy.

Figure 9.5. Rockfall and slump (upper arrow) along the cliffs at Sea Ranch. Photo by Lauret Savoy.

to the north, with precipitous slide-prone cliffs over 1,000 feet in elevation plunging to the sea (fig. 9.7). Most of this area is undeveloped and privately owned, and the steepness of the slopes prevents access to the shore. Southward from the Russian River to Bodega Head, a distance of 11 miles, the shoreline is less mountainous, but still quite rugged and beautiful. The Sonoma Coast state beaches, including several small beaches such as Shell Beach, Wright Beach, and Portuguese Beach, line the shore, in addition to the more extensive Salmon Creek Beach to the north

of Bodega Head. Land use is primarily recreational, as most of this shoreline is state beach. Two small bluff-edge subdivisions are located at Ocean View near Duncans Point and along Gleason Beach. South of Bodega Head to Tomales Bay rugged cliffs and numerous offshore rocks characterize the coastline. This land is almost entirely undeveloped with the exception of a small residential community at Dillon Beach in Marin County. The rugged coastline southward to the Bolinas Peninsula lies within Point Reyes National Seashore.

Geology. A few miles south of Fort Ross the San Andreas Fault extends offshore and reaches land again just east of Bodega Head. Where the fault crosses the coast near Fort Ross, the character of the rock and slopes changes abruptly. Sandstones and mudstones are found to the north of the fault. To the southwest occur the grassy, yet steep slopes disrupted by numerous landslides characteristic of the inherently unstable Franciscan Formation. Any development in this region should also consider the possibility of earthquake-induced damage.

Erosion and protection. Landsliding within the Franciscan is quite hazardous to coastal development, with slides varying in both size and activity; some creep slowly, others fail rapidly. Many slumps occurred during storms in 1983. Although most of the cliff-top homes at the Gleason Beach subdivision were protected by concrete seawalls of various designs, erosion of the bluff face was common (fig. 9.8). At least one house was left partially unsupported as the underlying rain-saturated material slipped away (fig. 9.9). The neighboring Pacific View development did not receive much damage because the bluff there is fronted by a wide, protective beach. Homes situated on the cliff edge at Dillon Beach, however, are threatened by erosion.

Bolinas—case study of an unplanned development

Located approximately 25 miles north of San Francisco, the Bolinas Peninsula forms the southernmost extension of the larger Point Reyes Peninsula. Duxbury Point forms the southwestern tip of the Bolinas Peninsula. The reef extending south from the point disperses waves entering Bolinas Bay from northerly through westerly directions. The bay, like other crescent-shaped bays along the coast, such as Half Moon Bay, is exposed to severe southerly winter storms. Most of Bolinas Peninsula is a broad, flat plateau or terrace that is bordered by steep seacliffs 140 to 200 feet high.

Bolinas Beach, which fronts the southeast-facing bluffs of the peninsula, has been a major recreational area in Marin County for over 100 years. Originally, Bolinas and neighboring Stinson Beach were part of the Baulenes Rancho land grant, which was first settled and used as a cattle ranch in the 1830s. In 1880 the first summer home subdivision was developed in what is presently the town of Bolinas. Since that time seacliff erosion has been a never-ending hazard.

Geologic setting. The Bolinas Peninsula is made up of 2 distinct geologic formations that are separated by the western edge of the San Andreas Fault zone. Mudstone lies to the west of the fault, whereas weak siltstones and sandstones occur to the east of the fault. The San Andreas Fault separates the Point Reyes Peninsula from the rest of Marin County, creating the long linear depressions at Tomales Bay, Olema Valley, and Bolinas Lagoon. An earthquake of large magnitude in this general region could easily trigger cliff failure.

Erosion and protection. Development is limited along the segment north of Duxbury Point to residences along the bluffs near the point. Farther to the north the mesa is primarily open space

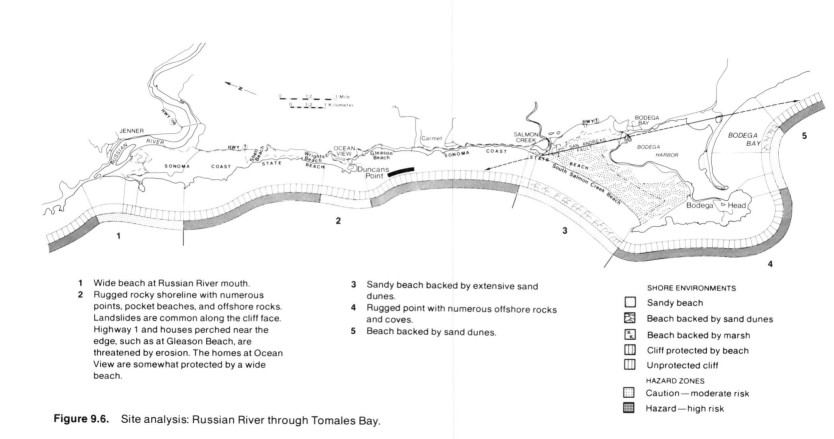

1 Wide beach at Russian River mouth.
2 Rugged rocky shoreline with numerous points, pocket beaches, and offshore rocks. Landslides are common along the cliff face. Highway 1 and houses perched near the edge, such as at Gleason Beach, are threatened by erosion. The homes at Ocean View are somewhat protected by a wide beach.

3 Sandy beach backed by extensive sand dunes.
4 Rugged point with numerous offshore rocks and coves.
5 Beach backed by sand dunes.

SHORE ENVIRONMENTS

☐ Sandy beach
▨ Beach backed by sand dunes
▨ Beach backed by marsh
▥ Cliff protected by beach
▥ Unprotected cliff

HAZARD ZONES

▦ Caution—moderate risk
▦ Hazard—high risk

Figure 9.6. Site analysis: Russian River through Tomales Bay.

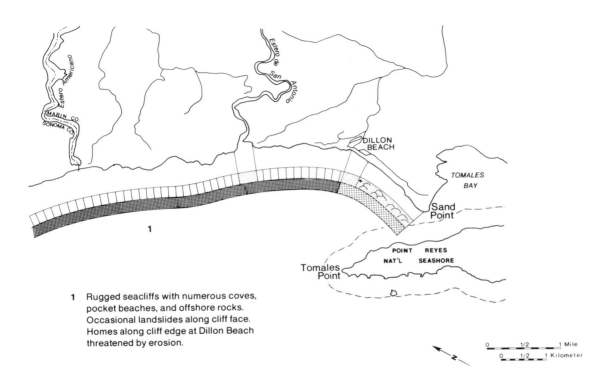

1   Rugged seacliffs with numerous coves,
pocket beaches, and offshore rocks.
Occasional landslides along cliff face.
Homes along cliff edge at Dillon Beach
threatened by erosion.

Figure 9.7. Massive landslide in steep cliffs along Highway 1 to the north of the Russian River. Photo by Lauret Savoy.

Figure 9.8. Erosion of loose, rain-saturated cliff material at Gleason Beach. Photo by Lauret Savoy.

Figure 9.9. Slumping of weak, water-saturated cliff material has partially undermined this house at Gleason Beach. Photo by Lauret Savoy.

with the exception of the site of an old RCA transmitting station presently leased by the Commonweal Association. The entire seacliff appears to be a continuous plane of landslides. Resistant, wave-truncated outcrops of the shale at beach level create a narrow shingle-covered tidal platform (fig. 9.10). Duxbury Reef is a partially submerged offshore extension of this rock platform. As is common throughout the area, most retreat occurs during the wet winter months. Groundwater seeps can be seen along the base of the seacliff during winter and spring. Most of this water originates from rainfall percolating through the fractured bedrock. Heavy rains also tend to cause gullying on the cliff face.

An RCA engineer recorded the relative positions of the terrace edge and cliff base on a base map every few years from 1913 until 1950. A survey of the present position of the seacliff indicates that along the few sections where the mudstone is less disrupted and supports more vegetation, the edge has receded an average of 0.4 to 7 inches per year since 1913. This average rate increases southward to over 36 inches per year where the mudstone is more fractured and less competent.

Between Bolinas and Duxbury points, the narrow, wave-cut platform and beach are often inundated by waves. Along this west-facing shore, attacking waves cause landslides by undercutting the base of the cliffs. Failure is facilitated by increased subsurface flow of water and saturation due to septic effluent from cliff-top homes as well as winter rainfall. Along this reach, the average rate of recession along the cliff base ranges from 6 to 24 inches per year.

Figure 9.10. Wave action has formed this nearly flat bench along Agate Beach to the north of Duxbury Point. Photo by Lauret Savoy.

and septic system leakage, as residences on Ocean Parkway and Terrace Avenue rely on individual septic tanks.

A combination of the inherent weaknesses of the bedrock, groundwater seepage, and wave attack causes rapid recession of the cliff. Seacliff failure occurs in a variety of forms, from flows of loose earth to massive landsliding. East of Duxbury Point, landslides have been extensive and destructive enough to damage portions of Ocean Parkway and remove several homes from their foundations (fig. 9.11).

The cliffs just west of the entrance to Bolinas Lagoon support the most densely developed residential area in Bolinas, with 20

Figure 9.11. Extensive landsliding on the cliffs has destroyed portions of Ocean Parkway. Photo by Lauret Savoy.

Numerous homes line the cliff edge directly to the northeast of Duxbury Point. Since the area was initially subdivided in 1927, many of these ocean-front lots and the fronting Ocean Parkway have been either partially removed or damaged by cliff erosion. Between Duxbury Point and Terrace Avenue, the mudstone is deeply weathered and exhibits extensive fracturing and shearing, particularly near where the San Andreas Fault slices through the bluff. The instability of the cliff is evident in the numerous landslides. Seeps commonly occur along the seacliff base during the winter and spring. This water probably originates from rainfall houses located on the brow of the cliff, some of which are pre-cariously perched on the face. Fronting these bluffs is the most heavily used stretch of beach. With each winter the position of the homes, road, and utilities located at the terrace edge becomes increasingly precarious. This is not a new problem—the hazards resulting from wave attack and erosion have threatened Bolinas for over 100 years.

The materials forming the seacliffs are easily erodible and sus-ceptible to failure, particularly during winter months. Cliff re-treat commonly has occurred in a variety of forms, from debris flows to large rock falls (fig. 9.12). The first bulkheads and groins were emplaced in the 1880s beneath the cliffs backing the beach, in an attempt to halt cliff erosion and stabilize the beach. These early wooden structures were regularly maintained by the com-munity for many years, and provided a beach of sufficient dimen-sions to shield the cliff base from wave attack. Permanent and seasonal dwellings were constructed on the beach during this time. During the harsh winter of 1912–13 storm waves destroyed many of the protective structures in addition to the homes. The groins and seawalls were subsequently rebuilt and well main-tained until the 1930s. By 1942, however, severe storms had dam-aged the structures to the point of uselessness. Within 3 years most of the beach sand had been removed, exposing bedrock. In 1947 the community constructed a reinforced concrete groin near the mouth of Bolinas Lagoon, and within a year the beach sand was restored. The groin functioned well until the storm waves of the past several winters severely damaged it and substantially

Figure 9.12. Large rockfall at base of cliffs to the west of the en-trance to Bolinas Lagoon. Photo by Lauret Savoy.

Figure 9.13. Severely damaged concrete groin as it appeared in 1983. Photo by Lauret Savoy.

Figure 9.14. Extensive bluff erosion that occurred during recent winters left this Bolinas house unsupported. Photo by Lauret Savoy.

decreased its effectiveness (fig. 9.13). The few remaining bulkheads are damaged and scattered. Although a few individual property owners have placed riprap and other protective structures below their cliff property, much of the bluff remains unprotected. The combination of groin failure and the lack of protection at the cliff base has resulted in increasing damage to structures on the terrace edge every winter. During the January 1982 storms, significant cliff failure occurred. Four homes along Terrace Avenue above Bolinas Beach were irreparably damaged and subsequently condemned. With the winter storms of 1982–83, 1 cliff-top home was completely destroyed while several others suffered damage (fig. 9.14). The effects of cliff erosion have been nothing short of disastrous in this community. Had the geologic

conditions been evaluated prior to development, much of the erosion "problem" could have been forestalled through better planning and setbacks.

Erosion has proceeded at average rates of 6 to 20 inches per year since 1939. An 1882 plot of the Terrace Avenue area shows the road at one location as then 88 to 127 feet inland of the cliff edge; the same road is presently at the edge of the cliff.

Stinson Beach (fig. 9.15)

Stinson Beach lies on a narrow sand spit that fronts Bolinas Lagoon. The beach is exposed to the full brunt of winter southerlies. It is no surprise that waves commonly inundate the beach.

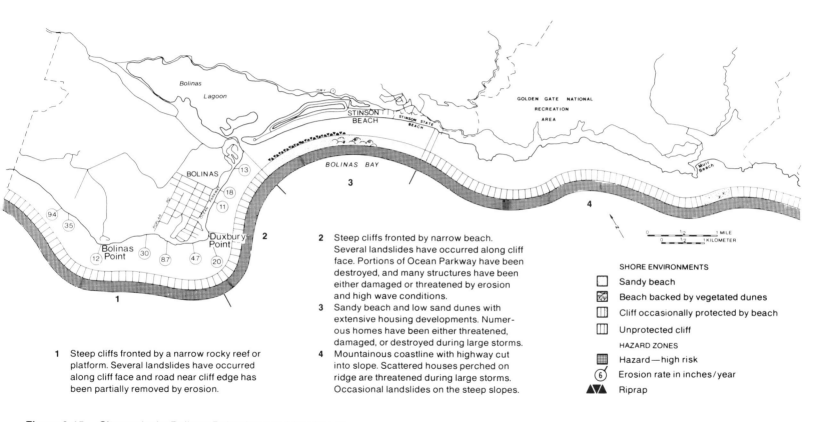

1 Steep cliffs fronted by a narrow rocky reef or platform. Several landslides have occurred along cliff face and road near cliff edge has been partially removed by erosion.

2 Steep cliffs fronted by narrow beach. Several landslides have occurred along cliff face. Portions of Ocean Parkway have been destroyed, and many structures have been either damaged or threatened by erosion and high wave conditions.

3 Sandy beach and low sand dunes with extensive housing developments. Numerous homes have been either threatened, damaged, or destroyed during large storms.

4 Mountainous coastline with highway cut into slope. Scattered houses perched on ridge are threatened during large storms. Occasional landslides on the steep slopes.

SHORE ENVIRONMENTS

☐ Sandy beach

▨ Beach backed by vegetated dunes

⊞ Cliff occasionally protected by beach

⊞ Unprotected cliff

HAZARD ZONES

▦ Hazard—high risk

⑥ Erosion rate in inches/year

▲▼▲ Riprap

Figure 9.15. Site analysis: Bolinas Point through Muir Beach.

Originally subdivided in 1906, the spit contains 2 separate developments—Stinson Beach Village at its eastern terminus, and the newer, exclusive Seadrift subdivision nearer to the tip. Houses in both of the developments are situated either directly on the beach or on the sensitive dunes behind the beach.

The result of unwise environmental planning can be seen in the damaging and costly effects of storms over the last several years. Storm waves and high tides during the 1977–78 winter endangered 9 beach-front homes at Seadrift as 10 to 90 feet of protective sand dunes were eroded. After first attempting to protect the beach from further erosion with sandbags, residents installed 600 feet of Longard Tubing (polyethylene tubes filled with sand slurry) along the beach. In time this tubing was undermined and torn by debris. Subsequently, riprap was emplaced at great cost (fig. 9.16). Total protection and repair costs reached $191,000, much of this at public expense. The county assessor estimated that each of the ocean-front lots lost approximately $50,000 each in market value.

The storms of 1982–83 were particularly disastrous to Stinson Beach Village. One home was completely washed out to sea, approximately 15 homes were destroyed, and over 50 houses were damaged (fig. 9.17). Estimated damage to these homes approached $1 million. The lesson learned here is painfully clear. Before any construction on vulnerable beach and dune areas is allowed, the hazards of erosion and wave wash should be carefully considered.

Figure 9.16. Extensive riprap was emplaced during the winter of 1982–83 to protect the beach homes at Seadrift. Photo by Lauret Savoy.

Figure 9.17. Recent winter storm waves irreparably damaged this Stinson Beach house. Photo by Lauret Savoy.

10. The San Francisco coastline

Kim Fulton and Lauret Savoy

When one thinks of San Francisco's waterfront, the extensive bay and Fisherman's Wharf often come to mind rather than the city's coastline. This coastline extends for an 8-mile length from the Golden Gate Bridge to the San Mateo County line (fig. 10.1). Steep, frequently crumbling cliffs fronted by narrow beaches characterize the shore between the Golden Gate and Seal Rocks. Most of this area is federal and state property (military and beach facilities), and the only private development occurs along a short stretch where expensive homes line the edge of the bluffs near Bakers and Phelan state beaches. The long, sandy Ocean and Fort Funston beaches, which are part of the Golden Gate National Recreation Area, lie to the south of Seal Rocks. The Esplanade, which is protected by an extensive seawall, and the Great Highway run along the shoreline behind the beach.

History and development

For over 200 years after European explorers first sailed along the northern California coast, San Francisco Bay and the peninsula

Figure 10.1. Location map for the San Francisco area coastline.

remained unknown. In 1769 an overland expedition led by Don Gaspar de Portola accidentally discovered the bay and peninsula while searching for Monterey Bay. During the next several years the bay area was further explored, and Captain Juan Bautista de Anza chose sites for the mission and presidio on the peninsula in 1776. The original mission was named San Francisco de Asis for the patron of the Franciscan order, and the presidio served as the northernmost such post in Alta California until it fell into disrepair. These remained the only major settlements on the peninsula until the 1830s.

By the early nineteenth century, with the growth of maritime trade in the Pacific, Yerba Buena Cove on the bayside of the peninsula had become a major port of call for ships from various countries, including Russia and the United States. The town of Yerba Buena grew around this bayside port, and in 1847, after the Bear Flag Revolt, the town was officially renamed San Francisco. Not a trace of Yerba Buena Cove can be seen today. The cove was later filled in with the rapid expansion of San Francisco, and if you follow Market Street to the Ferry Building, you would stand on what was once the center of the cove.

Through the remainder of the nineteenth century and into the twentieth century, the city of San Francisco grew around Yerba Buena Cove and Telegraph Hill. During this time the western side of the peninsula, with the exception of the military posts and the popular Golden Gate Park and Ocean Beach, remained what many residents considered dreary, barren sand hills.

Erosion and protection (with emphasis on private property)

Golden Gate to Point Lobos (Cliff House) (fig. 10.2)

The spectacular rocky bluffs between the Golden Gate and the Cliff House are made up largely of serpentine, which gives them a greenish color. Some parts of these cliffs contain relatively solid rocks, with few cracks, making them quite resistant to wave attack. These erode occasionally when large blocks break off and fall into the surf, but in general they have relatively low average rates of retreat. Other sections of cliff contain serpentine that has been broken up and weathered. These areas are prime locations for landslides, especially where water runs over or through the cliffs. Figure 10.3 shows several larger slides are on public land in the area, but they can occur anywhere that the cliffs are made up of fractured and weathered serpentine.

Other houses, near Bakers Beach, are sitting on comparatively low, sandy bluffs, which erode rapidly when hit by waves. This beach is relatively protected from big surf, but several homeowners have found it necessary to construct low, wooden seawalls. Despite these, and numerous retaining walls higher up on the slope, some house foundations are threatened here, primarily by slow slumping of the loose cliff materials (fig. 10.4). Such slumps may be triggered as much by surface and subsurface water as by wave erosion.

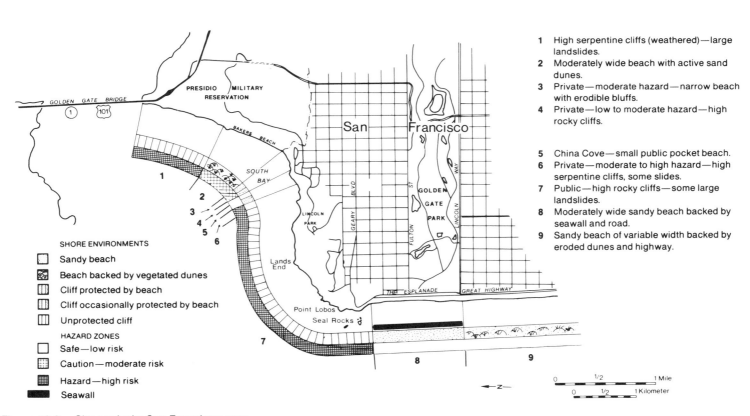

1 High serpentine cliffs (weathered)—large landslides.
2 Moderately wide beach with active sand dunes.
3 Private—moderate hazard—narrow beach with erodible bluffs.
4 Private—low to moderate hazard—high rocky cliffs.

5 China Cove—small public pocket beach.
6 Private—moderate to high hazard—high serpentine cliffs, some slides.
7 Public—high rocky cliffs—some large landslides.
8 Moderately wide sandy beach backed by seawall and road.
9 Sandy beach of variable width backed by eroded dunes and highway.

SHORE ENVIRONMENTS

☐ Sandy beach
▨ Beach backed by vegetated dunes
▥ Cliff protected by beach
▥ Cliff occasionally protected by beach
▥ Unprotected cliff

HAZARD ZONES

☐ Safe—low risk
▦ Caution—moderate risk
▨ Hazard—high risk
■ Seawall

Figure 10.2. Site analysis: San Francisco area.

Figure 10.3. A landslide occurs in the foreground next to expensive homes built on the coastal bluff. Photo by Kim Fulton.

Figure 10.4. Damaged concrete and wooden retaining walls on eroding bluffs near Bakers Beach. Photo by Kim Fulton.

The Cliff House to Fort Funston (fig. 10.2)

This continuous stretch of sandy beach and dunes is entirely in public ownership. A massive concrete seawall has protected the northern portion since 1929. This immense structure, almost half a mile long, and extending 20 feet above and below sea level, cost almost $600,000 when it was first built. These days, the cost of all that concrete might be over $6 million. Although it has experienced some tremendous storms since then, this wall remains in relatively good condition.

South of the wall is a region of actively migrating sand dunes, which have occasionally threatened to bury the Great Highway. This wide strip of asphalt runs along the dunes, seaward of any permanent buildings. The south end of the Great Highway has been periodically endangered by erosion of the dunes on which it is built.

Even farther south, toward the San Francisco County line, the high sand cliffs of Fort Funston provide beautiful views for park-goers but are relatively unstable. The Ocean Shore Railroad ran along these bluffs from 1907 until 1920. In the 1930s a major road was built along the same route. By 1957 this highway was closed because it cost so much to maintain. Those stretches that have not been destroyed by slides can still be seen, somehow hanging onto this near-vertical slope, far above the waves.

11. San Francisco to Año Nuevo

Kenneth R. Lajoie and Scott A. Mathieson

During the last major ice age, about 20,000–15,000 years ago, a large quantity of water from the oceans was stored on the continents as glacial ice, and, as a consequence, sea level was about 300 feet lower than it is today. During that time, the coastline of the area now called San Mateo County was situated 7–25 miles west of its present location. As the large continental glaciers melted at the end of the ice age, sea level rapidly rose and reached its present elevation about 5,000 years ago. Therefore, all the distinctive coastal landforms (seacliffs, seastacks, beaches, and dunes) found along the San Mateo coast today were formed by coastal processes (wave erosion, littoral drift, and wind transport) over only the past 5,000 years, a very short period of time in geologic terms. We create coastal hazards by building structures that can be damaged or destroyed by these and related natural processes (slope failure and flooding). Prudent land use demands that we understand the various natural processes active in the coastal zone and then recognize the potential adverse consequences of ignoring or changing them.

Regional setting

The Pacific coastline of San Mateo County extends about 56 miles from the city and county of San Francisco on the north to Santa Cruz County on the south (fig. 11.1). In map view, this roughly north-south trending rocky coastline consists of 9 geometrically distinct segments (fig. 11.1) that reflect local variations in coastal processes and geologic conditions, primarily topography and rock type. The distinctive character of each coastal segment generally reflects the local balance between the erosional power of the nearshore marine processes and the erosional resistance of the cliffs, bluffs, or beaches exposed to wave attack along the shoreline. Overall, this coastline is sand-deficient, and only locally do broad beaches protect the sea cliffs from wave erosion.

In most of the relatively straight or linear coastal segments (Daly City, Pacifica, and Half Moon Bay), sedimentary rocks with little or no erosional resistance are exposed to wave attack along the shoreline. In these areas dynamic factors, such as the size, direction, and frequency of storm waves and ocean swells, control coastal configuration and produce the straight or only gently curved coastlines. Under natural conditions the rate of seacliff erosion is generally very low along these coastlines because a dynamic balance is attained in which sand derived primarily from local erosion forms permanent beaches that protect the vulnerable cliffs from all but the largest storm waves. However, the potential for rapid beach and seacliff erosion in these areas is very high, especially if natural wave patterns are altered

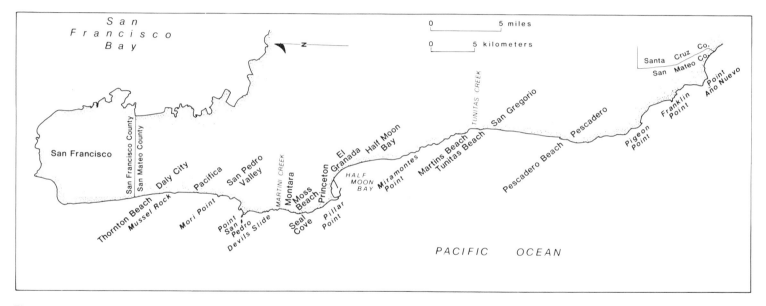

Figure 11.1   Location map for the coastline between San Francisco and Año Nuevo.

suddenly, as by construction of a breakwater or jetty such as at Half Moon Bay.

In most of the irregular cliffed coastal segments (Devils Slide, Montara, and Pescadero to Franklin Point) rocks highly resistant to erosion are exposed to wave attack along the shoreline. Here, sea cliff erosion is minimal, and even drastic changes in wave patterns would have little effect on coastal configuration.

In the remaining coastal segments (Point San Pedro, Moss Beach–Pillar Point, Miramontes Point–Pescadero, and Point Año Nuevo), rocks with moderate erosional resistance are exposed to wave attack along much of the shoreline. In these areas, which are characterized by high, vertical seacliffs with numerous sea caves and arches, the coastal features are variable. Generally, small, rocky headlands separate small pocket beaches. Here, geologic conditions and rates of seacliff erosion are also variable. Locally, natural erosion rates are the highest recorded along the San Mateo coast. The response of the steep seacliffs to changes in wave patterns in these areas is unpredictable, but could be severe locally.

Slope failure, though not uniquely a coastal process, contributes significantly to the formation of many coastal landforms and is a major contributor to geologic hazards along the San Mateo coast. Large, deep-seated coastal landslides, which form only where weak or highly fractured rocks are exposed in high coastal bluffs, occur at Thornton Beach, Mussel Rock, Devils Slide, Seal Cove, and Tunitas Creek (fig. 11.1). Minor landslides and blockfalls occur along steep seacliffs that are retreating at slow-to-intermediate rates under persistent wave attack. These conditions exist at Pacifica, Moss Beach–Pillar Point, Miramontes Point–Pescadero State Beach, and the south shore of Point Año Nuevo. Sudden roof collapse of sea caves occurs occasionally along the vertical seacliffs between Miramontes Point and Tunitas Creek and along the south shore of Año Nuevo. The only parts of the San Mateo County coastline with no slope failures are the high, stable cliffs just south of Devils Slide and the low, stable bluffs between Pescadero State Beach and Franklin Point.

Seacliff erosion and related slope failures are the primary geologic processes related to hazards and risks of land use along the coastline of San Mateo County. Historical maps and photographs, and old structures such as early railroad and highway grades, provide most of the information for measuring long-term (80–140 years) rates of cliff erosion in this area. We report rates of seacliff erosion simply as inches per year on the accompanying strip maps, but we must be careful in how we interpret or use these data. Cliff erosion is very irregular in both space and time, so we should not expect the yearly increment of erosion along any particular segment of coastline to equal the value we report here. For example, if we report an erosion rate of 12 inches per year we do not mean to imply that the cliff has eroded back 12 inches every single year. More likely, one storm every 10 years removes 120 inches (10 feet) of cliff, or one storm every 50 years removes 600 inches (50 feet). Also, one small section of coast may be cut back a certain amount one year and an adjoining section cut back the same amount a few years later. Over the long term, these episodic increments average out to the rates we report here; in a few places, where historical data are sufficiently detailed, we can document this episodic nature of cliff erosion (Seal Cove, Princeton, and El Granada). The heavy winter storms of 1981–82 and especially 1982–83 caused severe cliff erosion in many parts of coastal San Mateo County that had little or no erosion over the past 100 years. We discuss these data in the text, but exclude them from the erosion rates listed on the strip maps.

We document landslide movement by observing damage to railroad grades, highway grades, and buildings. Even though slope failure is not a uniquely coastal process, it is ultimately caused by wave erosion along coastal bluffs, and is a major factor in assessing land-use hazards and risks in coastal San Mateo County.

Following is a brief history of human occupation and cultural land use in coastal San Mateo County that illustrates the nature and quality of historical data used to assess coastal hazards and risks in this area.

Cultural history and data sources

The original inhabitants of the central California coastal region probably arrived before sea level reached its present elevation about 5,000 years ago. Therefore, many of their older coastal habitation sites either lie below present sea level or have been destroyed by seacliff erosion over the past 5,000 years. However, numerous younger sites, primarily shell middens (refuse heaps), occur along the San Mateo coastline; several are presently being destroyed by wave erosion, while others are being covered or exposed by beach and dune activity. Unfortunately, none of the archeological data long the San Mateo Coast yield long-term rates of coastal erosion, but one site in nearby northern Santa Cruz County indicates that seacliff retreat was extremely rapid just after sea level reached its present elevation about 5,000 years ago, and has been relatively slow ever since, roughly the past 4,000 years; some of the cliffed coastlines in central California may have retreated landward as much as 1 mile to their present locations between 5,000 and 4,000 years ago.

The early European explorers of the California coast (1543–1776) left no detailed charts or maps of the San Mateo coastline. However, the written records of some of the early expeditions contain some useful information. For example, on 3 January 1603 the Spanish explorer Vizcaino sailed past the prominent marine headland at the southern tip of San Mateo County and named it Punta de Año Nuevo (New Year's Point). The written account of this voyage makes no mention of the island that lies off the point today, which suggests that it was connected to the mainland in 1603. Thus, Año Nuevo Island may have been formed by rapid seacliff erosion sometime after 1603.

During the time of the Spanish missions and the Mexican land grants in central California (1791–1822), several maps of the San Mateo coast were made, but none is sufficiently detailed to provide useful information on coastal changes. Two years after gold was discovered at Sutter's Mill (1848) in the Sierran foothills, California became the thirty-first state of the United States. The population of central California expanded rapidly in the 1850s, and detailed navigational charts were needed to guide the growing maritime traffic that supplied the gold fields through the port of San Francisco. In response to this navigational need, the U.S. government commissioned a series of hydrographic charts and topographic maps of the California coast at a scale of 1:10,000. The 7 detailed topographic maps (1853–66) in this series that cover the San Mateo Coast provide an excellent means of assessing coastal changes and for making quantitative measurements of long-term rates of seacliff erosion. These early hand-drawn maps are of such high quality that comparative studies are limited more by lack of detail on the highly generalized modern topographic maps (1:24,000) than by minor topographic inaccuracies on the older maps (fig. 11.15). All detailed erosion data derived from more recent sources (aerial photographs, subdivision maps, etc.) merely add short-term temporal data to the general erosion patterns and long-term erosion rates derived from comparisons between the original and more recent topographic maps.

During the late 1800s coastal San Mateo County was sparsely populated, mainly because it was physically isolated by the Santa Cruz Mountains that form the topographic backbone of the San Francisco peninsula; most of the area is still sparsely populated for the same reason. The mountainous barriers to overland transportation were so great that lumber and farm produce from this area were shipped to nearby San Francisco on small coastal schooners that were loaded by wire cables from long piers or from staging areas on high coastal bluffs. The scant remains of a few of these cargo-loading facilities provide some information on seacliff erosion.

In 1905 the Ocean Shore Railroad Company was incorporated in San Francisco to build a railroad along the coastline of San Mateo and Santa Cruz counties that would connect San Francisco and Santa Cruz. This railroad was never completed, but the railbed along the entire proposed route was excavated and rail was laid between San Francisco and Tunitas Creek, about halfway down the San Mateo coast. This section of the railroad was used for freight and passenger service from 1908 to 1921. The Ocean Shore Railroad Company went bankrupt in 1921, due in large measure to high costs of building and maintaining the railbed across the steep, unstable coastal bluffs south of Thornton Beach and near Point San Pedro in the northern part of San Mateo County. The abandoned railbed across the bluffs between Thornton Beach and Mussel Rock was converted to highway use and carried automotive traffic from the late 1930s to 1957, when landslides caused by a local earthquake closed the grade. One short section of railbed across the stable bluffs south of Point San Pedro was converted to highway use in 1939 and is still occupied by state Highway 1 today. Parts of this old railroad grade, especially those along the steep, unstable coastal bluffs in the northern part of the county, are still visible today and provide excellent data for measuring coastal erosion and assessing slope stability.

In the aftermath of the San Francisco earthquake and fire of 1906 there was widespread belief that many residents would leave the city out of fear that a similar catastrophe would happen again. This belief was incorrect, but it, and the access afforded by the new Ocean Shore Railroad, led to speculative land development along the San Mateo coast. Between 1906 and 1908 several new coastal communities along the rail line's route were incorporated and subdivided. Most of these communities never developed, but in some of them streets were excavated and concrete sidewalks were installed. These streets and sidewalks, and the detailed plot maps of some of these old subdivisions, provide valuable reference marks for quantitatively measuring local rates of seacliff retreat over the past 75 years.

Old photographs (1895–1950) from early geologic reports, business documents, and historical archives provide detailed information on seacliff retreat at several localities along the San Mateo coast. The first aerial surveys of this area were flown in the late 1920s, and comparisons between these and several sets of later aerial photographs provide time-lapse analysis of coastal changes over the past 60 years.

As mentioned previously, most of coastal San Mateo County is sparsely populated, and the few developed areas in the northern part of the area have been built up in only the last 30 years. This recent development took place with little regard for potential hazards. Consequently, much of our data on short-term erosion rates and landslide movement are derived from recently built structures that are presently being damaged or threatened. Also,

some of the most rapid seacliff erosion along this coast was artificially induced by construction of the breakwater at Half Moon Bay in 1959. Ironically, these recent sources of erosional information best illustrate the coastal hazards that could be avoided with prudent land-use policies.

In 1975 the California Coastal Zone Conservation Commission zoned the coastline of San Mateo County as part of the geotechnical element of its statewide comprehensive coastal zone management plan. This commission used a threefold stability classification (low, moderate, and high stability) for the coastline based on historical erosion rates and the inherent stability of the rocks exposed in the seacliffs (fig. 11.2). The coastal commission adopted land-use policies for coastal bluffs based on this stability classification (fig. 11.2). The commission placed the greatest land-use restrictions on those areas classified as unstable, and the least restrictions on those areas classified as stable (fig. 11.2). San Mateo County adopted the state commission's classifications and policies for the natural hazards element of its general land-use plan. The stability/risk zones outlined in the present study differ only slightly from those adopted and used by the California Coastal Zone Conservation Commission and San Mateo County. The high, moderate, and low hazard/risk categories in the present classification are roughly equivalent to the stable, intermediate, and unstable categories in the previous classification; the present classification simply refines and updates the previous classification. In this summary report we cannot comprehensively document the historical and geological information on which the

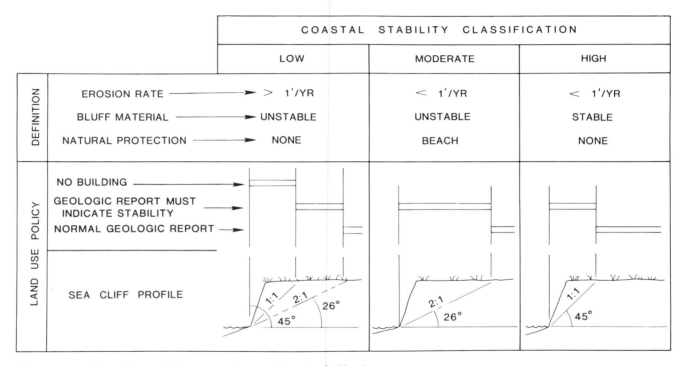

Figure 11.2. Shoreline stability categories used by the California State Coastal Zone Conservation Commission and by the County of San Mateo to regulate land use along the San Mateo County coastline.

coastal hazards classification is based. However, the photographs and maps that accompany the text illustrate the nature and quality of the historical data.

In the following section we briefly describe the geology, summarize the long-term (1850–present) and short-term (1982–83) seacliff erosion rates, and discuss the geologic hazards in each of the 9 coastal segments outlined in figure 11.1. This information is summarized diagrammatically on the strip maps of the San Mateo County coastline (figs. 11.3, 11.14, and 11.24).

Coastal hazards and risks

County line–Mussel Rock (fig. 11.3)

The straight 3.1-mile coastline between the San Mateo County line on the north and Mussel Rock on the south, essentially the corporate limits of Daly City, consists of a narrow, flat beach backed by high (125–730 feet), steeply sloping bluffs cut into weak sands and gravels (figs. 11.4, 11.5, and 11.6). A large block landslide forms a prominent step in the bluff face at Thornton Beach State Park in the northernmost part of this coastal segment (fig. 11.3), and a large debris landslide occupies a broad, bowl-shaped topographic depression in the southernmost part where the San Andreas Fault intersects the coastline just north of Mussel Rock (fig. 11.5). A large, deep gully evenly bisects the otherwise continuous bluff face between these 2 active landslides.

The abandoned grade of the Ocean Shore Railroad (1906–21) and state Highway 1 (1939–57) forms a prominent bench about 150 feet above sea level across the unstable bluff face and the active landslides. Residential subdivisions built between 1956 and 1960 occupy the rolling uplands at the crest of the bluff and near the head or top of the landslides.

The narrow, flat beach along this coastline does not protect the base of the bluff from seasonal wave attack. Therefore, wave erosion oversteepens the lower part of the bluff and triggers numerous landslides and debris flows that are slowly destroying the old railroad grade. Because the debris removed from the bluff face by wave erosion is derived from such a large area, the rate of actual cliff retreat is very slow. Therefore, most of the homes along the bluff top are not threatened directly by wave erosion. However, wave erosion contributes to the movement and enlargement of the 2 large landslides at Thornton Beach and at Mussel Rock, and the resultant landward expansion of each landslide has damaged streets and homes since the area was developed in the late 1950s (figs. 11.4 and 11.6). Several homes have been removed from these areas to prevent their complete destruction, and many others are presently being damaged and threatened by continued landslide movement.

During heavy storms in January 1983 wave erosion and groundwater saturation caused the large slump block of Thornton Beach State Park to move. At the southern end of the landslide, a drop of 21 feet destroyed the access road to a parking lot situated on the landslide itself. Recent headward expansion of the slide's northern end threatens Skyline Boulevard near the San Mateo County line. During the 1983 storms wave erosion destroyed 2

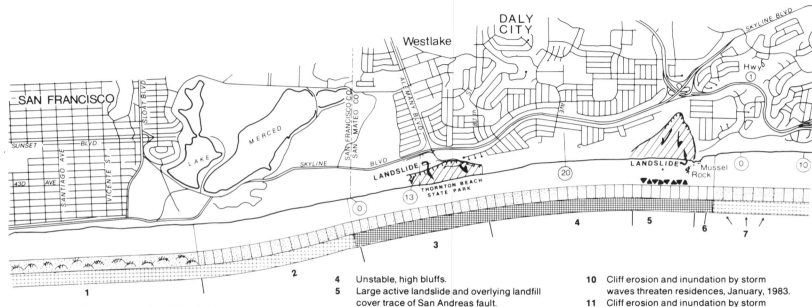

1. Sandy beach of variable width backed by eroded dunes and highway.
2. Beach backed by high, erodible bluffs.
3. Large landslide. Smaller landslides damage and threaten highway, access road to state beach, streets, and residences, 1970–1983. Several residences at top of bluff relocated 1973–1976.
4. Unstable, high bluffs.
5. Large active landslide and overlying landfill cover trace of San Andreas fault.
6. Cliff erosion threatens road.
7. Active erosion of large gullies.
8. Toe of seacliff eroded 33 feet by storm waves, January, 1983.
9. Seacliff protection and riprap below mobile home park periodically damaged by storm waves, 1970–1982.
10. Cliff erosion and inundation by storm waves threaten residences, January, 1983.
11. Cliff erosion and inundation by storm waves threaten streets and residences 1970–1983. Riprap covering low cliffs below street built 1982–1983.
12. Marked erosion of beach, low cliff, and dunes by storm waves, January, 1983.
13. Cliff erosion by storm waves, January, 1983. Residence threatened.
14. Unstable coastal bluffs. No useful facilities threatened by landslides.

Figure 11.3. Site analysis: San Francisco through Shelter Cove.

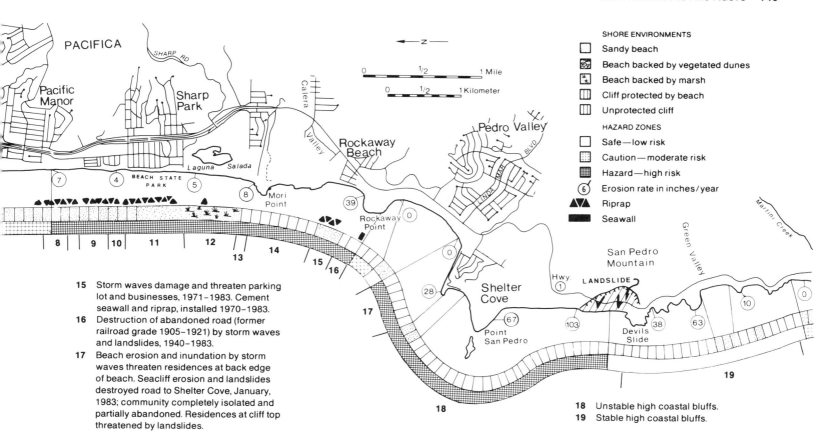

SHORE ENVIRONMENTS

- Sandy beach
- Beach backed by vegetated dunes
- Beach backed by marsh
- Cliff protected by beach
- Unprotected cliff

HAZARD ZONES

- Safe—low risk
- Caution—moderate risk
- Hazard—high risk
- ⑥ Erosion rate in inches/year
- ▲▼▲ Riprap
- ▬ Seawall

PACIFICA

Pacific Manor

SHARP RD

Sharp Park

Calera Valley

Laguna Salada

BEACH STATE PARK

Rockaway Beach

Pedro Valley

LINDA MAR BLVD

Mori Point

Rockaway Point

Shelter Cove

San Pedro Mountain

Green Valley

Martini Creek

Hwy. 1

LANDSLIDE

Devils Slide

Point San Pedro

15 Storm waves damage and threaten parking lot and businesses, 1971–1983. Cement seawall and riprap, installed 1970–1983.

16 Destruction of abandoned road (former railroad grade 1905–1921) by storm waves and landslides, 1940–1983.

17 Beach erosion and inundation by storm waves threaten residences at back edge of beach. Seacliff erosion and landslides destroyed road to Shelter Cove, January, 1983; community completely isolated and partially abandoned. Residences at cliff top threatened by landslides.

18 Unstable high coastal bluffs.
19 Stable high coastal bluffs.

Figure 11.4. Thornton Beach State Park in Daly City about 0.9 mile south of the northern San Mateo County line. The bluffs are cut into weak sandstones. The large step in the bluff face to the left is a large landslide slump block. The narrow bench to the right is the abandoned grade of the Ocean Shore Railroad (1906–21) and later State Highway 1 (1939–57), which was closed permanently in 1957 by slope failures triggered by a local earthquake. Several homes at the top of the bluff have been damaged by slope failures; 6 have been removed to avoid complete destruction and another is presently abandoned. Landslide movement caused by heavy rains and wave erosion at the base of the slope destroyed the access road to the state park during the winter of 1983. U.S. Geological Survey photo by K. Lajoie, March 1971.

Figure 11.5. Daly City bluffs about 0.9 mile south of northern San Mateo County line. Abandoned railroad and highway grade across the bluff face is gradually being destroyed by slope failures caused by wave erosion. U.S. Geological Survey photo by K. Lajoie, March 1971.

Figure 11.6. Daly City bluffs just north of Mussel Rock. Large, irregular depression in center and to left is a landslide that covers the trace of the San Andreas fault. The graded area in the lower part of the landslide is a sanitary landfill protected from wave erosion by riprap. Upslope expansion of the landslide has damaged several homes (built 1957–59) in upper right. The 2 homes at the extreme upper left were abandoned in early 1984, and at least 2 additional homes on either side of those already abandoned are close to this same fate. The vacant lots in the upper right center are where homes have been removed to avoid complete destruction. The house shown next to the empty land was recently abandoned; the street pavement in front of this house is presently breaking up; and 1 house across the street from those already removed is being pulled away from its foundation. Wave erosion at toe of slope contributes to slide movement. U.S. Geological Survey photo by K. Lajoie, March 1971.

pedestrian access routes from the parking lot to the state beach at the toe of this landslide.

Between 1866 and 1956, the head of the large gully bisecting the bluffs in the central part of this coastal segment eroded inland at an average rate of 20 inches per year. Severe erosion in this gully caused by heavy rains during the winter of 1981–82 threatened several homes in this area. An extensive landfill was emplaced in 1983, which stabilized this potentially serious condition.

In summary, the primary coastal hazard in this area is movement and headward or landward expansion of 2 large landslides that threaten numerous homes along the crest of high, unstable coastal bluff. Wave erosion contributes to the movement of these and numerous smaller landslides in the area.

Mussel Rock–Point San Pedro (fig. 11.3)

The 5.6-mile coastline between Mussel Rock on the north and Point San Pedro on the south, which corresponds to the corporate boundaries of the City of Pacifica (fig. 11.1), consists of a low, narrow coastal plain in its northern half and 2 small coastal valleys flanked by narrow ridges in its southern half. This low coastal area is backed by high, rolling hills incised by deep, narrow canyons. The northern coastal plain slopes gently southward from an elevation of 160 feet near Mussel Rock to below sea level at Laguna Salada, a natural lagoon north of Mori Point. The straight shoreline along the seaward margin of this narrow plain consists of a flat, sandy beach backed by low, vertical cliffs

Figure 11.7. Pacifica. About 1.2 miles south of Mussel Rock. Low cliffs cut into weak, fairly loose terrace sands and gravels. Broad beach protects vulnerable cliffs from all but largest storm waves. U.S. Geological Survey photo by K. Lajoie, March 1971.

Figure 11.8. Sharp Park, Pacifica. 0.7 miles north of Laguna Salada. Mobile home park, 1971. Note minor damage to seawall. See figure 11.9 for damage done during storms in January 1983. U.S. Geological Survey photo by K. Lajoie, March 1971.

cut into loosely consolidated sands and gravels (figs. 11.7, 11.8, and 11.9).

This area, which includes the local communities of Pacific Manor and Sharp Park, is densely populated and has numerous homes and businesses built close to the edge of the low, exposed seacliffs (figs. 11.7–11.11). Over the past 20 years a few riprap seawalls were built along the base of the cliffs to protect cliff-top property from periodic but generally minor wave erosion. However, during the severe storms in January 1983 wave erosion destroyed a seawall, an access road, and several mobile home

foundations at a mobile home development in Sharp Park (fig. 11.9). Twenty-three mobile homes were moved to prevent further damage. The toe of the cliff was cut back at least 100 feet at the northern end of this development. Just north of the mobile home park, a large 3-story building that was almost undercut by cliff erosion during these storms was also moved to prevent damage. South of the mobile home park, severe cliff erosion undercut and threatened several permanent homes during the 1983 storms (fig. 11.10); most of these homes are now protected by extensive riprap revetments.

Figure 11.9.   Sharp Park, Pacifica. Same locality as figure 11.8. Road and foundations damaged by waves during storms in January 1983. Toe of seacliff receded up to 100 feet in 1 storm. Mobile homes removed to prevent destruction. See figure 11.8. U.S. Geological Survey photo by K. Lajoie, April 1983.

Along the low cliffs of Beach State Park in southern Sharp Park, beach and cliff erosion threatened a public road during the January 1983 storms. Riprap was rapidly emplaced to curtail the damage to the road and to homes along its landward side (fig. 11.11). Just south of this area, heavy surf severely eroded the broad beach and inundated the windbreak along the spit at Laguna Salada during the 1983 storms. Several trees were toppled and beach sand was washed over a large part of the golf course around the lagoon. At the southern end of the spit just

Figure 11.10.   Sharp Park, Pacifica. These homes built in the late 1950s are adjacent to the trailer park shown in figures 11.8 and 11.9. Low seacliffs cut into weak terrace sands and gravels. Wave erosion was caused during heavy storms in January 1983. U.S. Geological Survey photo by K. Lajoie, April 1983.

Figure 11.11. Sharp Park, Pacifica. Approximately 0.3 miles north of Laguna Salada. Riprap (1982–83) protecting low cliffs cut into weak terrace sands and gravels. Since the late 1960s periodic wave erosion has threatened public roads and private property. During heavy storms in January 1983 waves washed across the road and into yards on the east side of the road. U.S. Geological Survey photo by K. Lajoie, April 1983.

north of Mori Point, beach erosion exposed a low cliff to wave attack in 1983. Here seacliff retreat, which amounted to at least 80 feet in 1983, now threatens a house and garage; yet for at least 100 years prior to 1983 there was virtually no cliff erosion in this area.

Calera Valley and San Pedro Valley (fig. 11.13) south of Mori Point are graded to sea level and terminate at the shoreline in broad pocket beaches flanked by rocky headlands. Dense commercial development in the community of Rockaway Beach extends down to the surf zone at the mouth of Calera Valley. In San Pedro Valley dense residential and commercial development is restricted to the east side of state Highway 1, which runs along the back edge of the beach, well behind the surf zone. However, a few homes and businesses lie on the beach berm west of the highway. At the south end of this beach and 0.4 miles to the west at Shelter Cove, several homes are built along the back edge of steep, narrow beaches. The abandoned rock-cut grade of the Ocean Shore Railroad forms a prominent bench around Rockaway Point between Calera Valley and San Pedro Valley (fig. 11.12) and along the steep north-facing bluffs of Point San Pedro.

Over the past 15 years wave erosion has periodically threatened and damaged a parking lot on the seaward side of a restaurant at Rockaway Beach. A low riprap seawall now protects this and more recent commercial development on the berm of this small pocket beach. However, waves have cut deeply into the low cliffs south of this seawall in the past few years.

Over the past 60 years erosion and landslides have destroyed

Figure 11.12. Rockaway Point. Narrow bench around point is remnant of Ocean Shore Railroad grade. A seawall is seen protecting the most rapidly eroding part of the railroad grade; date of wall construction is not known. U.S. Geological Survey photo by K. Lajoie, March 1971.

much of the abandoned railroad grade around Rockaway Point, but the rocky shoreline below the grade has not receded measurably during the same period (fig. 11.12).

During the heavy storms in January 1983 beach erosion and wave inundation threatened the homes at the back edge of the beaches at San Pedro Valley and Shelter Cove. During these storms, cliff erosion caused a major slope failure that destroyed a section of the access road (old railroad grade) to Shelter Cove and threatened homes at the top of the steep cliff above the road. The community of Shelter Cove is now completely isolated, and several homes there have been abandoned.

In summary, the main hazard in the coastal segment between Mussel Rock and Point San Pedro is the wave erosion that threatens and periodically damages structures along the low cliffs in Sharp Park, Rockaway Beach, the southern part of San Pedro Valley, and Shelter Cove.

San Pedro Mountain (fig. 11.13)

The 2.8-mile coastline between Point San Pedro on the north and Martini Creek on the south consists of steep (40–50 degree) coastal bluffs up to 900 feet high that plunge precipitously into the ocean along the western flank of San Pedro Mountain (fig. 11.2). The bluff face in the northern half of this coastal segment is composed of deformed and pervasively fractured sandstones and shales and is traversed by rock-cut remnants of the Ocean Shore Railroad grade 200–300 feet above sea level. The rock-cut grade of state Highway 1, built in 1939, traverses the southern part of this bluff face above the railroad grade, but joins it in a deep notch behind the small headland called Devils Slide (fig. 11.13A).

Numerous debris chutes scar this broad, unstable slope and a large, complex landslide occurs at its southern end, just north of Devils Slide. Wave erosion removes landslide debris from the base of these bluffs and oversteepens the lower part of the slopes by cutting into the fractured bedrock (fig. 11.13A). Over the long term, wave activity is responsible for the overall slope instability in this area, but over the short term it is not sufficient to cause specific slope failures. It is the cumulative effect of many small slope failures that is slowly destroying the old railroad grade, and it is periodic movement on the large landslide that is damaging the existing highway (fig. 11.13B). Most of these slope failures are caused by groundwater infiltration during heavy rains, not by wave erosion during individual storms.

Figure 11.13. (A) San Pedro Mountain. Unstable bluffs cut into highly fractured sandstone and shale; granitic rocks exposed in lower right. Remnants of Ocean Shore Railroad grade on left. State Highway 1 is closed periodically by slope failures on slopes to right. U.S. Geological Survey photo by E. Brabb, March 1970. (B) Same locality as figure 11.13A. Road damaged by landslide movement in January 1983. This section of Highway 1 is periodically closed due to slope failure caused mainly by heavy rains, but contributed to by wave erosion at base of slope. U.S. Geological Survey photo by K. Lajoie, April 1983.

Historically, San Pedro Mountain and its steep coastal bluffs restricted access from the north to the central and southern parts of coastal San Mateo County. As mentioned earlier, the Ocean Shore Railroad penetrated this barrier for 13 years before its bankruptcy in 1921. The present-day highway is closed periodically because of slope failures, especially where the roadbed crosses the active landslide just north of Devils Slide. The California Department of Transportation plans to permanently repair or relocate this stretch of highway in the near future to insure reliable access to the residential communities to the south.

The steep, irregular bluffs in the southern half of this coastal segment, between Devils Slide and Martini Creek, are composed of deeply weathered but stable granitic rocks. Here, Highway 1 follows the rock-cut grade of the Ocean Shore Railroad except where it lies on an artificial fill across Green Valley. We can document no historical changes along these precipitous but stable bluffs. However, rapid runoff from heavy rains in the winter of 1981–82 washed out the fill and highway grade across Green Valley.

In summary, the main hazards in the coastal segment between Point San Pedro and Martini Creek are the slope failures that periodically damage and close Highway 1 just north of Devils Slide.

Montara Valley–Pillar Point (fig. 11.14)

The 4.4-mile convex coastline between Montara Valley on the north and Pillar Point on the south consists of low, rolling hills truncated along their western flank by low-to-intermediate sea-cliffs. This area is sparsely to moderately developed with most of the population concentrated in the residential communities of Montara and Moss Beach. These 2 communities, and the smaller community of Seal Cove in the southern part of the area, were subdivided in 1908 but have acquired most of their population in only the past 20 years. In all 3 communities many homes and a few businesses have been built close to, and directly on, the sea-cliffs.

The geologic diversity of this coastal area is expressed by the variable nature of the shoreline; there are 3 distinct shoreline segments along this open coast. The straight 0.7-mile northern shoreline segment is a broad, sandy beach (Montara Beach) backed by low cliffs cut into loose sediments that cover the floor of Montara Valley. These sediments offer little resistance to wave erosion, but the long-term rate of cliff retreat is very low because the broad beach protects the cliffs from all but the largest storm waves. However, in January 1983 storm waves cut into these cliffs and threatened a restaurant built close to the cliff top near the south end of the beach. A large riprap wall now protects this structure from further cliff erosion. Runoff from heavy rains during the winter of 1982–83 washed out the artificial fills beneath Highway 1 where it crosses 2 deep gullies near the north end of the beach.

The 1.3-mile central coastal segment consists of a rocky shoreline backed by irregular cliffs cut into highly jointed, but stable

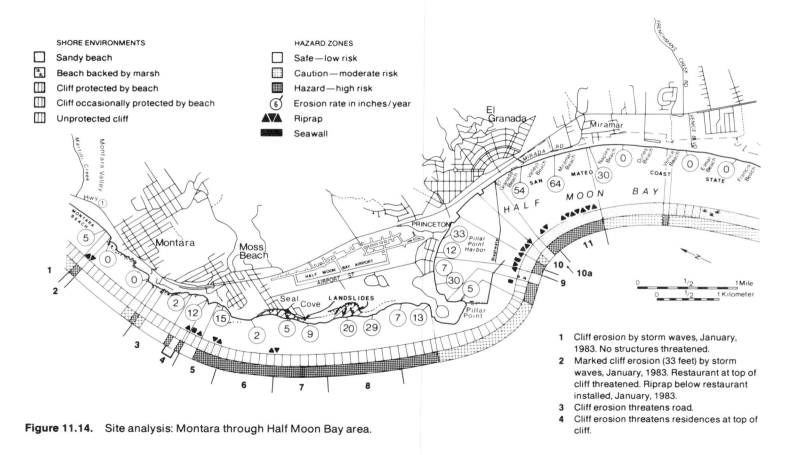

SHORE ENVIRONMENTS

- Sandy beach
- Beach backed by marsh
- Cliff protected by beach
- Cliff occasionally protected by beach
- Unprotected cliff

HAZARD ZONES

- Safe—low risk
- Caution—moderate risk
- Hazard—high risk
- ⑥ Erosion rate in inches/year
- ▲▼ Riprap
- Seawall

Figure 11.14. Site analysis: Montara through Half Moon Bay area.

1 Cliff erosion by storm waves, January, 1983. No structures threatened.

2 Marked cliff erosion (33 feet) by storm waves, January, 1983. Restaurant at top of cliff threatened. Riprap below restaurant installed, January, 1983.

3 Cliff erosion threatens road.

4 Cliff erosion threatens residences at top of cliff.

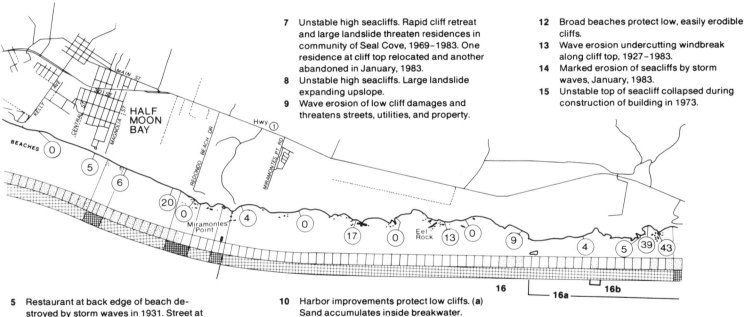

7 Unstable high seacliffs. Rapid cliff retreat and large landslide threaten residences in community of Seal Cove, 1969–1983. One residence at cliff top relocated and another abandoned in January, 1983.

8 Unstable high seacliffs. Large landslide expanding upslope.

9 Wave erosion of low cliff damages and threatens streets, utilities, and property.

12 Broad beaches protect low, easily erodible cliffs.

13 Wave erosion undercutting windbreak along cliff top, 1927–1983.

14 Marked erosion of seacliffs by storm waves, January, 1983.

15 Unstable top of seacliff collapsed during construction of building in 1973.

5 Restaurant at back edge of beach destroyed by storm waves in 1931. Street at top of cliff destroyed by storm wave erosion in 1957. Residences at top of cliff threatened by cliff erosion, 1957–1983. Riprap installed April, 1983.

6 Windbreak at cliff top undercut by wave erosion and landslides 1970–1983.

10 Harbor improvements protect low cliffs. (a) Sand accumulates inside breakwater.

11 Greatly accelerated seacliff erosion caused by reflection of storm waves off breakwater built in 1959. Cliff erosion and inundation by storm waves damage and threaten roads, utilities, and property. Riprap emplaced, 1977–1983.

16 Shoreline undeveloped. Minor seacliff erosion. Groundwater seeps destabilize cliff face locally. (a) Sea caves at base of high, vertical seacliffs. (b) Large amphitheater developed by wave erosion and landsliding between 1861 and 1927.

granitic rocks that are overlain by nonresistant sand and gravel. The granitic rocks are highly resistant to wave erosion, so seacliff retreat is minimal. However, the overlying deposits locally yield to wave attack during heavy storms. A few homes in the southern part of this area were threatened, and one was badly damaged, by erosion of these soft deposits during heavy winter storms in 1982 and 1983.

The 2.4-mile southern coastal segment consists of narrow, rocky beaches backed by steep seacliffs 30–130 feet high cut mainly into mudstone, which is covered by weak sand and gravel. The southern end of the linear ridge into which these cliffs are cut is the prominent headland called Pillar Point. Two broad landslide depressions indent the seaward side of this narrow ridge, and the community of Seal Cove lies partly within the northern landslide depression (fig. 11.15).

At the northwestern end of this ridge at Moss Beach, cliff retreat due to wave erosion in the low seacliffs has been very rapid. At one point the cliff has retreated about 150 feet in 120 years, which yields a long-term average erosion rate of 15 inches per year (figs. 11.15 and 11.16). In 1971 the planning department rejected permit applications to develop 3 residential lots above the cliff at this locality because of the potential erosion hazard. A restaurant built at the back of this beach in 1905 was destroyed by storm waves in 1957. Subsequently, wave erosion destroyed the public road at the top of the cliff behind the restaurant, and now threatens several homes close to the edge of the cliff. Severe cliff erosion during the heavy storms in January 1983 prompted

2 property owners to construct large riprap walls to protect their homes and remaining property.

Because the unstable seacliffs south of Moss Beach are not protected by beaches, long-term rates of seacliff retreat are fairly high, up to 8 inches per year over a 95-year period (fig. 11.15). Several residential lots and 3 homes near the top of the cliffs in Seal Cove have been damaged within the past 10 years by slope failures; 1 home was moved in 1983 to avoid complete destruction (fig. 11.17). However, even though cliff erosion is a serious problem in Seal Cove, the primary hazard is movement and headward expansion of the large landslide on which part of the community is built. A few homes have been badly damaged by this slope failure, and several more are threatened. The large landslide south of Seal Cove is also active and is expanding northward; here blocks up to 200 feet wide are sliding seaward. Wave erosion at the base of the steep cliffs and groundwater infiltration from above contribute to the pervasive slope instability in this area. The engineering department of San Mateo County has identified high-hazard zones along the top of the seacliffs in this area and has restricted development in the most unstable zones as a safety measure.

In summary, the hazards in the coastal segment between Montara Valley and Pillar Point are restricted to the southern part of the area and are related to rapid seacliff erosion and resultant slope failures. Existing local land-use restrictions should prevent the serious coastal hazards in the southern part of Moss Beach and in Seal Cove from becoming worse.

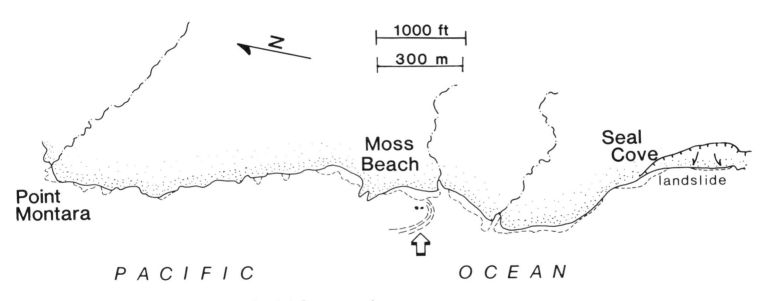

Figure 11.15. Moss Beach. 1866 Coast and Geodetic Survey map of area. Note U-shaped reefs near base of seacliff (see arrow). The cliff face (solid line) is now 180 feet away from large rocks in center of reef. These rocks were only 30 feet away from the cliff face in 1866.

Figure 11.16. Moss Beach looking south-southeast. The cliff face at the left is estimated to have eroded back 150 feet between 1866 and 1971. To the extreme lower left is a portion of the cliff face that has been coated with gunnite to reduce erosion and thus provide protection for a cliff-side home. U.S. Geological Survey photo by K. Lajoie, April 1983.

Figure 11.17. Seal Cove with Pillar Point in background. Unstable seacliffs composed of mudstone and covered by loose sands and gravels. Cliffs retreating at average rates up to 0.5 feet per year over the past 80 years. House at cliff top was damaged by slope failure and moved in 1983 to avoid complete destruction. U.S. Geological Survey photo by K. Lajoie, June 1982.

Half Moon Bay (fig. 11.14)

Half Moon Bay is an embayment between Pillar Point on the north and Miramontes Point on the south (fig. 11.18). The 6.5-mile smoothly arcuate shoreline of this open embayment is a broad, sandy beach backed by a narrow coastal terrace.

The northern half of Half Moon Bay is moderately developed with most of the population concentrated in the old but rapidly growing communities of El Granada and Half Moon Bay. At Princeton an expanding marina and harbor facility provides berths for numerous recreational boats and commercial fishing vessels. The area south of the town of Half Moon Bay is sparsely developed. Here, agriculture is the primary land use.

Half Moon Bay is constrained by headlands on the north (Pillar Point) and south (Miramontes Point). Between these rocky headlands, loose nonresistant sands and gravels occur at the shoreline and are exposed to wave erosion. The prevailing swell along the central California coast is from the northwest, and the waves bend or refract around the Pillar Point headland. The refracted waves erode the nonresistant deposits between the confining headlands into a simple, gradually tightening curve (log spiral curve) that evenly distributes wave energy along the shoreline. In this configuration the crests of the waves are parallel to the beach in all places. Sand periodically eroded from the soft cliffs accumulates to form a broad permanent beach that protects the cliffs from all but the largest storm waves, which usually approach the coastline from the southwest. Under these conditions the rate of seacliff retreat is very low, even though the cliffs themselves offer little or no resistance to wave attack.

These equilibrium conditions prevailed in Half Moon Bay prior to the construction of the Pillar Point breakwater at the northern end of the bay in 1959. Between 1866 and 1959 the only significant change in Half Moon Bay occurred along the northern, tightly curved portion of the shoreline (figs. 11.18 and 11.19). The small amount of cliff retreat recorded here was probably caused by erosional retreat of the controlling headland at Pillar Point. Along the straight shoreline in the southern part of Half Moon Bay, cliff erosion was and still is localized and very sporadic. Near the town of Half Moon Bay a windbreak along the undeveloped cliff top is being undercut by periodic storm waves. Here and 1 mile to the south, near Miramontes Point, waves cut deeply into the soft cliffs during the heavy storms in January 1983.

The riprap breakwater built across the northern tip of Half Moon Bay in 1959 disrupted the equilibrium wave pattern and focused wave energy at the low cliffs south of the breakwater. Here, where a county road was destroyed and state Highway 1 threatened, rates of seacliff retreat jumped from 3 inches per year to as high as 80 inches per year (figs. 11.20 and 11.21). Periodic dumping of concrete and riprap failed to check this erosion (fig. 11.22). More substantial riprap emplaced after the heavy storms in 1982 and 1983 will probably protect Highway 1 where it passes within 30 feet of the present shoreline near the eastern end of the breakwater. The accelerated cliff erosion ex-

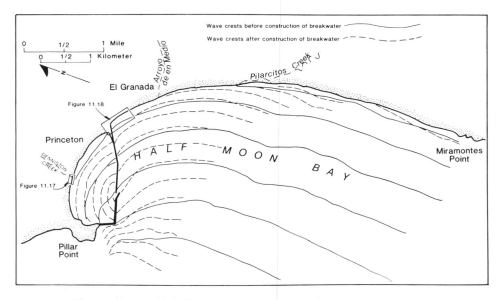

Figure 11.18. Half Moon Bay between Pillar Point and Miramontes Point. Wave crests originating from the northwest are shown refracting (bending) around Pillar Point before the construction of the harbor breakwater (solid curving lines) and after construction of the breakwater (dashed lines inside and outside the harbor). This bending causes the waves to have a head-on impact on the cliffs adjacent to the east breakwater. Serious erosion has occurred in these unconsolidated bluff materials since 1956–60. The locations of figures 11.19 and 11.20 have also been included.

tends 1 mile south of the breakwater to the community of Miramar where another part of the county road, several homes and businesses, and an apartment building are threatened (fig. 11.23). Riprap emplaced between 1978 and 1983 offers only temporary protection for these exposed structures.

Surprisingly, construction of the breakwater in 1959 did not eliminate erosion of the low, unprotected seacliffs inside the enclosed harbor. Roads, utilities, and private property in the harbor community of Princeton have been damaged and are still threatened by seacliff retreat that averages 33 inches per year in places (fig. 11.19). Extensive harbor improvements now protect all but the most quickly eroding part of the low cliffs (at the foot of Broadway Avenue) in this area.

In summary, the main coastal hazard in Half Moon Bay is the accelerated cliff erosion caused by the construction of the breakwater at Princeton Harbor between 1956 and 1960.

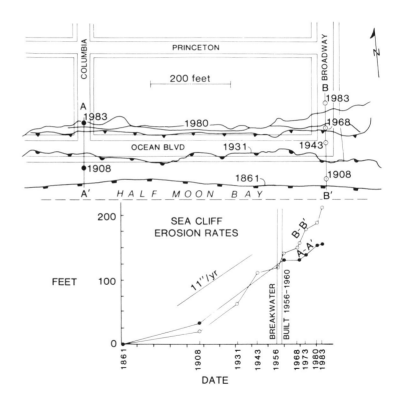

Figure 11.19. Seacliff erosion at Princeton, inside Pillar Point Harbor, Half Moon Bay. The rate of erosion did not change after construction of the Pillar Point Harbor breakwater between 1956 and 1960. The propagation pattern of the wave produced within the harbor (see fig. 11.18) has focused wave energy in the Broadway Avenue area since completion of the breakwater. During the heavy storms of the past 3 years 24 feet (96 inches per year) of cliff erosion has occurred at this point.

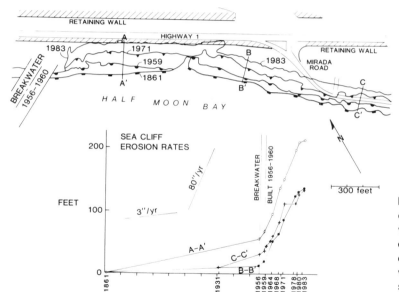

Figure 11.20. Seacliff erosion at El Granada in Half Moon Bay adjacent to and extending to the south of the Pillar Point Harbor breakwater. Erosion at El Granada State Beach increased markedly after construction of the breakwater between 1956 and 1960. Under natural conditions prior to 1956 these easily eroded cliffs were protected by wide beaches. The breakwater interrupted the littoral drift of sand southward from the Princeton cliffs and focused wave energy, which has resulted in severe cliff erosion.

Figure 11.21. This is one example of significant erosion that has occurred since completion of the Pillar Point Harbor breakwater. The man is standing on the west abutment of a bridge constructed in approximately 1914. There was never more than 10 feet of cliff seaward of this bridge at any time during the last 120 years. The effects of the breakwater were not noted here until the late 1960s, at which time previously insignificant erosion rates of 1–2 inches per year increased to 1–3 feet per year, and then between 1971 and 1972 became enormous at 16 feet per year. U.S. Geological Survey photo by K. Lajoie, March 1980.

Figure 11.22. El Granada in Half Moon Bay, February 1976. Mirada Road was damaged and is now completely destroyed by wave erosion after breakwater was built 0.5 miles to the north in 1959. The riprap on the beach was placed at the base of the cliff in the late 1960s and remains precisely where placed; the cliff on the other hand has retreated approximately 90 feet since that time. See figure 11.19 for erosion rates. U.S. Geological Survey photo by K. Lajoie, March 1980.

Figure 11.23. Miramar in Half Moon Bay, February 1976. Prior to 1959 these low cliffs were covered by dunes. Construction of a breakwater 0.9 miles to the north in 1959 disturbed wave patterns in Half Moon Bay, which resulted in beach and cliff erosion. By 1984 the cliff face had eroded back to the center of this building built in 1972. The trees seen here were never farther than 10 feet from the cliff face (during their lifetime or the past 75 years). Both of these trees were toppled in January 1983. Erosion here will likely increase drastically during the next winters, similar to 1982 and 1983. U.S. Geological Survey photograph by K. Lajoie, March 1980.

Miramontes Point–Tunitas Creek (figs. 11.14 and 11.24)

The 6.2-mile coastline between Miramontes Point on the north and Tunitas Creek on the south consists of steep-to-overhanging sea cliffs (70–160 feet high) that back small pocket beaches separated locally by sharp, rocky headlands. These precipitous cliffs truncate and form the southwestern margin of a narrow coastal terrace that is backed by rolling hills with broad, open valleys. The streams from these valleys cross the flat coastal terrace in deeply incised ravines that locally plunge over the steep seacliffs as small seasonal waterfalls.

This sparsely populated coastal area is traversed from north to south by state Highway 1, which runs along the landward margin of the coastal terrace, and by the abandoned grade of the Ocean Shore Railroad, which runs midway between the highway and the steep seacliffs. Artificial fills supporting the railroad grade across swales and gullies provide convenient empoundment for several stock and irrigation ponds. A rock-cut portion of the grade crosses the sheer cliff face north of Tunitas Creek. Agriculture is the primary land use in this area. Residential and recreational development is restricted to several summer homes on the high cliffs near the mouth of Tunitas Creek and at the back of Martins Beach, a shallow pocket beach about 1.6 miles north of Tunitas Creek. A residential golf course community is presently being constructed at Miramontes Point. Access to most of the isolated pocket beaches along this coastline is extremely limited and dangerous.

The sandstones and mudstones that form the seacliffs in this area are cut locally by numerous vertical joints that weaken the rock. Waves erode deep sea caves along these joints where they intersect the shoreline. Resistant zones in these rocks form a wide variety of arches, sharp headlands, and offshore stacks along this entire coastline. A veneer of sand and gravel up to 65 feet thick covers the older sandstone and mudstone in this area (fig. 11.25). Groundwater from precipitation and irrigation percolates downward through these permeable sands and gravels, flows laterally along the sandstone and mudstone, and emerges locally as springs in the vertical seacliffs. These springs contribute to cliff erosion by weakening the jointed rocks exposed in the cliff face.

The rates of seacliff erosion in this area are quite variable. Overall, changes in the seacliffs have been minor over the past 80 to 120 years, but locally some changes have been very dramatic. For example, about 0.6 miles north of Martins Beach, a large U-shaped cove has formed since 1861 when the first detailed map of the area was made. Erosion may have been initiated by collapse of a sea cave similar to the ones just south of this site. A long, narrow headland at the north end of Martins Beach has been reduced to an isolated stack over the past 120 years. On the other extreme, the abandoned rock-cut grade of the Ocean Shore Railroad across the vertical cliff face north of Tunitas Creek has not been destroyed by cliff retreat over the past 80 years.

On the shorter time scale, beach erosion and wave inundation have periodically threatened the summer homes at the back of Martins Beach over the past 15 years. Construction activity for a golf course clubhouse at the top of the seacliff at Miramontes Point caused a large section of cliff face to collapse; the proposed clubhouse was moved to a new site. These 2 examples not withstanding, the coastal hazards and risks in this coastal segment are very low, due primarily to sparse development and limited access to the shoreline. However, potential coastal hazards related mainly to unpredictable seacliff retreat are locally great.

Tunitas Creek–San Gregorio–Pescadero (fig. 11.24)

The straight 6.2-mile coastline between Tunitas Creek on the north and Pescadero Creek on the south consists of high coastal bluffs and vertical sea cliffs cut into the western flanks of high, rolling hills that are dissected by deep, narrow canyons and slightly broader valleys. Seasonal sand beaches extend along most of this coastline during the summer months. At the mouth of Pescadero Creek a permanent spit capped by dunes separates tidal lagoons and marshes from the open ocean.

This hilly coastal area is virtually undeveloped; sheep and cattle grazing are the primary land uses. State Highway 1, which locally runs along the crest of high vertical seacliffs, traverses the area from north to south. The only other permanent structures near the coastline are a few summer homes near the mouth of the Tunitas Creek. Access roads from the eastern part of the county intersect the coastal highway near the mouths of Tunitas, San Gregorio, and Pescadero creeks. Parking lots and paths to state

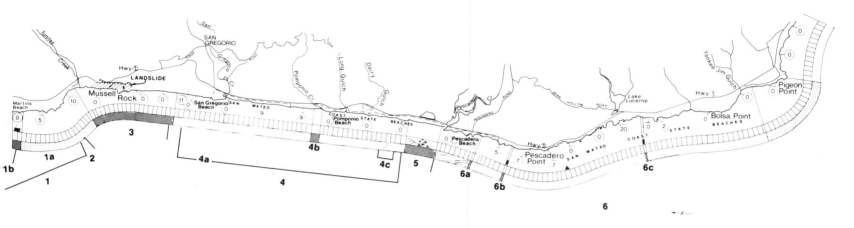

1   Shoreline undeveloped. Minor seacliff erosion. Groundwater seeps destabilize cliff face locally. (**a**) Sea caves at base of high, vertical seacliffs. (**b**) Beach erosion and storm waves periodically threaten private road and residences at back of Martins Beach.

2   Wave erosion at base of seacliff, 1982–1983. Residences at top of cliff not presently threatened.

3   Unstable, high coastal bluffs. Landslides gradually destroying abandoned railroad grade. Wave erosion destabilizes slopes below railroad grade. Large landslides cause some highway damage.

4   Shoreline undeveloped. Present open space land use not threatened by seacliff erosion. (**a**) Hazardous rockfalls from high vertical seacliffs onto public beaches. (**b**) Parking lot threatened by seacliff erosion, January, 1983. (**c**) Shallow sea caves.

5   Marked beach and dune erosion, January, 1983. Continued erosion would threaten highway.

6   Very stable irregular shoreline. Low sea-cliffs. Undeveloped. (**a**) Slow cliff erosion threatens highway. (**b**) Cliff erosion threatens residence. (**c**) Cliff erosion threatens residence.

Figure 11.24.   Site analysis: Martins Beach through Greyhound Rock area.

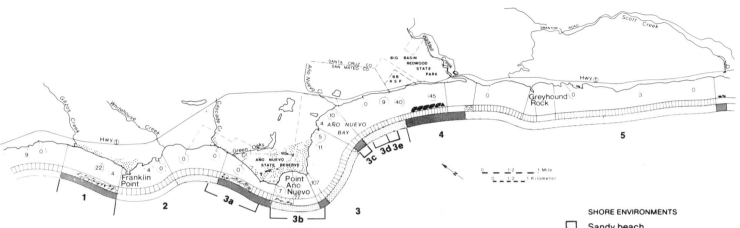

1 Marked beach and dune erosion by storm waves, January, 1983. Undeveloped shoreline; no immediate hazard.

2 Stable shoreline. Low seacliffs. Undeveloped.

3 Undeveloped shoreline. Open space and agricultural land uses not threatened by seacliff erosion. (**a**) Marked beach and dune erosion by storm waves, January, 1983. Undeveloped. No immediate hazard.

(**b**) Historical data suggest Año Nuevo Island may have been connected to mainland in 1603. Erosion rate of 9 feet/year for Point Año Nuevo. (**c**) Sea caves. (**d**) Sea caves. Large sea cave collapsed in April, 1971. (**e**) Marked beach and seacliff erosion by storm waves, January, 1983.

4 Highway One endangered by wave action during storms. About 2500 feet of road now protected by riprap; 2000 feet of this emplaced when highway was undercut during 1983 storms.

5 Steep mudstone cliffs 150 feet high with occasional pocket beaches. Measured erosion rates in recent years are low.

SHORE ENVIRONMENTS

☐ Sandy beach

▨ Beach backed by sand dunes

▨ Beach backed by vegetated dunes

▨ Beach backed by marsh

▥ Cliff protected by beach

▥ Unprotected cliff

HAZARD ZONES

☐ Safe—low risk

▨ Caution—moderate risk

▨ Hazard—high risk

⑥ Erosion rate in inches/year

▲▲ Riprap

■ Seawall

Figure 11.25. About 0.6 miles south of Martins Beach. Sea caves form where mudstone is weakened along vertical joints. U.S. Geological Survey photo by K. Lajoie, February 1976.

recreational beaches are situated near the mouths of San Gregorio, Pomponio, and Pescadero creeks. Sandstones and mudstones form the seacliffs along this entire segment of coastline. Locally, a thin veneer of terrace sand and gravel overlies these older rocks.

The rocks that form the 500-foot irregular coastal bluffs in the northern part of this area, between Tunitas and San Gregorio creeks, are sheared and fractured leaving these bluffs unstable and scarred by deep gullies and large landslides. Movement on 1 landslide damaged the coastal highway in the 1950s. The old rock-cut Ocean Shore Railroad grade (built in 1905-1908 but never used) that traverses these bluffs about 100 feet above the beach is being gradually destroyed by gullying and by small landslides caused by wave erosion at the base of the slope. This process was greatly accelerated by severe cliff erosion during the storms of January 1983.

The rocks exposed in the seacliffs between San Gregorio Creek and Pescadero Creek in the southern part of this coastal segment are fairly competent and are not highly deformed. Therefore, the seacliffs in this area are nearly vertical and locally overhanging broad, shallow sea caves that form along widely spaced joints in the rock. Seacliff retreat in this area is very slow and poses no immediate threat to the coastal highway. However, occasional block falls from the vertical cliffs are a hazard to the increasing number of people who use the beaches below for recreational purposes.

In January 1983 heavy storm waves cut deeply into the beach berm and dune ridge on the spit forming the lagoon near the mouth of Pescadero Creek. Continued erosion of this spit would threaten the coastal highway that runs along its landward edge.

In summary, there are no coastal hazards in this area because of the generally stable seacliffs and the lack of urban development. Any future development proposals should however carefully consider any potential erosion hazards.

Pescadero Creek–Franklin Point (fig. 11.24)

The irregular 11.2-mile coastline between Pescadero Creek on the north and Cascade Creek on the south consists of low, rocky cliffs and infrequent small pocket beaches backed by low, terraced hills. Sandstones, mudstones, and boulder conglomerates, which are highly resistant to wave erosion, form the cliffs along most of this open coastline. A thin layer of terrace sand and gravel overlies these older rocks and forms the upper part of the cliff face in some areas. At Franklin Point, near the southern end of this coastal segment, sand dunes now stabilized by vegetation overlie the terrace deposits.

Agriculture is the primary land use in this sparsely populated coastal area. In the central part of the area several permanent homes have recently been built west of state Highway 1, which traverses the region close to the shoreline.

Because the rocks exposed along the shoreline in this area are highly resistant to wave erosion, the long-term rates of seacliff retreat are virtually nil (figs. 11.26A and 11.26B). This shoreline probably has not eroded significantly over the past 5,000 years.

Figure 11.26A. Low, rocky cliffs about 3.2 miles south of Pescadero Creek, May 1905. U.S. Geological Survey photo by R. Arnold.

Figure 11.26B. This exposed shoreline has changed very little in 66 years (same locality as fig. 11.26A, November 1971). U.S. Geological Survey photo by K. Lajoie, November 1971.

However, wave surge during the severe storms in January 1983 locally eroded the weak terrace sands and gravels from the upper parts of the seacliffs, and threatened 2 residences and a short section of the coastal highway in the northern part of the area. Wave erosion during these storms also cut deeply into the sand dunes on the other shore of Franklin Point. These minor problems notwithstanding, actual and potential coastal hazards and risks in this coastal segment are very low due to the stable seacliffs and sparse development.

Point Año Nuevo (fig. 11.24)

Point Año Nuevo is the broad headland extending 1.7 miles from the foot of the rugged Santa Cruz Mountains near the southern boundary of San Mateo County. The flat surface of Point Año Nuevo is an old wave-cut platform or terrace. A stabilized and extensively quarried dune field covers the seaward half of this low headland. The entire Año Nuevo area is sparsely populated. Historically, agriculture was the principal land use in this area, but recently Año Nuevo Island and the western and southern parts of the headland were designated an open space and wildlife preserve by the State of California. Consequently, recreational uses such as hiking and picnicing have increased greatly in the last 10 years. State Highway 1 lies inland along the base of the Santa Cruz Mountains and runs close to the seacliffs only south of Año Nuevo where the mountains intersect the coastline.

The 1.0-mile northern shoreline of Point Año Nuevo is a straight, sand/gravel beach backed by low cliffs cut mainly into stabilized dunes. The 0.8-mile southwest-facing shoreline that forms the headland itself is a curved, sandy beach between 2 rocky points. Año Nuevo Island lies about 2,300 feet offshore from the southern point. The irregular 1.0-mile southern shoreline of Point Año Nuevo consists of low vertical-to-overhanging cliffs. Here, small headlands are cut into chert (a siliceous or glassy sedimentary rock used prehistorically by the original Indian population for making arrowheads and other tools), sandstone, and mudstone. Large sea caves, small arches, and a prominent seastack occur along this southeast-facing shoreline. A straight shoreline backed by high, vertical cliffs cut mainly into sandstones and mudstones and fronted by seasonal sandy beaches extends 1.2 miles southeastward from the base of Point Año Nuevo to the county line.

Because the shoreline of this coastal segment is completely undeveloped and will most likely remain that way, there are no major coastal hazards in the area. However, a few examples of dramatic coastal changes along this remote coastline are worth mentioning.

As mentioned earlier in this chapter, the written record of Vizcaino's 1603 voyage makes no mention of an island off the headland; therefore, the present Año Nuevo Island may have been part of the mainland at that time (fig. 11.27). The island was most likely formed fairly recently when southward enlargement of a circular embayment between the island and the northern point cut through a narrow peninsula composed of weak rocks that connected the island to the southern point. The rapid retreat of the southern point since 1853 is consistent with the recent separation of the island (fig. 11.27).

Because of southward littoral drift along the southern San Mateo County coastline, sand accumulates to form sandy beaches along the northern shorelines of Franklin Point and Point Año Nuevo. Under natural conditions, the prevailing northwest winds would blow this sand across these 2 low headlands in migrating dunes that eventually cascaded over the southern cliffs (fig. 11.28A). Dune migration was virtually halted on both headlands by natural and introduced vegetation in the first half of this century. Also, sand quarrying operations depleted the dune field on Point Año Nuevo in the 1950s. With their supply of sand cut off by dune stabilization, the broad, sandy beaches along both the north and south shores of Point Año Nuevo (fig. 11.28A) have virtually disappeared (fig. 11.28B). Surprisingly, we can document no increase in cliff erosion related directly to the recent and drastic depletion of these beaches. However, large waves cut deeply into the unprotected dunes on the north shore of Año Nuevo during the severe storms of January 1983. The once-vegetated cliffs north of the mouth of Año Nuevo Creek have also

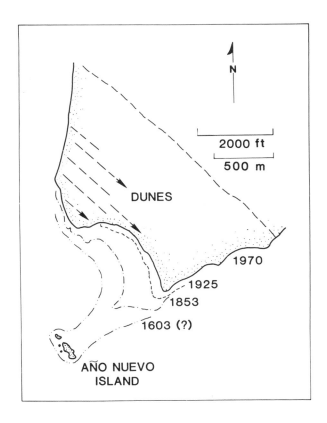

Figure 11.27.   Point Año Nuevo. The island was probably separated from the mainland after 1603 by wave erosion. Dashed lines are indicated dates. Arrows are alignments of dunes, which indicate prevailing winds.

A

Figure 11.28A. About 0.3 miles east of Point Año Nuevo, May 1905. Under natural conditions dunes derived from the beach on the north shore of the point cascaded over these low seacliffs to form broad beaches. U.S. Geological Survey photo by R. Arnold.

Figure 11.28B. Same locality as figure 11.28A, November 1971. Extensive quarrying operations in the 1950s and natural and artificially introduced vegetation have stabilized the dunes on this low headland. U.S. Geological Survey photo by K. Lajoie.

B

been eroded back significantly over the past 5 years. It is possible that the sudden roof collapse of a large sea cave on the south shore in 1971 may have been caused by increased wave erosion related to the recent disappearance of beaches in this area.

Summary and conclusions

The primary existing coastal hazards in San Mateo County include slope failures at Thornton Beach, Mussel Rock, and Devils Slide. Upslope expansion of the large landslides at Thornton Beach, Mussel Rock and Seal Cove have damaged and continue to threaten homes and streets at the top of the high coastal bluffs. Movement on the large landslide at Devils Slide periodically closes state Highway 1. Wave erosion at the toe of the high, unstable bluffs originally caused and subsequently maintained the slope instability in these 4 areas.

Wave erosion and inundation is a serious, long-term hazard at Sharp Park, Rockaway Beach, Seal Cove, and Half Moon Bay. Of these hazards, only the one at Half Moon Bay was caused by human activity, in this case the construction of the breakwater between 1956 and 1960. Large waves accompanying the heavy storms in the winters of 1982 and 1983 eroded many low-lying coastal bluffs that had not been threatened in historic times. Most of this damage took place on the north side of headlands indicating that the damaging waves came from the northwest; in the past most coastal damage was done by waves from the southwest.

South of Miramontes Point coastal hazards are presently minimal, mainly because of the low population density and sparse development. However, between Pescadero Beach and Franklin Point the low, exposed coastline has few hazards primarily because of the resistant nature of the rock exposed to wave erosion in the sea cliffs.

The hazard zones delineated on the accompanying maps modify and update the coastal stability categories established 10 years ago by the California State Coastal Zone Conservation Commission. It is hoped these revisions will contribute to the commission's pioneering work in minimizing coastal hazards in San Mateo County.

12. Año Nuevo to the Monterey Peninsula

Gary Griggs

History

The first inhabitants of the central California coast arrived well over 10,000 years ago. Archeological evidence, including radiocarbon dating, suggests that the initial immigrants were hunters, but that about 7,500 years ago a strong dependence on shoreline resources and seed gathering developed. Ancient kitchen middens (shell mounds), which clearly record both the diet and the waste disposal practices of the original inhabitants, are widespread from Año Nuevo south to Santa Cruz, in the midbay region, and also on the Monterey Peninsula. In addition to access to a rich intertidal area for food, the Indians sought two of the things we still search for today: freshwater and protection from the wind.

The first Europeans to reach the area, a Spanish expedition led by the Portuguese explorer Cabrillo, entered Monterey Bay in 1543. Sixty years later in 1602, Sebastian Vizcaino sailed into Monterey Bay and described it as "a harbor sheltered from all winds." Although it may have been somewhat protected from northwest winds during that voyage 380 years ago, the coastline of the bay has been repeatedly attacked and damaged over the years by severe storms from the west and southwest. Vizcaino

passed Año Nuevo Point on 3 January 1603 and gave it its present name (Punta de Año Nuevo). The early marine and overland expeditions along this coastline provided clear evidence for the rate of coastal retreat and shoreline changes at Año Nuevo Point. It seems apparent from the travel accounts of Father Crespi (1769), Father Palou (1776), and Captain Vancouver (1798) that the island was a peninsula in the not-too-distant past. As late as 1857 at least one detailed and official map shows no island, only a point connected by a sandbar. Now that the point here has been breached, erosion of the low terrace is proceeding extremely quickly, averaging about 9 feet per year for the past hundred years (fig. 3.8).

By the mid-1800s a variety of settlers had been attracted to the central coast from Año Nuevo to Monterey by the region's natural resources: fertile flatland and water for agriculture and grazing, redwood for lumber, and limestone for cement. Although these same natural resources are mainstays of the coastal economy today, central California's inland population was also attracted to the coast as a vacation area by the late 1800s. Hotels and resorts were followed by more and more permanent residents who built their cliff-top and ocean-view homes. Within the last 15 years a number of large-scale ocean-front developments have been proposed and built along much of inner Monterey Bay. These often consist of vacation or second homes. Vacant ocean-front real estate within or adjacent to urban areas, despite the hazards, commands astronomical prices. In June 1983 a beach-front vacant lot along Beach Drive in Rio Del Mar was for sale at $250,000 in an area partially inundated by high tides and storm waves a little over 4 months earlier. People will pay a lot for their ocean-front property, and often they have to pay again to protect it.

Año Nuevo to Natural Bridges State Park (figs. 12.1 and 12.2)

Physical setting and land use. The 25 miles of coastline between Año Nuevo and Natural Bridges State Beach is undeveloped for the most part. A sequence of up to 5 elevated marine terraces or benches used principally for agriculture (brussel sprouts and artichokes) and grazing flank the coastline throughout much of the area.

The small community of Davenport, which is built around a large cement plant, is the only town in the area. Captain John Davenport, a whaler from New England, established a whaling and lumber shipping settlement at Davenport Landing, about a mile and a half north of the present town, in 1867. A 450-foot-long wharf was soon constructed and was used as late as 1905. The sea, however, eventually claimed this pier, as well as the inner portion of another one constructed in 1934 at Davenport for shipping cement by sea (fig. 12.3). The cement is now shipped by trucks and rail.

A few older farmhouses scattered along the sheltered coastal valleys are now inhabited by the fieldworkers. Most of the coastal land is now owned by several large firms. The state recently purchased one of these large holdings for a state park— the

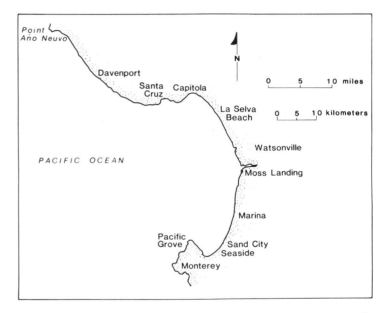

Figure 12.1. Location map for the coastline between Point Año Nuevo and the Monterey Peninsula.

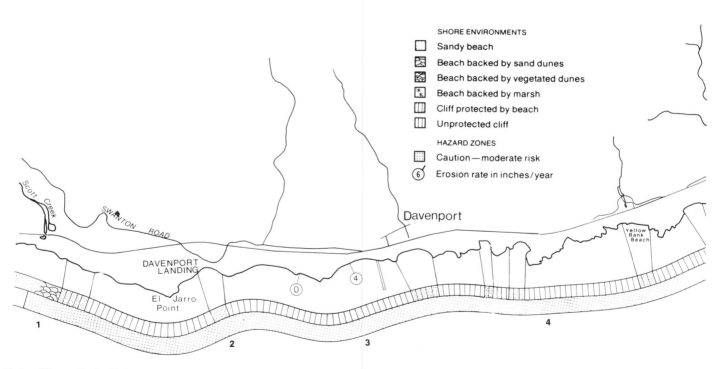

SHORE ENVIRONMENTS

Sandy beach

Beach backed by sand dunes

Beach backed by vegetated dunes

Beach backed by marsh

Cliff protected by beach

Unprotected cliff

HAZARD ZONES

Caution—moderate risk

⑥ Erosion rate in inches/year

Davenport

Scott Creek

SWANTON ROAD

DAVENPORT LANDING

El Jarro Point

Yellow Bank Beach

Figure 12.2. Site analysis: El Jarro Point through Needle Rock Point.

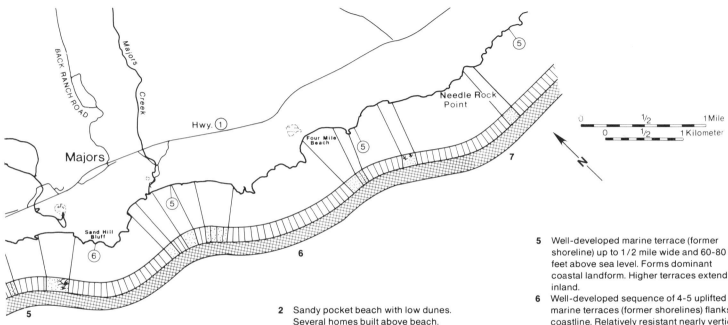

5 Well-developed marine terrace (former shoreline) up to 1/2 mile wide and 60-80 feet above sea level. Forms dominant coastal landform. Higher terraces extend inland.

6 Well-developed sequence of 4-5 uplifted marine terraces (former shorelines) flanks coastline. Relatively resistant nearly vertical mudstone cliffs periodically broken by sandy pocket beaches which form at stream mouths. Erosion along sandstone layers and faults has produced a scenic coastline. Cave or arch collapse produces rapid retreat at specific locations.

7 Coastal land from Wilder Creek to Majors Creek is now part of State Park system.

2 Sandy pocket beach with low dunes. Several homes built above beach.

3 Near vertical mudstone cliffs with small pocket beaches at mouths of coastal streams. Railroad embankments extend across stream valleys. Pier formerly used for loading cement onto ships now partially collapsed.

4 Railroad track undercut by wave action during major storm in 1973.

1 Sandy beach at mouth of Scott Creek flanked by sand dunes. Off road vehicle activity has destroyed vegetation such that sand blows onto Highway 1. Wave erosion has endangered highway at north end of beach, leading to riprap emplacement.

Figure 12.3. Remnants of a pier at Davenport formerly used to load cement onto ships. Photo by Gary Griggs.

Wilder ranch, immediately north of Santa Cruz—and plans for public use are now being developed. One of the major environmental factors controlling land use along the north coast of Santa Cruz is the strong wind that blows persistently across the terraces almost every afternoon from about March to October.

Geologic conditions. The bedrock exposed in the cliffs, which are 30 to 200 feet high along nearly this entire reach of coast, is a mudstone that is moderately resistant to erosion. More rapid erosion in zones of weakness within the mudstone, such as at faults, joints, and fractures, as well as at sandstone layers, has produced a variety of scenic arches, tunnels, and caves. A close inspection of a particular stretch of cliff where an arch or tunnel exists will usually reveal the zone of rock weakness responsible for the differential erosion (fig. 12.4).

A low rock bench or shore platform 50 to 100 feet wide (well developed at the north end of Natural Bridges State Beach) commonly occurs at the base of the seacliff and buffers against wave attack. Most of the year wave energy is expended against the outer edge of this bench or platform rather than against the seacliff (fig. 12.5). During major storms and high tides, however, water from breaking waves will reach across this platform to the base of the seacliff.

The nearly vertical seacliffs are broken a number of times by the sandy pocket beaches that form at the mouths of coastal streams. Beaches at the mouths of Waddell and Scott creeks are good examples. Some of the larger beaches are backed by active dunes and lagoons. For the most part, however, the dunes can-

Figure 12.4. A tunnel has been cut into mudstone cliffs north of Santa Cruz due to erosion along a sandstone dike (arrow). Photo by Gary Griggs.

Figure 12.5. A shore platform or shelf along the coast north of Santa Cruz protects the cliffs from direct wave attack. Photo by Gary Griggs.

not migrate due to the steep cliffs at the downwind (southwest) end of the pocket beaches. A small dune field that developed at the mouth of Scott Creek was stabilized by vegetation until recent years. The use of motorcycles and 4-wheel drive vehicles has virtually eliminated the vegetation, so that the sand dunes are now migrating onto Highway 1 and have created a constant maintenance problem (fig. 12.6).

Another problem location exists between Waddell Creek and the San Mateo County line, where high, steep cliffs flank the

Figure 12.6. Sand dunes encroach onto Highway 1 at the mouth of Scott Creek, providing a constant maintenance problem. Photo by Gary Griggs.

Figure 12.7. Highway 1 has been maintained along the base of the steep Waddell Bluffs only through the emplacement of riprap during 1983. Photo by Gary Griggs.

inland side of Highway 1. These actively eroding bluffs mark the former coastline and historically formed a barrier to transportation between Santa Cruz and San Francisco (fig. 12.7). The Ocean Shore Railway was built both southward and northward but was never able to bridge this gap. Travelers made the connection for years on a Stanley Steamer that crossed the beach below the bluffs at low tide. In 1947 the state removed some 1 million cubic yards of loose rock at the base of the cliffs and constructed the present Highway 1. Weathering of the cliffs leads to

continual sloughing of both small chips as well as large rocks. As the maintenance ditch at the base of the cliff becomes filled, rocks may begin to enter the roadway where they endanger motorists.

Cliff erosion and protection. Considerable wave energy is expended against the exposed stretch of seacliffs along the north coast of Santa Cruz County. Yet average erosion rates in most places are moderately low (3–6 inches per year or less), due to the erosional resistance of the mudstone and also the presence of the shore platform. Periodically an arch or sea cave collapses producing rapid retreat (up to 75 feet in one case) nearly instantaneously.

The cliffs from Pt. Año Nuevo to Waddell Creek, however, are eroding far more rapidly for several reasons. The San Gregorio Fault Zone separates the flat coastal terraces, which form the Año Nuevo Point area, from the mountains at this location. On the seaward side of the fault much of the cliff consists of poorly consolidated or weak material with little erosional resistance. As mentioned earlier, the Año Nuevo Point area has been retreating at about 9 feet per year for well over 100 years. This is probably the area of most rapid, long-term cliff retreat along the entire California coast. The low, unstable cliffs flanking Año Nuevo Bay are also rapidly eroding, with well-documented rates of 7 feet per year over the last 4 years.

Because the uplifted terrace land is used primarily for agriculture and grazing, the slow erosion of the seacliffs has produced no significant hazard or damage in this area. The greatest threat is to Highway 1 and the railroad between Santa Cruz and Davenport. Combined stream runoff and wave action in 1973 did undercut about 30 feet of railroad track just south of Davenport, leaving it dangling in midair. Riprap was subsequently emplaced and track replaced. Farther north along the steep bluffs just above Waddell Creek, a 600-foot stretch of riprap was emplaced during construction in 1947 to protect a section of Highway 1 exposed to wave action. During the winter 1983 storms, waves removed all of the loose rock and fill protecting about 2,000 feet of highway at the southern end of the bluffs. For years the rock debris coming off the cliffs was simply taken annually from the protective ditch along the roadway and dumped on the opposite side of the road into the surf zone. The coastal commission in recent years, however, has opposed this ocean disposal of the rock. Thus even some modest degree of "natural" protection was lost. Only through the installation of 24 thousand tons of rock riprap was Highway 1 saved in the winter of 1983 (fig. 12.7). With the exception of these 2 areas, however, the remainder of this coastline is unprotected. No immediate erosion threats exist, primarily due to the lack of development.

Natural Bridges to New Brighton Beach (figs. 12.2 and 12.8)

Physical setting and land use. The urbanized portion of Santa Cruz County begins at Natural Bridges and extends well into Monterey Bay. The 10 miles of coastline between Natural Bridges and New Brighton Beach consists primarily of cliffs that range in height from 25 to 75 feet and are often fronted by sandy beaches. The protection from northwesterly wave action offered by the headland of Lighthouse Point, the low relief at the mouths of the rivers and creeks, and the sand supplied by those streams all contribute to the formation of the normally wide, stable, protective beaches. These beaches and the protection offered around the northern margin of Monterey Bay (Santa Cruz to Capitola), combined with the excellent climate and closeness to San Francisco Bay area, have attracted tourists to the Santa Cruz area for over 100 years. Ocean-front hotels and motels, piers, casinos,

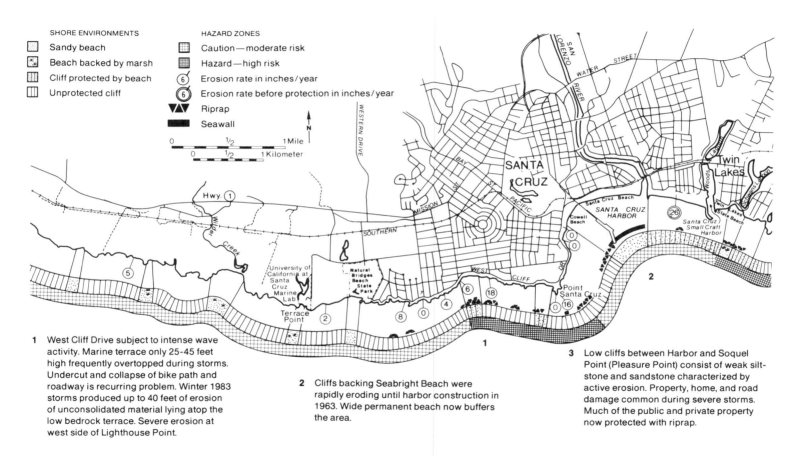

SHORE ENVIRONMENTS

- ▦ Sandy beach
- ▦ Beach backed by marsh
- ▦ Cliff protected by beach
- ▦ Unprotected cliff

HAZARD ZONES

- ▦ Caution—moderate risk
- ▦ Hazard—high risk
- ⑥ Erosion rate in inches/year
- ⑥ Erosion rate before protection in inches/year
- ▼▲▼ Riprap
- ▬ Seawall

1 West Cliff Drive subject to intense wave activity. Marine terrace only 25-45 feet high frequently overtopped during storms. Undercut and collapse of bike path and roadway is recurring problem. Winter 1983 storms produced up to 40 feet of erosion of unconsolidated material lying atop the low bedrock terrace. Severe erosion at west side of Lighthouse Point.

2 Cliffs backing Seabright Beach were rapidly eroding until harbor construction in 1963. Wide permanent beach now buffers the area.

3 Low cliffs between Harbor and Soquel Point (Pleasure Point) consist of weak siltstone and sandstone characterized by active erosion. Property, home, and road damage common during severe storms. Much of the public and private property now protected with riprap.

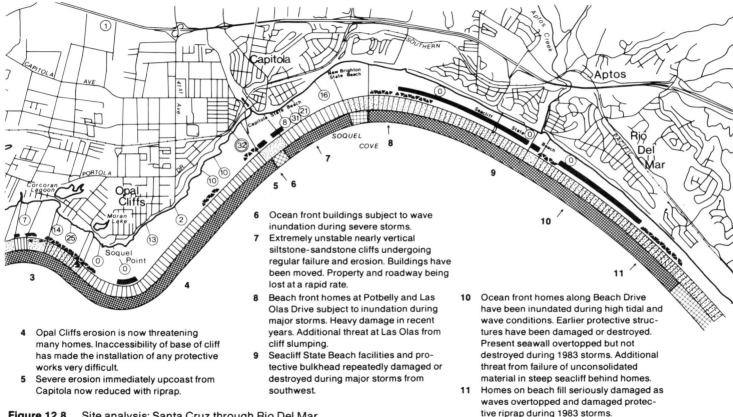

6 Ocean front buildings subject to wave inundation during severe storms.

7 Extremely unstable nearly vertical siltstone-sandstone cliffs undergoing regular failure and erosion. Buildings have been moved. Property and roadway being lost at a rapid rate.

8 Beach front homes at Potbelly and Las Olas Drive subject to inundation during major storms. Heavy damage in recent years. Additional threat at Las Olas from cliff slumping.

9 Seacliff State Beach facilities and protective bulkhead repeatedly damaged or destroyed during major storms from southwest.

10 Ocean front homes along Beach Drive have been inundated during high tidal and wave conditions. Earlier protective structures have been damaged or destroyed. Present seawall overtopped but not destroyed during 1983 storms. Additional threat from failure of unconsolidated material in steep seacliff behind homes.

11 Homes on beach fill seriously damaged as waves overtopped and damaged protective riprap during 1983 storms.

4 Opal Cliffs erosion is now threatening many homes. Inaccessibility of base of cliff has made the installation of any protective works very difficult.

5 Severe erosion immediately upcoast from Capitola now reduced with riprap.

Figure 12.8.   Site analysis: Santa Cruz through Rio Del Mar.

and the boardwalk have come and gone with fires and storms over the years. Old Victorian houses and many newer homes exist side by side along almost the entire cliff edge throughout this area. A few ocean-front lots still exist, and these command very high prices when sold, despite their limited life spans.

Geological conditions. The mudstone that makes up the cliffs along the north coast of the county extends a short distance into the city of Santa Cruz and underlies the West Cliff Drive area up to Almar Avenue (about three-fourths of a mile north of Lighthouse Point). At this point the rock type changes from the Santa Cruz Mudstone to the Purisima Formation, a sedimentary rock consisting of sandstones and siltstones with occasional layers or lenses composed almost entirely of mollusc shells. These fossil beds are particularly well exposed in the cliffs between Capitola and New Brighton Beach and are commonly seen as isolated fragments lining the walkways and front porches of countless beach cottages.

Much of both the east and west sides of Santa Cruz has been built on the first or lowest marine terrace, a flat benchland up to several miles wide. This terrace, which was formerly at sea level, provides evidence that the landscape is young and still rising out of the sea. Where the terrace is well exposed above the beaches along West Cliff Drive, the flat bedrock bench contains the well-preserved shells of boring clams. Uplift of the terrace was not uniform, however. In the East Cliff Drive area between Corcoran Lagoon and Moran Lake, the bedrock of the terrace is nearly at sea level, leaving only the loose sand and gravel deposits sitting atop the terrace to resist wave action. To the east the terrace climbs quickly, and from Opal Cliffs to Capitola the bedrock has been raised to heights of over 50 feet. This varying relief has produced widely varying erosion rates, with the lower areas usually eroding more rapidly.

Cliff erosion and protection. The coastal area from Santa Cruz to Capitola has been studied in detail because of both the relatively rapid rate of erosion and the impact of the cliff retreat on developed property. Good aerial photographic and map coverage extends back over half a century and provides a clear record of the coastline changes.

Three factors are of key importance in affecting the erosion rates in this area: (1) the abrupt geological transition to the soft sandstones and siltstones of the Purisima Formation, (2) localized weaknesses within the sedimentary rocks, and (3) the impact of human activities, in particular the placement of various engineering structures along the coast.

For the most part the long-term retreat rates of the mudstone cliffs extending from Natural Bridges to Almar Avenue are relatively low (4 to 8 inches per year or less on the average.) Photographs of some areas show no recognizable changes in over 50 years. When erosion does occur, however, it is usually episodic rather than gradual (fig. 12.9). For example, 2 of Natural Bridge's 3 arches collapsed during large storms. The major threats along West Cliff Drive are to the ocean-front roadway and bicycle path. There is only a single home built at the cliff edge on the seaward side of the road in the West Cliff area.

Figure 12.9. (A) Natural Bridge on the Wilder Ranch about 1890;
(B) same location in 1970. 1970 photo by Gary Griggs.

As soon as the weaker rock of the Purisima Formation appears, erosion rates increase markedly and the coastline changes direction as a result. Although inner Monterey Bay provides the weaker rocks making up the cliffs with considerable protection against the dominant northwesterly waves, these cliffs still erode faster than the exposed cliffs to the northwest. In part this is because they are simply soft and uncemented. Additionally, however, the northern bay provides no real protection for less frequent but not uncommon storm waves from the west or southwest. Because of the relatively rapid cliff erosion throughout this area over the years, considerable effort and expense have gone into trying to protect the cliffs through various means. In fact, of the 10 miles of coastline in this particular area, about 4.5 miles are protected one way or another. Although the effectiveness of the protection measures varies considerably, it is clear that they have significantly reduced or at least temporarily halted cliff erosion. Many of the rates listed on the map therefore represent erosion conditions prior to the emplacement of any riprap or seawall.

Rock or riprap was placed along much of West Cliff Drive about 1965; some areas were reprotected in 1981. During the winter of 1983, however, the combined high tides and extreme waves struck hard at the low bluffs and produced considerable retreat and damage to both the bike path and roadway (fig. 12.10). To illustrate the episodic nature of retreat and document the severity of the 1983 storm, we can look at one particular site, the end of Woodrow Avenue. From 1931 to 1982 the edge of the

Figure 12.10. Storm waves and high tides during the winter of 1983 destroyed the sidewalk and bicycle path at many locations along West Cliff Drive in Santa Cruz. Photo by Gary Griggs.

bluff had retreated about 25 feet, for an average of 6 inches per year. During the winter of 1983, however, over 18 feet of retreat took place over several months, which triples the "long-term" rate to almost 19 inches per year. This is why average erosion rates, particularly those based on a short time interval that may not include major storms, need to be used with caution.

The jetties of the Santa Cruz small craft harbor completed in 1965 have had a pronounced effect on coastal processes and erosion both immediately upcoast and for a considerable distance downcoast. The west jetty formed a major barrier to littoral drift coming down the coast and, as a result, created a wide stable beach (Seabright Beach) in an area previously undergoing rapid erosion (see fig. 5.22). A large portion of the 300,000 cubic yards of annual littoral drift fills the harbor entrance and requires annual dredging, now with a price tag of about $.5 million each year (table 5.1). The sand trapped behind the west jetty and in the harbor mouth has not been available on a year-round basis to downcoast beaches. As a result, cliff erosion rates increased significantly for a distance of about 4 miles downcoast, all the way to New Brighton Beach. Erosion rates in the weak layers of the Purisima Formation and along joints and fractures had already created an area of rapid cliff retreat; harbor construction added to the problem. As a result, most cliff-top homeowners have now had riprap emplaced to protect their homes. Costs and effectiveness have varied depending upon the amount, structure, and size of rock used. Many homeowners have had to replace or reinforce their protection after settlement and collapse. Following the 1983 storms, one property owner spent $100,000 to re-construct a massive rock barrier (fig. 1.2). This is clearly a high tax to pay for an unobstructed ocean view.

Immediately after harbor construction, the wide, normally year-round beach at Capitola disappeared. Consequently, the waves began to attack the ocean front structures and parking area. In order to alleviate this situation and to provide a beach for the summer tourists on which the community depends, Capitola eventually constructed a 250-foot-long groin at the down-coast end of their beach and brought in about two thousand truckloads of local quarry sand to build a beach. Capitola's sandy beach has returned and provides a protective buffer except during periods of severe storm waves from the west or southwest. This is exactly what happened during the winters of 1978 and 1983 when waves washed debris over a low seawall and into the downtown area (fig. 5.21).

Average annual erosion rates in this area prior to harbor construction and riprap installation varied between several inches and 1–2 feet. Some unprotected areas are still retreating at rates as high as 20 to 30 inches per year. One particular problem area is the stretch of high cliffs immediately east of Capitola. Apartments have been undercut, houses have been moved, roads have fallen in and entire parcels have disappeared (fig. 12.11). There is no permanent protective beach and the rock weaknesses particularly favor the failure of large blocks of rock (fig. 12.12). The city of Capitola is presently evaluating its options for protecting this rapidly retreating area.

Figure 12.11. These apartments at Capitola have been progressively undercut by continued cliff failure. The duplex on the right was moved off the site in the late 1970s due to continued erosion. Photo by Gary Griggs.

Figure 12.12.   Cliffs at Capitola are retreating rapidly through large-scale collapse of massive blocks of bedrock. Photo by Gary Griggs.

New Brighton Beach to the Pajaro River (figs. 12.8 and 12.13)

Physical setting and land use. The coast from New Brighton Beach to the Pajaro River constitutes southern Santa Cruz County and lies within the usually protected inner portion of Monterey Bay. The coastline follows a very gentle, smooth curve throughout this entire reach which tells us a good deal about its geologic development (see the following section). Wide, sandy beaches border this stretch of coast and protect the flanking seacliffs almost permanently. From New Brighton to La Selva Beach an uplifted marine terrace is the dominant coastal landform. A very steep cliff, approximately 100 feet high, forms the seaward edge of the terrace. To the south the terrace disappears, and a field of recent and active sand dunes mantles the low-lying area and dominates the topography.

Although both the seacliff and the beach itself have been heavily developed in the north (Rio Del Mar and Aptos Seascape, for example), much of the southern segment has historically remained in agriculture and park lands. Over the last 15 years, however, with the rush to the coast and the increased popularity of retirement or vacation homes, more and more developments of this sort have appeared on the drawing boards and, in a number of places, on the cliffs or sand dunes themselves. Major portions of the beaches have been set aside as state parks (New Brighton, Seacliff, Manresa, and Sunset state beaches). This has, however, not guaranteed protection either from the ocean waves or from encroaching development. Improved areas of Seacliff State Beach have been regularly damaged or destroyed during major storms, and homes and condominiums have been built on many of the cliffs and dunes overlooking the state beaches.

Geological conditions. The terrace and seacliff between New Brighton and Aptos Creek are composed of the moderately resistant sandstones and siltstones of the Purisima Formation. Although these rocks are identical to those making up the cliffs in the Santa Cruz–to–Capitola area, the cliffs backing the northern portion of the inner bay are more stable and vegetated because they are protected from wave action and erosion by wide, sandy beaches.

South of Aptos Creek the underlying material exposed in the bluffs is primarily weak, poorly consolidated sands that, for the most part, represent ancient sand dunes. This material is less stable than the bedrock to the north and can erode very quickly where it is subject to rainfall, streamflow, or wave action. It is also prone to slumping and sliding and, therefore, should be considered potentially unstable in any hillside or bluff area. Unfortunately, the loose and unstable nature of these sands has not been given much consideration either in the development of the cliffs themselves, or in beach-front construction at the base of these cliffs, along Beach Drive in Rio Del Mar, for example.

The Sunset Beach–Pajaro Dunes area forms the southern Santa Cruz County coastline. The dunes here are active, meaning that they are still connected to their beach sand sources and undergo periodic erosion under severe storm conditions, followed by subsequent buildup.

Coastal erosion and protection. The interior of northern Monterey Bay presents a clear example of the problems associated with constructing permanent structures on the beach. A wide, sandy beach, which is in equilibrium or balance with the predominant northwesterly waves, normally flanks this entire stretch of coast. These waves undergo considerable refraction or bending as they enter the bay, and in doing so lose much of their energy. The historic record, however, shows the repeated impact of storm waves from the southwest that remove much of the sand, carry large logs across the beach, and often reach the base of the sea cliff itself. The very presence of beach sand and stranded driftwood logs at the base of the seacliff are clear testimony to past storm activity. This backbeach area, now intensively developed, is equivalent to a river's floodplain. The question to ask yourself as a potential homeowner is not *if* it will be inundated, but how often, and to what depth?

For a distance of nearly 3 miles from Pot Belly Beach (just south of New Brighton State Beach) to Aptos Seascape, extensive public and private development has taken place on the backbeach area (fig. 12.14). Dozens of private homes, in addition to a recreational vehicle campground (Seacliff State Beach), a roadway, restrooms, and a major sewer line have been built on or buried beneath the beach. Storm damage in 1978 and 1983 was extensive in this area. A look at the historical record, however

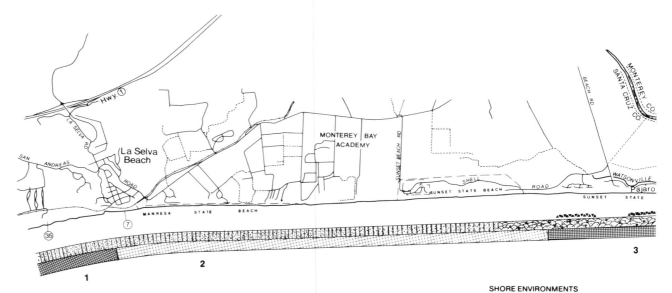

1   Cliffs consist of poorly consolidated sand and silt, subject to gullying or rapid erosion when protective beach removed.
2   Coastal land used primarily for agriculture and other lower density uses. Low bluffs consist of weak sands and silts. Subject to erosion.

3   Condominiums and homes built directly on active dunes subject to undercutting and erosion during severe wave and tidal conditions. 1983 storms eroded front face of dunes threatening dozens of structures. Several million dollars in riprap emplaced.

SHORE ENVIRONMENTS

☐   Sandy beach
▨   Beach backed by vegetated dunes
▥   Cliff protected by beach

HAZARD ZONES

▦   Caution — moderate risk
▦   Hazard — high risk
⑥   Erosion rate in inches / year
▲▲   Riprap
▰   Seawall

Figure 12.13.   Site analysis: La Selva Beach through Moss Landing.

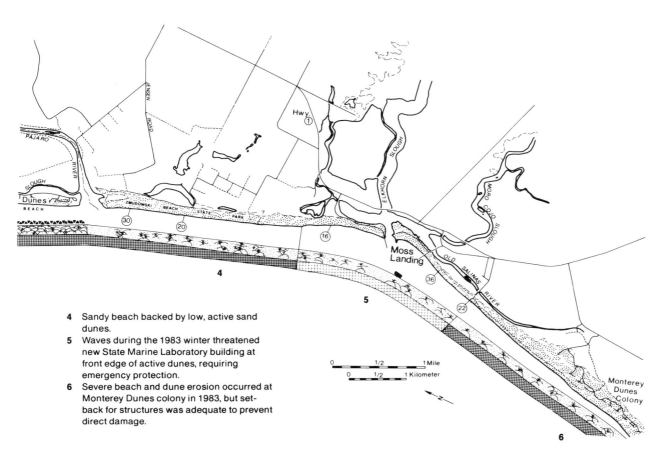

4 Sandy beach backed by low, active sand
dunes.
5 Waves during the 1983 winter threatened
new State Marine Laboratory building at
front edge of active dunes, requiring
emergency protection.
6 Severe beach and dune erosion occurred at
Monterey Dunes colony in 1983, but setback for structures was adequate to prevent
direct damage.

shows that these storms are not new to this portion of northern Monterey Bay.

The history of recurrent storm damage at Seacliff State Beach, for example, has been well documented. Ten times in 58 years, or about once every 6 years on the average, seawalls and bulkheads on the beach at Seacliff have been destroyed or heavily damaged. After extensive damage in 1939 and 1940, the seawall was rebuilt. Storms in the winter of 1941 destroyed it again (fig. 5.12). Following extensive damage to a piling and timber bulkhead in 1978 and again in 1980, a new bulkhead 2,600 feet long, identical to the previous structure, was reconstructed along with the campground. This new structure was built at a cost of $1.7 million and was supposed to last 20 years. In late January 1983, within 2 months of its completion and dedication, storm waves and high tides overtopped the bulkhead (fig. 5.11). Large logs battered the timbers and pilings loose, and about 700 feet of the bulkhead was once again destroyed. The parking lot, recreational vehicle camp sites, and restrooms were heavily damaged. Logs, sand, and debris were carried over and through the battered bulkhead to the seacliff. Damage costs have been estimated at $740,000.

Private development at Pot Belly Beach and Las Olas Drive immediately upcoast of Seacliff, and houses along Beach Drive just downcoast, have suffered in the same storms. Nearly 100 homes have been built over the years directly on the beach, some on pilings or piers, some on slabs. Many of these homes have suffered serious damage, some repeatedly, particularly during the severe winters of the past 5 years. Decks and stairways to the sand have been removed. Windows, doors and entire walls facing the ocean have been ripped out by wave action, leaving furniture afloat in seawater. In 1983, 2 houses on shallow pier foundations along Beach Drive were destroyed. They collapsed onto the beach when 6 feet of sand was scoured away, undermining their piers (fig. 5.2).

At Aptos Seascape, just to the south, 21 homes were built on fill above the beach in 1969 and protected by riprap. Political pressure led to approval of the project despite objections from the county planning department about the difficulty of guaranteeing

Figure 12.14. Along much of northern Monterey Bay homes have been built directly on the backbeach only a few feet above sea level. Photo by Gary Griggs.

partially collapsed (fig. 12.15). Damage estimates at Seascape range from $2 to $2.5 million.

In addition, about half of the riprap was destroyed as sand was scoured from beneath it and the rock tumbled onto the beach. In June 1983, less than 6 months later, county approval was given to a new seawall (fig. 12.16) that would extend 1,000 feet and cost $2.9 million; this amounts to $2,900 per front foot for protection, or over $100,000 per property owner! A few miles downcoast another expensive seawall was being rapidly built in November 1983 with hopes of forestalling future bluff erosion (fig. 12.17).

Without question there is a definite hazard associated with beach-front living in this area. Whether we like history or not, the record, the risks, and the costs are clear. No one should naively purchase beach-front property or a home without carefully looking at the impact of winter storms, particularly those since 1978.

The seacliffs behind the beach, which have been our concern throughout most of the remainder of Santa Cruz County, are a far safer environment than the beach along inner Monterey Bay. One important human factor has minimized the problem of wave erosion at the base of the cliff. The nearly continual row of homes along the beach itself has protected the cliff: waves must first scour away the sand, then remove or wash through the homes, before finally reaching the cliff. The aerial photo record of the past 45 or 50 years indicates cliff erosion of this sort has either been negligible or minor. The beach provides enough protection to keep bluff erosion due to scouring at the base from becoming a major problem. The winter of 1983 did produce some failure of

Figure 12.15. Home fronts at Aptos Seascape were heavily damaged during 1983 storms. Much of riprap ultimately failed. Photo by Gary Griggs.

protection from wave action. The homes were initially required to be set back 20 feet from the top of the riprap. In time this was reduced to 10 feet and then eliminated altogether. During the first southwesterly storms of 1983 in late January, 13-to-20-foot waves riding atop tides of 6.6 feet overtopped the protective wall. Nineteen of the 21 homes received major damage. In most cases the waves broke through the windows, doors, and house fronts facing the ocean and washed through the homes. One house

Figure 12.16. A new seawall was built in late 1983 at Aptos Seascape at a cost of $2.9 million, or $2,900 per front foot. Photo by Gary Griggs.

Figure 12.17. A steel pile and timber bulkhead was built in the early winter of 1983 to protect condominiums at Sand Dollar Beach. Photo by Gary Griggs.

this sort in the loose, sandy cliffs immediately downcoast from Aptos Seascape.

The cliff failure that has taken place in this area has resulted from surface runoff creating gullies, or intense rainfall that has produced slumping. The rainstorms of early January 1982 generated widespread failure in the loose cliffs above Beach Drive. Two homes at the base of the cliff were totally destroyed, while many others received heavy damage. Property owners atop the cliff lost backyards and some stood in danger of losing their homes. The cliffs have been heavily developed, and although photographic measurements indicate at least 40 years of stability,

the 1982 rainstorm changed that history in at least one area. A safe setback and the routing of all runoff back away from the cliff are wise precautions.

The sand dunes at the southern end of the Santa Cruz County coastline present a different set of concerns. The only development in the area, Pajaro Dunes, consists of 396 condominiums, 24 townhouses, and 145 single-family dwellings. All units are built on the active sand dunes with 66 houses and a number of the townhouses and condominiums built directly on the foredune above the beach. The pattern in this area over the past 50 to 75 years has been one of dune erosion or removal during severe storms, followed by gradual buildup of sand, or accretion. This is known as a dynamic equilibrium. Unfortunately, the condominiums and homes do not shift with the dunes and herein lies the problem. Since development was initiated in 1969, 3 winters of severe storms from the southwest (1969, 1978, and 1983) have eroded portions of the front face of the dune. The January 1983 storms cut back the dunes up to 40 feet, and left a near vertical cut 15 to 18 feet high that came right to the foundations of many of the homes (fig. 12.18). Only the emergency emplacement of thousands of tons of rock saved these expensive homes from disaster (fig. 12.19). At the end of the storm season, riprap and a revetment had been emplaced along the seaward frontage of this development at a cost of several million dollars (fig. 12.20). Any resemblance to a natural dune environment, however, had disappeared.

Although northern Monterey Bay is protected to some degree, as the early explorers noted, it is at the mercy of the southwesterly storms that have repeatedly damaged ocean-front developments throughout the area. Caution is advised prior to any investment in a coastal home or property in this area. Although many sites are safe and without problems, many others only survive because of massive and repeated efforts in costly protective works.

Figure 12.18. Storm waves and high tides in 1983 cut back the dunes severely at Pajaro Dunes. Photo by Gerry Weber.

Figure 12.19. The emergency dumping of thousands of tons of riprap in 1983 saved Pajaro Dunes from additional dune erosion. Photo by Rogers Johnson.

Figure 12.20. In December 1983 Pajaro Dunes was temporarily protected but the dunes are badly eroded. Photo by Gary Griggs.

Pajaro River to the Monterey Peninsula (figs. 12.13 and 12.21)

Physical setting and land use. The coastline from the Pajaro River to the Monterey Peninsula consists of a broad coastal lowland fronted by wide, sandy beaches. A wide belt of dunes, commonly rising as high as 100 feet, flanks the beaches. The broad beaches and dunes are the result of large quantities of sand brought to Monterey Bay over thousands of years by the Salinas and Pajaro rivers, as well as littoral drift from the north. The combination of abundant sand, an extensive low area behind the beach in contrast to cliffs, and a dominant wind direction from the northwest has enabled these dunes to form and migrate.

Between the mouths of the Salinas and Pajaro rivers there are a number of lagoons or sloughs that are remnants of former estuaries. These brackish water bodies (particularly Elkhorn Slough) are important refuges for aquatic birds and other wildlife.

Rivers along this coast have very little if any runoff during much of the year, so that waves and longshore currents commonly block their mouths with sandbars. During the early part of the century the mouth of the Salinas River flowed northward in the lower part of its course, paralleling the shoreline and only separated from the sea by a narrow stretch of dunes (fig. 12.22). The river actually discharged into Elkhorn Slough until about 1910 when it broke through the dunes at approximately its present location. Ultimately a dike was constructed to prevent the river's northward flow into its old channel.

Immediately offshore from the mouth of Elkhorn Slough lies the head of Monterey Submarine Canyon, one of the world's deepest and largest underwater canyons. This feature has affected the coastline here in a number of ways discussed in the next section.

Most of the low ocean-front land from the Pajaro River mouth to Monterey is undeveloped. Much of the beach is set aside as state parks (Zmudowski State Beach, Salinas River State Beach, Moss Landing State Beach, Monterey State Beach) or is part of the Fort Ord Military Reservation. As a result, most of the beaches and dunes are used for recreation while the land behind the dunes is agricultural or open space. Some local development on the dunes occurs at Moss Landing, in particular the State

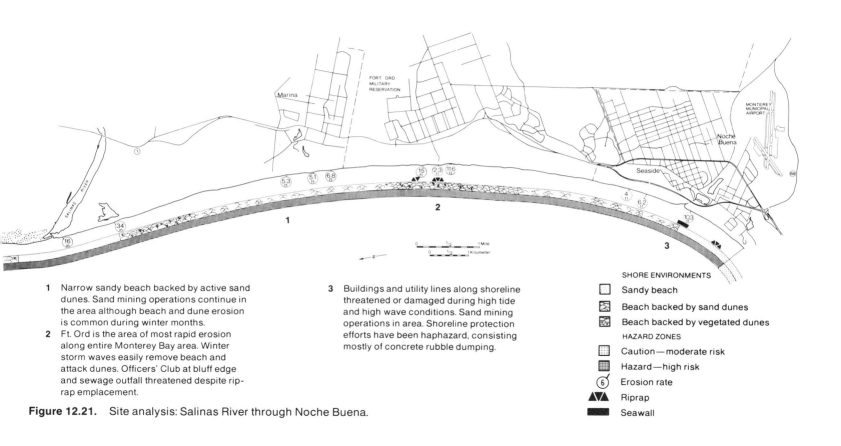

Figure 12.21. Site analysis: Salinas River through Noche Buena.

1 Narrow sandy beach backed by active sand dunes. Sand mining operations continue in the area although beach and dune erosion is common during winter months.

2 Ft. Ord is the area of most rapid erosion along entire Monterey Bay area. Winter storm waves easily remove beach and attack dunes. Officers' Club at bluff edge and sewage outfall threatened despite rip-rap emplacement.

3 Buildings and utility lines along shoreline threatened or damaged during high tide and high wave conditions. Sand mining operations in area. Shoreline protection efforts have been haphazard, consisting mostly of concrete rubble dumping.

SHORE ENVIRONMENTS

▫ Sandy beach

▨ Beach backed by sand dunes

▨ Beach backed by vegetated dunes

HAZARD ZONES

▦ Caution—moderate risk

▦ Hazard—high risk

⑥ Erosion rate

▲▼▲ Riprap

▬ Seawall

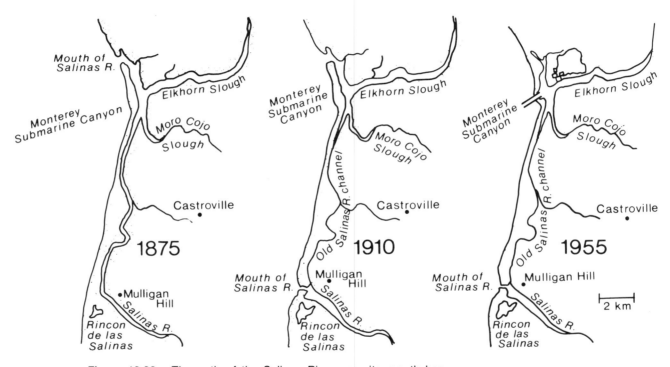

Figure 12.22. The path of the Salinas River near its mouth has changed significantly over the years. Redrawn from B. Gordon, *Monterey Bay Area Natural History and Cultural Imprints* (Pacific Grove: Boxwood Press, 1979).

University Marine Laboratory buildings. Just north of the Salinas River mouth the Monterey dunes colony occupies a 1.25-mile-long strip of active coastal dunes. Between 1973 and 1976 120 homes in groups of 2 or 3 were built on the dunes. To the south the Fort Ord Officers Club, the Holiday Inn, and some ocean-front condominiums just north of Monterey, and several sand mining operations and sewage treatment plants are the only structures in or near the ocean front. As a result, erosion and storm problems directly affecting structures have been somewhat limited.

Geologic conditions. The beaches and dunes form the dominant coastal geologic features in this area, although the Salinas River and Monterey Submarine Canyon are important features as well. The ocean-front dunes are geologically quite young (less than about 3,000–5,000 years) and are still active. The migration of the dunes inland can be seen clearly along Highway 1 from Marina to Sand City, where drifting sand must regularly be removed from the roadway. Further landward, principally beneath the Fort Ord area, are older dunes that extend several miles inland and are now stabilized by vegetation.

Monterey Submarine Canyon, over 3,000 feet deep, is also a key element in the sediment transport system within the bay. Just as inland transport of sand in dunes represents a permanent loss to the beach, so does underwater transport of sand into submarine canyons. The canyon head itself extends inshore almost to the mouth of Elkhorn Slough. When the Salinas River emptied into the slough, much of its sediment probably ended up in the canyon. While the shift of the river mouth to the south resulted in far greater volumes of sand reaching the bay, the construction of the 2 jetties at Moss Landing in 1946 is known to have channeled some littoral drift offshore into the canyon where it is permanently lost. In fact, one of the jetties and the Moss Landing pier partially collapsed due to slumping at the canyon head.

Another major loss of sand in southern Monterey Bay is due to 2 sand mining operations in the Marina and Sand City areas. About 300,000 to 350,000 cubic yards of beach sand is removed annually. The sand produced from southern Monterey Bay is unique in the world for its amber color, hardness, particle shape, and wide range of usable grain sizes. This particular sand has a variety of industrial uses including sandblasting and abrasives, water filtration, and stucco manufacture. A critical issue in a recent sand dredging permit renewal was whether or not the shoreline of southern Monterey Bay is eroding, and if so, how fast and in what areas? Of considerable significance is whether or not the sand removal operations significantly affect coastal erosion.

Coastal erosion and protection. The generally smooth outline of inner Monterey Bay combined with the wide sandy beaches and active dunes suggests an equilibrium coast, or one more or less in balance with the forces or processes acting upon it. In other words, sand is brought to the coast by rivers, and longshore currents or littoral drift sort the material out and move the sand alongshore. Some of the sand disappears into the head of Monterey Submarine Canyon, some is blown off the beach onto sand dunes, and some is removed by mining operations never to see the beach again.

This, in fact, may be the long-term picture, but the historical and photographic record combined with recent observations indicate that the shoreline is dynamic; the beaches narrow and then reform, the dunes are cut back and are subsequently rebuilt. Of major concern is the amount of change involved in these cycles or processes and how this affects our development and proposed development. What happens during major storms or are there long-term changes that can be documented?

Although aerial photographs from 1937 to 1980 show erosion alternating with accretion on the coast near Moss Landing, for example, the overall trend is one of gradual beach accretion and progradation or outbuilding of vegetation. Measured rates of buildup averaged from 1937 to 1983 are about 2 feet per year.

Despite this overall pattern of buildup, however, severe erosion can occur in any individual year. An addition to the Moss Landing Marine Labs was proposed in 1982 in the dune area. The geologic consultants were convinced that the area was quite stable and that the structure would need no seawall or other protection. Immediately after the foundation was poured in December 1982, however, storm waves removed 17 feet of the dunes and threatened the foundation, requiring emergency riprap to be brought in. With the subsequent storms of early 1983 a permanent low concrete seawall was designed and emplaced for future protection (fig. 12.23). The seawall does not protect the south side

Figure 12.23.   A new addition to the Moss Landing Marine Laboratories was begun in December 1982 on the backbeach where dunes were removed. A seawall was required before the structure was completed. Photo by Gary Griggs.

Figure 12.24.   The Officers Club at Fort Ord in 1980. Erosion of the loose sand has now brought the edge of the cliff to within a few feet of the structure. Photo by Harold Field.

of the structure, where runup from storm waves can be expected to reach the buildings.

At Monterey Dunes Colony, just north of the present Salinas River mouth, the coastline has both retreated and advanced over the years. The older aerial photographs of the area (1937) reveal an older, presumably storm-cut scarp; somewhat subdued, it now lies landward of many of the present structures. By 1937 the dune field had been built seaward 300 to 400 feet from the old scarp and was sparsely vegetated. In subsequent years vegetation expanded, and from 1956 until 1978 the area remained quite stable. Between 1978 and 1983, however, storms cut back the dunes as much as 100 feet in places. During the winter of 1983, the protective offshore bar was breached at several points, and this allowed larger waves to break closer to the beach. The wave runup combined with the high tides and storm surge led to considerable dune erosion. Fortunately the structures at Monterey Dunes Colony were built farther back from the front edge of the dunes than those at Pajaro Dunes, and thus no structures were damaged or undermined. Nonetheless, the evidence of the older storm scarp in the dunes and the winter storms of the last several years have led the dune homeowners to begin evaluating protection alternatives.

To the south, Marina State Beach and the Fort Ord Military Reservation are areas of moderate-to-high erosion rates. The Marina area is also the site of a sand mining operation. The construction of the freeway through the area in the 1970s and increased recreational use of the Marina Beach and dune areas have markedly decreased vegetative cover on the coastal dunes. As a result, blowouts in the dunes, or locations where vegetation has been removed and wind erosion is concentrated, have become more common. Average erosion rates in the Marina State Beach area between 1937 and 1983 averaged 5 to almost 7 feet per year; about a third of the total retreat has occurred since 1978. The Marina Sewage Treatment Plant experienced dramatic storm wave damage during the 1982-83 winter when a 40-foot section of the emergency bypass outfall line was undercut and collapsed. The storms of January 1978 removed some 50 to 80 feet of dunes and bluff at Monterey Sand Company's Marina sand mining operation, causing severe damage. A sand mining bunker, once located safely inland, toppled onto the beach during that storm.

Erosion at Fort Ord is the most severe in the southern Monterey Bay, despite the military's protective measures. Erosion here is dramatically episodic. The Fort Ord Officers Club has reportedly been abandoned as too hazardous for continued use due to its precarious cliff-top location. The lower portion of the bluff was protected with riprap prior to the 1983 storms (fig. 12.24). Winter waves, however, broke over the rock and also undermined the loose sand, causing it to slump onto the beach. Continuing wave attack led to additional cliff failure and buried much of the riprap. By early 1984 the edge of the eroding bluff was within 15 to 20 feet of the Officers Club and a new protection project was underway.

Perhaps the most spectacular erosion in the area is that at the

Figure 12.25.   Sewer outfall line at Fort Ord in 1978. Note end of outfall at left side of photo. (See fig. 12.26 for comparison.) Photo by Harold Field.

Figure 12.26.   Fort Ord outfall line shown in figure 12.25 in January 1983. The supporting piers are 35 feet apart. Note erosion that has taken place between 1978 and 1983. Photo by Harold Field.

sewage outfall line. This large-diameter pipe was initially placed atop massive concrete piers (set 35 feet apart) and buried below beach level in 1962. The outfall line is now 20 feet above beach level and 6 piers are exposed, indicating a loss of 175 feet of beach in 21 years (fig. 12.26). Remnants of riprap emplaced after the 1978 storms are evident and clearly show the extent of 1983 erosion (fig. 12.25); as much as 40 feet of bluff was lost during 1983.

The Sand City area lies at the southern end of the inner Monterey Bay dune field and has experienced high erosion rates, particularly since 1978. Two sand companies remove sand from the beach within Sand City. Both companies in addition to the city itself have built seawalls or dumped concrete rubble on the beach in an attempt to arrest dune erosion. In 1972 the Department of Navigation and Ocean Development prepared a feasibility study for a groin and public beach project here that was never initiated. Erosion rates along several transects in the Sand City area indicate average retreat rates of 4 to 10 feet per year with much of this taking place since 1978. A large Holiday Inn was constructed right on the beach just north of Monterey in 1968. In order to protect the structure a massive concrete seawall was built completely across the front and along the north side of the hotel. It has served its function well and shows no sign of deterioration. It is clear, however, that any sand removal from the beach tends to eliminate the natural buffer system and thereby increases the potential for storm-induced erosion. It should also be clear that any construction along the shoreline at this location could be endangered quickly by rapid retreat of the dunes. South of the Holiday Inn all of the utility lines run beneath the beach in front of a large apartment complex. Beach and dune erosion in 1983 broke the water line, which had to be rerouted, and threatened the sewer and electrical lines. Continuing dune erosion by Janu-

ary 1984 had come to within 14 feet of the pilings supporting the apartments (fig. 12.27). Five thousand tons of rock were being brought in to provide temporary winter protection for the ocean-front apartments.

Immediately north of the apartments, where new development is being proposed, a broken pipeline indicates at least 37 feet of 1983 dune retreat. An important question to ask at this time is whether the rapid rates of recent coastline erosion between Fort Ord and Monterey are short-term fluctuations, or whether they are the net result of a reduction in the sand supply, at least in part due to sand mining.

Figure 12.27. Apartment complex on the sand dunes at Monterey threatened by erosion in 1983. Five thousand tons of rock have now been emplaced in an effort to protect these buildings. Photo by Gary Griggs.

Summary

The sandy beaches and dunes of the Monterey Bay area have been popular summer recreational areas for nearly a century, but the coastline is a far different place during the winter months. Although the beaches, dunes, and cliffs are regularly attacked by winter storm waves and high tides, these same environments have also been intensively developed, particularly since the late 1960s. The hazards of ocean-front living are now more clearly recognized and understood, as are the high costs of protecting or attempting to protect this property. The damages and erosion from the storms of 1978–83, in addition to longer-term historic coastline changes, must be carefully evaluated prior to approval of any additional ocean-front development.

13. The outer coast: Point Pinos to Point Buchon

Jef Parsons

The first known description of the central California coast (fig. 13.1) by Europeans is from Juan Rodriguez Cabrillo's voyage of discovery in 1542. Cabrillo's ships rounded the Cabo de Galera (Point Conception) and sailed north into the outer coast. They found a bold coast without shelter behind which stood a long mountain range or sierra. To these mountains they gave the name the Sierras de Martin and the cape at which they ended, Cabo de Martin. A large storm forced them out to sea. Land was once again sighted at Punta de Reyes (Point Reyes), and Cabrillo headed back south along the coast, seeking shelter in a large enseñada (Monterey Bay). On Saturday, 18 November, the ships sailed south along the central coast.

The original ship's diary, translated by H. Wagner in 1928, reads:

> All the coast passed this day is very bold, there is a great swell and the land is very high. There are mountains which seem to reach the heavens and the sea beats on them, sailing along close to land, it appears as though they would fall on the ships. They are covered with snow to the summits, so they named them the Sierra Nevadas. At their beginning there is a cape which projects into the sea which they named

Cabo de Nieve (Cypress Point). Whenever the wind blew from the northwest the weather was clearer.

Again at Cabo de Martin, they described the downcoast areas as of good quality and inhabited by Indians.

Most of the names along the coast have since changed (many were given by Vizcaino in 1602), but the character of the land has seldom been so well described. The ruggedness and isolation of this coast has until very recently kept civilization at its borders. In 1769 Monterey became the first capital of California and the Mission San Carlos de Borremea the first Spanish development on the outer coast. The Spanish imprint on the land was small, and it was not until the early American period that conflict with the ocean's fury began to be noted.

Coastal schooner traffic and commerce increased tenfold in the 1860s as whaling stations, dairy farms, and mining developed in the coastal communities. Numerous landings about Estero Bay and the San Simeon in San Luis Obispo County as well as at Carmel on the Monterey Peninsula served the rapid influx of settlers to the region. The great swells noticed by Cabrillo began sending ships onto the nearshore rocks, while piers along the coast were repeatedly damaged.

The character of development changed on the coast after the completion of the Southern Pacific Railroad route through the Salinas Valley. By the 1920s subdivisions had begun on the coastal terraces both in the Monterey area and in San Luis Obispo County about Estero Bay and Cambria. Built close to the shore, these areas soon were threatened by shoreline erosion and the

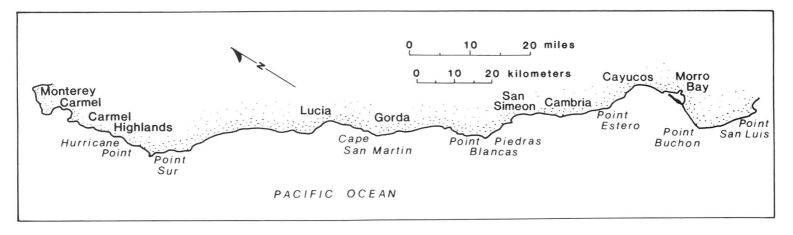

Figure 13.1.   Location map for the coastline between Monterey and
Morro Bay.

hazards of wave inundation and debris. These problems continue to worsen today, particularly in the Cayucos area and along state Highway 1.

Coastal hazards in the central coast are characterized by a combination of high wave energy and complex geology. The off-shore area for much of the region drops off rapidly into the deep ocean. The dominant wind direction is nearly parallel to the shore, being from the northwest, but winter storms are highly varied and much damage occurs during frequent and violent southwesters. Winds of 75 miles per hour are often reported out at sea during these storms.

Wide, sandy beaches exist in the southern area around Estero Bay, but most of the outer coast loses its sand to offshore submarine canyons. Small pocket beaches can be found in protected spots or where there is a local sand supply such as a stream or eroding cliffs. The cliffs themselves are of varying geology, some with resistant materials, others that erode relatively rapidly.

These characteristics make the central coast a unique area of great scenic value. Much of the wild Big Sur area has been a part of the Los Padres National Forest since 1908, and such attractions as Point Lobos State Reserve, Julia Pfeiffer Burns State Park, and the Hearst Castle are visited by thousands of people each year. Although recreation is the primary use of the central coast, thriving communities on the Monterey Peninsula and around Estero Bay continue to expand, and new development continues to locate along the ocean bluffs, particularly near Cambria.

The Monterey Peninsula (figs. 13.2 and 13.3)

The Monterey Peninsula is a large granite projection of the Santa Lucia mountains that slope down to Monterey Bay in a series of steps or coastal terraces. The cities of Monterey and Carmel-by-the-Sea are the 2 areas central to modern development. Originally the Spanish military garrison and the ecclesiastical capital of Padre Junipero Serra, the 2 towns have been a focus for twentieth-century development. The Pacific Improvement Company, owned by the famous Big Four (Huntington, Stanford, Hopkins, and Crocker) of the Central Pacific Railroad, built the Del Monte Hotel on the outer coast as a retreat for the Victorian-age social elite. Seventeen Mile Drive was developed and the area promoted by Del Monte Properties in the 1920s. Expensive homes have since been constructed along this scenic coast. Some of these homes have been built on the sand dunes that were earlier mined for use in glass manufacture.

The granitic bedrock of the Monterey Peninsula has protected the area from shoreline erosion problems. The rock, which has the formal name of the Santa Lucia Granodiorite, is very resistant to erosion. Virtually no changes can be detected in photographs taken in the period of 1915–25 and retaken today (fig. 13.4). North of Cypress Point, however, the coast is low with a terrace less than 20 feet in elevation. Great waves can splash and throw large debris, such as boulders weighing several hundred pounds, onto the terrace. The mast of the Steamer *St. Paul*, wrecked at Point Joe in 1896, was thrown 50 feet across Seventeen Mile Drive in 1931. The same storm threw tons of debris

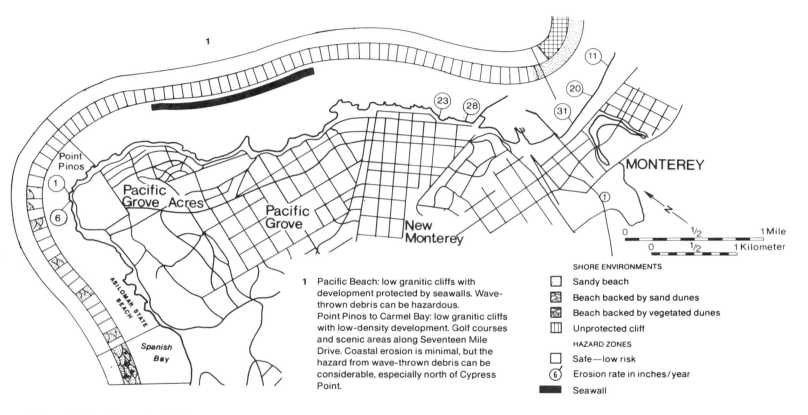

1   Pacific Beach: low granitic cliffs with development protected by seawalls. Wave-thrown debris can be hazardous.
Point Pinos to Carmel Bay: low granitic cliffs with low-density development. Golf courses and scenic areas along Seventeen Mile Drive. Coastal erosion is minimal, but the hazard from wave-thrown debris can be considerable, especially north of Cypress Point.

SHORE ENVIRONMENTS

▫ Sandy beach

▨ Beach backed by sand dunes

▨ Beach backed by vegetated dunes

▥ Unprotected cliff

HAZARD ZONES

▫ Safe—low risk

⑥ Erosion rate in inches/year

▬ Seawall

Figure 13.2.   Site analysis: Monterey area.

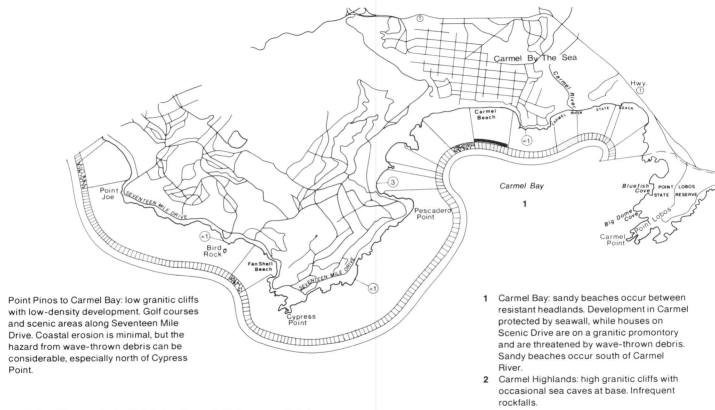

Point Pinos to Carmel Bay: low granitic cliffs with low-density development. Golf courses and scenic areas along Seventeen Mile Drive. Coastal erosion is minimal, but the hazard from wave-thrown debris can be considerable, especially north of Cypress Point.

1 Carmel Bay: sandy beaches occur between resistant headlands. Development in Carmel protected by seawall, while houses on Scenic Drive are on a granitic promontory and are threatened by wave-thrown debris. Sandy beaches occur south of Carmel River.

2 Carmel Highlands: high granitic cliffs with occasional sea caves at base. Infrequent rockfalls.

Figure 13.3. Site analysis: Point Joe through Soberanes Point.

across Seventeen Mile Drive at Moss Beach and Fan Shell Beach, closing the road in both places.

The southern part of Asilomar State Beach is known as Spanish Bay. The Del Monte Hotel operated a boardwalk and bathhouse area here in the 1920s. Waves damaged these facilities several times, and wave runup was observed all the way to the Southern Pacific Railroad tracks hundreds of yards inland. Stranded at Spanish Bay early in 1935, the schooner *Aurora* was removed from the beach the following December when large storm waves lowered the beach profile, removed about 6 feet of sand that had accumulated in the hull, and refloated the wreck. Waves then drove the hulk around Point Pinos, past the municipal wharf, and smashed it into the KDON radio station transmitter's wharf pilings. Only quick work by the fire department to lasso and tie the wreck down to the beach saved the wharf from being battered to pieces.

It has been nearly 40 years since such large waves have battered the coast. New development has occurred, and only careful planning can avoid such costly wave damage in the future.

South of Cypress Point the coast becomes a little higher, and wave runup is less of a hazard. Coastal terraces are cut in sedimentary rock that is softer than the resistant granitic rock around Carmel Bay. Waves are more capable of eroding these softer rocks, and as a result seawalls were built in the 1920s to protect the area along Carmel Beach. Houses built on promontories of granitic rock, as along Scenic Drive, are exposed to wave debris

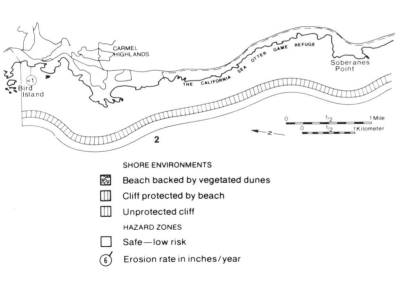

SHORE ENVIRONMENTS

Beach backed by vegetated dunes

Cliff protected by beach

Unprotected cliff

HAZARD ZONES

Safe—low risk

(6) Erosion rate in inches/year

Figure 13.4.   Homes and roadway protected from wave runup by stone seawall, Pacific Grove. Photo by Jef Parsons.

thrown into the air that has caused minor damage, such as broken windows.

The size of waves in the area is clear from historical accounts of waves that have completely overwashed Carmel River Beach to strike the cliffs behind. Off Seventeen Mile Drive, waves have been known to overwash Bird Rock completely. The area of large wave splash is well shown on the granitic rocks by a prominent dark algal zone usually about 20 to 25 feet above sea level. Debris can be thrown above this height.

Farther south by Carmel Highlands, tall, steep granitic cliffs greet the ocean. The granitic material erodes along fractures and joints. These joints have eroded over thousands of years into sea caves, the collapse of which can result in the loss of large blocks of overlying material. Erosion of this type can be seen at Point Lobos Reserve. At the end of Bluefish Cove near Guilemot Island a sea cave has been eroded along a joint. On the west side of Big Dome, a scar left by a large slab of granite that failed along a joint can be seen. The slab was about 100 feet high by 40 feet wide. Such block falls do not occur often in the granitic cliffs, but inspection of the cliff areas for such problems should be made by a geologist before construction takes place near the cliff face.

The Big Sur coast (figs. 13.3, 13.5, 13.6, 13.7, and 13.8)

South of Carmel the seacliffs grow in stature and grandeur. Often referred to as the "greatest meeting of land and sea," the Big Sur coast is rugged and often remote. In the 1880s, homesteaders gathered about the Rancho El Sur land grant in the Big Sur Valley were numerous enough to form a community served by a wagon road to Monterey and an anchorage at Point Sur. South of Post Hill only a horse trail followed the coast until construction of the Coast Highway from 1929–37. Many of the landholdings were consolidated in the 1920s as the urban millionaires, many from Hollywood, established ranches and retreats. The community of Big Sur itself became a legendary retreat for artists, writers, and others.

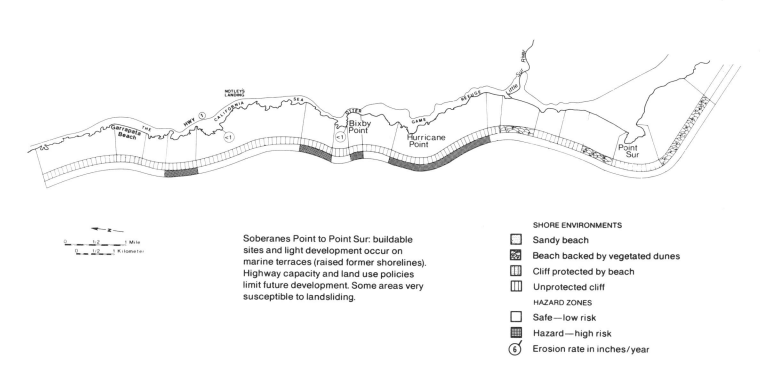

Soberanes Point to Point Sur: buildable
sites and light development occur on
marine terraces (raised former shorelines).
Highway capacity and land use policies
limit future development. Some areas very
susceptible to landsliding.

SHORE ENVIRONMENTS

Sandy beach

Beach backed by vegetated dunes

Cliff protected by beach

Unprotected cliff

HAZARD ZONES

Safe—low risk

Hazard—high risk

⑥ Erosion rate in inches/year

Figure 13.5. Site analysis: Garrapata Beach through Point Sur.

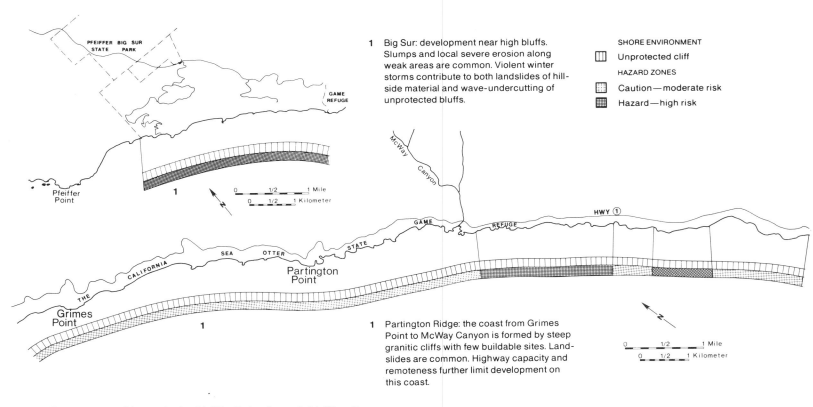

1 Big Sur: development near high bluffs. Slumps and local severe erosion along weak areas are common. Violent winter storms contribute to both landslides of hillside material and wave-undercutting of unprotected bluffs.

SHORE ENVIRONMENT

▥ Unprotected cliff

HAZARD ZONES

▦ Caution—moderate risk

▦ Hazard—high risk

1 Partington Ridge: the coast from Grimes Point to McWay Canyon is formed by steep granitic cliffs with few buildable sites. Landslides are common. Highway capacity and remoteness further limit development on this coast.

Figure 13.6.   Site analysis: Pfeiffer Point through McWay Canyon.

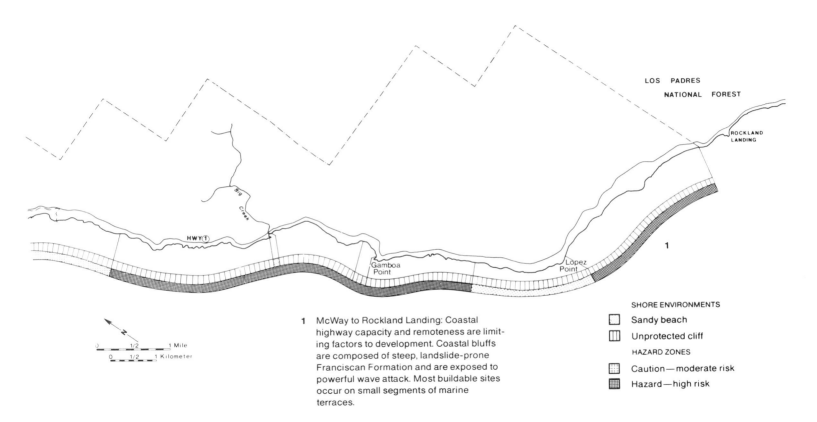

1 McWay to Rockland Landing: Coastal highway capacity and remoteness are limiting factors to development. Coastal bluffs are composed of steep, landslide-prone Franciscan Formation and are exposed to powerful wave attack. Most buildable sites occur on small segments of marine terraces.

SHORE ENVIRONMENTS

☐ Sandy beach

☐ Unprotected cliff

HAZARD ZONES

☐ Caution—moderate risk

☐ Hazard—high risk

Figure 13.7. Site analysis: Gamboa Point through Rockland Landing.

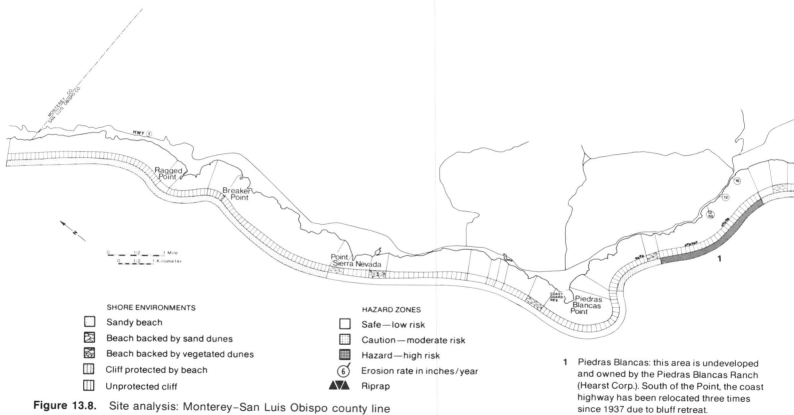

SHORE ENVIRONMENTS

- Sandy beach
- Beach backed by sand dunes
- Beach backed by vegetated dunes
- Cliff protected by beach
- Unprotected cliff

HAZARD ZONES

- Safe—low risk
- Caution—moderate risk
- Hazard—high risk
- (6) Erosion rate in inches/year
- Riprap

Figure 13.8. Site analysis: Monterey–San Luis Obispo county line through San Simeon Beach State Park.

1 Piedras Blancas: this area is undeveloped and owned by the Piedras Blancas Ranch (Hearst Corp.). South of the Point, the coast highway has been relocated three times since 1937 due to bluff retreat.

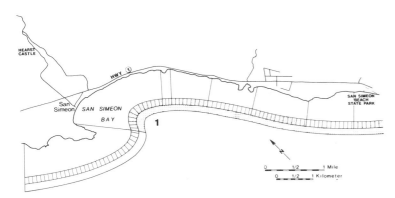

1 San Simeon: moderate erosion hazard in low bluffs consisting of Franciscan Formation rocks.

Opening of the Coast Highway in 1937 and the growth of recreation in the 1950s and 1960s have resulted in summer congestion and concern over maintaining the natural scenic integrity of the area. In spite of the highway congestion, minor development continues along the rugged coast. Private lands extend as far south as Lucia with development occurring either on the marine terrace remnants or on top of the coastal hills (fig. 13.9). A number of geologic problems affect both existing and future development.

Big Sur is characterized by both resistant and erosion-prone rocks. With the exception of the area near Point Sur, virtually no wide, sandy beaches exist. Steep cliffs generally greet the coast (figs. 13.10). North of Hurricane Point and also around Partington Point, the cliffs consist of granitic and metamorphic rocks that are very resistant to erosion. At Bixby Point the site of the Monterey Lime Company (1904–10) warehouse is still perched at the terrace edge, and the metamorphic rocks look the same as they did in the 1930s photographs by Big Sur resident Horace Lyon. Elsewhere, cliffs are composed of rocks of the Franciscan Formation. These areas often contain large, deep-seated landslides that affect several acres of land. Near Pfeiffer Beach and south of Esalen large sections of cliff consist of giant slides that are extremely unstable. During the winter of 1983 many sections of the Coast Highway near Lucia were destroyed as slide movement dropped the roadbed out. Such road closures are common on this coast and can be expected in the future (fig. 13.11).

Figure 13.9. Garrapata Beach. Limited development of single-family homes has taken place on the marine terrace during the 1970s. Photo by Jef Parsons.

Figure 13.10. The precipitous coastline at Lucia. Elevations rise from sea level to 5,000 feet in a little over 3 miles. Photo by Jef Parsons.

In 1983 1,000 feet of roadway was washed away at Hurricane Point while nearby areas of roadway were blocked by rock debris. The highway bridges may also be threatened in the near future. At Rocky Creek the south approach has been threatened by landslides, and a seawall was built in 1970 to protect the foot of the cliff from wave undercutting. This wall was damaged by large waves within several years of its construction. The cliffs south of Bixby Creek show scars from recent rockfalls in the harsh 1983 winter.

The long, steep slopes and violent winter storms also contribute to a thick veneer of rock debris material that is washed downslope by soil creep, slumping, and landsliding off the steep slopes. Along Big Sur the marine terraces are often buried by over 50 feet of rock debris. During the construction of the Pacific Coast Highway, a survey line 2,000 feet long moved seaward 2 feet and downward 8 feet in 2 years. The debris is seen clearly in the highway roadcuts, where it usually appears as a reddened soil filled with angular pieces of rock. Often, many of the small slides along the right-of-way originate in this loose material, due to the oversteepening of the slope by the highway cut.

The many geologic hazards in the Big Sur region require a qualified geologist to inspect any potential site before development. The ruggedness of the coast, as well as remoteness, have kept civilization at bay. Now visited by millions of visitors, scenic resources have also become an important planning issue. Limited water resources also limit new development, and the coast is likely to stay relatively undeveloped for some time to come.

The San Luis Obispo northern coast (figs. 13.8, 13.12, and 13.13)

The scenic Big Sur coast ends near the large promontory of Piedras Blancas (fig. 13.14). Extensive development has taken place south of here along the coast that Cabrillo described as of "good quality." Cayucos and Cambria saw an early period of development during the 1860s to 1880s at the height of the coastal schooner era. Landings were numerous, and piers were built at San Simeon in 1857 and at Cayucos in 1867. A pier at Cambria was destroyed by an early winter storm before it even had a chance to open in 1909. The current pier at San Simeon, built in 1953, and the Cayucos pier are occasionally damaged by large waves that completely overwash them. After the storms of 1983, 8 pilings at San Simeon and 30 pilings at Cayucos needed replacement.

Most of the accessible beach areas between San Simeon and Morro Bay are a part of the state park system, and development on the beaches is therefore precluded. Along the low terraces behind the beaches, however, residential subdivisions have been

Figure 13.11.   Landsliding poses a continual hazard to Highway 1 in the Big Sur area. The road was closed for 11 months in 1983 due to a number of large slides. This particular slide cost over $7 million to repair. Photo by Jef Parsons.

built on the bluff edges, and private property is critically threatened in areas. Most of the subdivisions were laid out in the 1920s or earlier. The decline of ocean shipping and the arrival of the railroad in San Luis Obispo were key factors in this change of land use. The local mining and lumber products could not compete with rail-supplied materials and they gave way to residential and tourism development. A postwar building boom in California has also contributed to the development now threatened by shoreline erosion. Bluff-top property is still being developed, and lots were for sale in 1983.

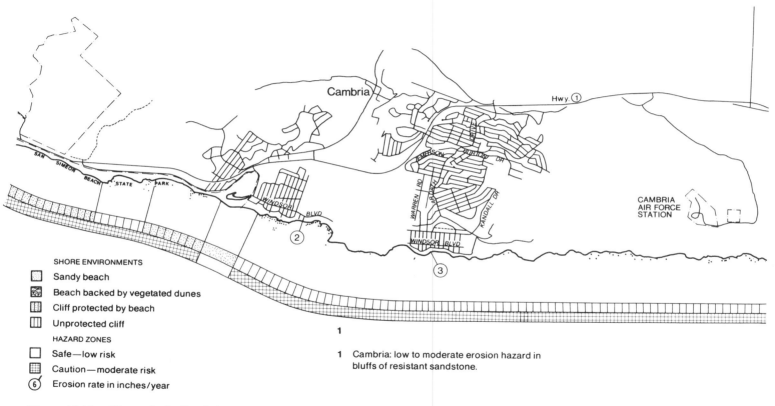

SHORE ENVIRONMENTS
- Sandy beach
- Beach backed by vegetated dunes
- Cliff protected by beach
- Unprotected cliff

HAZARD ZONES
- Safe—low risk
- Caution—moderate risk
- ⑥ Erosion rate in inches/year

1 Cambria: low to moderate erosion hazard in bluffs of resistant sandstone.

Figure 13.12.   Site analysis: Cambria area.

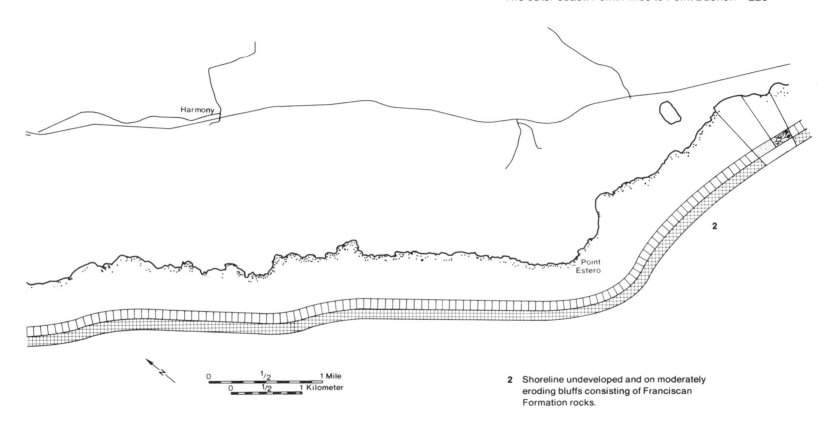

Harmony

Point
Estero

2

0   1/2   1 Mile
0   1/2   1 Kilometer

N

2 Shoreline undeveloped and on moderately
eroding bluffs consisting of Franciscan
Formation rocks.

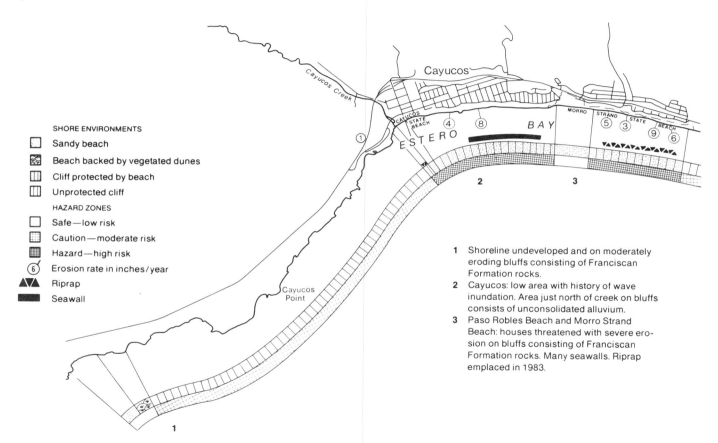

SHORE ENVIRONMENTS

- Sandy beach
- Beach backed by vegetated dunes
- Cliff protected by beach
- Unprotected cliff

HAZARD ZONES

- Safe—low risk
- Caution—moderate risk
- Hazard—high risk
- ⑥ Erosion rate in inches/year
- ▲▼▲ Riprap
- ▬ Seawall

1 Shoreline undeveloped and on moderately eroding bluffs consisting of Franciscan Formation rocks.
2 Cayucos: low area with history of wave inundation. Area just north of creek on bluffs consists of unconsolidated alluvium.
3 Paso Robles Beach and Morro Strand Beach: houses threatened with severe erosion on bluffs consisting of Franciscan Formation rocks. Many seawalls. Riprap emplaced in 1983.

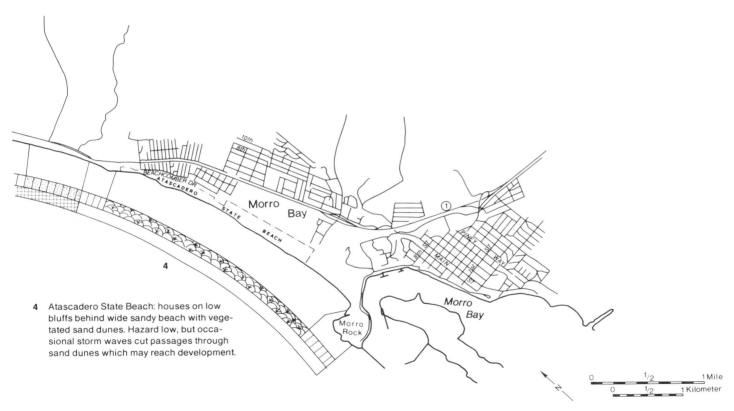

4   Atascadero State Beach: houses on low
    bluffs behind wide sandy beach with vege-
    tated sand dunes. Hazard low, but occa-
    sional storm waves cut passages through
    sand dunes which may reach development.

Figure 13.13.   Site analysis: Cayucos Point through Morro Bay.

Cambria and Cayucos

The Cambria and Cayucos shorelines are similar in many ways, but have very different risks of shoreline erosion. Both areas have bluffs about 20 feet high fronting the ocean. In the summer Cayucos has a wide, sandy beach, while Cambria has a smaller, more gravelly beach. During winter storms, however, both beaches virtually disappear and the waves strike directly against the bluffs. The bluffs are but the front of coastal terraces that have been covered with residential development since the 1950s and earlier. Both areas have about the same exposure to ocean waves, although Cayucos is a little bit better protected from northwesterly swells. Nevertheless, the Cayucos bluffs lost as much as 20 feet of land in 1983, while the Cambria area suffered little cliff retreat.

Understanding the complex geology of the area is important to predicting the relative hazard of shoreline erosion between Cambria and Cayucos. The entire coast from Piedras Blancas to Morro Bay lies within a part of the geologic province known as the Franciscan Formation. This geologic formation consists of a series of highly varied rock materials that have been mixed together into a very complex geologic unit. Within this unit are a series of distinct subunits that react to the force of incoming waves very differently. These rock types are usually exposed in the lower half of the coastal bluffs and support the coastal terraces. Where the rocks are weak and erode quickly, the bluffs retreat rapidly. Where the rocks are more resistant, the bluffs are more stable.

The coast around Cambria is composed of a rock unit called

Figure 13.14. Arroyo Laguna Vista Point on a low marine terrace near Piedras Blancas. The parking area is located on previous highway rights-of-way, much of which has been eroded. Photo by Jef Parsons.

the Cambria Slab. The Cambria Slab is composed of sandstones with layers of siltstone. The sandy nature of this rock is responsible for the sandy, porous soils about Cambria upon which the extensive forests of Monterey Pine grow (one of the three locations in the world to which the Monterey Pine is endemic.) The resistant sandstone layers allow this rock to withstand the force of winter waves rather well, and as a result the coastal terraces at Cambria retreat rather slowly. Measurements from a variety of sources such as aerial photographs, subdivision survey lines, and historic photographs all indicate an average erosion rate for these bluffs at only 2–3 inches per year. Excessive erosion occurs where the siltstone layers are exposed in the bluffs. An occasional seawall can be found protecting these areas, but the problem tends to be very local and is not extensive in the Cambria area.

In Cayucos, on the other hand, much less resistant Franciscan melanges are found. Melange is a term applied to rock types that include numerous small blocks of rock, often surrounded by small zones of sheared or crushed rock that is structurally weak and erodes easily. Some of the blocks are also composed of rock that is highly sheared and fractured, particularly the blocks composed of serpentine. The crushed and broken zones are quickly eroded by wave action, leading to collapse of the entire block and very rapid bluff retreat. The bluffs along Pacific Avenue and Studio Drive are characterized by these melange units and contain extensive blocks of sheared serpentine (fig. 13.15). Erosion by waves has exposed these resistant blocks, which then act as natural seawalls protecting the bluff—for a while. In several

Figure 13.15. Cayucos. Homes along Pacific Avenue and Studio Drive are constructed on rapidly eroding bluffs in the Franciscan Formation. Photo by Jef Parsons.

instances during 1983 these seemingly resistant blocks were breached by the intense storm waves, exposing heavily sheared and fractured rocks that were removed rapidly by the aggressive waves. Over 20 feet of bluff was removed in some of these spots during storms in January and February of 1983 alone, threatening homes built 30–40 feet from the previous bluff edge. Erosion rates are highly variable along this coast with measurements averaging 6–10 inches per year. As a result of these erosion rates, many seawalls have been built to protect homes along both Pacific Avenue and Studio Drive. Most of Studio Drive is now protected

Figure 13.16. Emergency riprap was placed in front of homes along Scenic Drive in March 1983. Oil Company pier in background was heavily damaged in January and February 1983. Photo by Jef Parsons.

by emergency riprap emplaced in March 1983. Rapid bluff erosion and its associated problems can be expected to continue in this area in the future (figs. 13.15 and 13.16).

Different units of the Franciscan melange occur upcoast from downtown Cayucos, as well as at San Simeon farther up the coast (fig. 13.17). These units do not include the small blocks of highly sheared rock found along Pacific Avenue and Studio Drive. The high silt content of the rocks, however, contributes to moderately high erosion rates (about midway between the rates of the Cambria Slab and the Pacific Avenue area). More important here, and a problem wherever Franciscan melange occurs, is the high percentage of clays in the unconsolidated terrace deposits that usually form the upper 10 feet of the bluffs around Cayucos. These clays are impervious to water and cause perched or elevated water tables. Where these empty out on the bluff face, they cause increased loss of soil as the saturated soil layers slump. Such problems can greatly increase the rate of bluff retreat. Homeowners should pay careful attention to on-site drainage.

Downtown Cayucos has yet another coastal hazard. The waterfront area is built upon the ancient valley of Cayucos Creek. This valley was filled in with unconsolidated sediment by the old stream, and shoreline erosion is a particular problem in this area. A recent condominium project off Lucern Drive just upcoast of the downtown area was threatened by bluff retreat soon after construction due to the poor resistance of this material in the bluffs. Some protection has been provided by riprap, but this must be periodically replaced. The downtown waterfront area itself is protected by a low seawall. In 1983 and several times in the recent past waves have overwashed the seawall, inundating the downtown area (fig. 13.18). Stores have been seriously damaged, and debris has been spread across Ocean Boulevard.

Morro Bay and vicinity (fig. 13.13)

Beginning at Atascadero State Beach and continuing south past Morro Bay to Montana de Oro State Park, the coast is fronted by a set of large sand dunes. These dunes provide protection for development on the terrace materials behind. The coastal terrace is warped downward at Morro Bay and depressed below sea level beneath the harbor. The low shore forms a trap for sand eroded from upcoast. Winds then transport the sand off the beach and build the dunes. Height of the dunes and width of the dune field increases where the terrace is lowest in elevation. At Atascadero Beach houses are built on a terrace about 15 feet high. The terrace is less than 10 feet high at Morro Bay High School and disappears completely beneath Morro Bay.

The sand accumulation north of the Morro Bay entrance has been significantly changed by the construction of the Pacific Gas and Electric Company's power plant and the harbor breakwater. Longshore drift north of Morro Bay has been interrupted, causing accumulation of sand along Atascadero Beach. In 50 years the coast has accreted, or extended seaward, about 250 feet near San Jacinto Avenue and almost 500 feet in front of the high

Figure 13.17.   Eroding bluffs of Franciscan material at San Simeon. Photo by Jef Parsons.

Figure 13.18.   Cayucos. High tides combined with storm waves during the winter of 1983 damaged stores and streets as they have in the past. Photo by Jef Parsons.

school. Before these changes Morro Rock was isolated from the coast, and the bay entrance was to the north of the rock. Now a tombolo (a sandbar) almost 2,000 feet in length has connected Morro Rock to the mainland, and the harbor entrance is to the south of the rock.

The addition of more dunes has increased protection of the Atascadero Beach subdivisions. Wave inundation had been a problem in the past, and the U.S. Army Corps of Engineers has considered a series of groins to increase the beach width here. The accretion of the shoreline may make this protection unnecessary, however. During 1983 waves cut back the dunes into steep faces. Extensive wave debris was found completely covering the foredunes. Wave runup was able to lap up onto the developed portion of the terrace through storm channels cut into the dunes, both at the high school and along Sandalwood Drive, but no damage was incurred. The storm channels occur where street runoff or other drainage outlets cut through the dunes. Construction plans should consider wave runup along sites with drainage into the dunes.

South of Morro Bay, coastal development is limited by Montana de Oro State Park and the industrial property downcoast of Point Buchon. Utilities and oil companies have begun developing the coastal areas that are isolated by Irish Hills (San Luis Range) from populated areas. Diablo Canyon Nuclear Power Plant and the 1978 proposed Rattlesnake Canyon Liquefied Natural Gas Terminal are two such projects on the small coastal terraces in this area. Offshore oil production may increase this development in the future. Erosion rates of 4–6 inches per year are found in areas of geology similar to that along the Irish Hills coast. Groundwater and impermeable clay horizons in the terrace materials can increase these rates through slumping, and proper site drainage should therefore be provided.

14. Morro Bay to Gaviota

Amalie Jo Orme, Kevin Mulligan, and Vatche Tchakerian

Point San Luis to Point Conception

The California coast between Point San Luis and Point Conception trends generally north-south, spanning some 72 miles of southern San Luis Obispo and western Santa Barbara counties (fig. 14.1). This coast embraces three natural units: (1) the 9-mile-long coast of San Luis Obispo Bay between Point San Luis and Pismo Beach, (2) the 44-mile-long Santa Maria coast from Pismo Beach to Point Pedernales, and (3) the 19-mile-long Jalama coast from Point Pedernales to Point Conception. The first 2 units represent the seaward edge of the onshore portion of the Santa Maria Basin. The third unit lies at the western end of the Santa Ynez Mountains, the most westerly landward component of the Transverse Ranges that extend 348 miles eastward across southern California to the north end of the Salton Trough. The first and third units are mostly cliffed, with marine terraces above the cliffs and pocket or fringing beaches below. The second unit is a long, sandy shoreline backed by extensive sand dunes, its continuity broken only by bold, rocky headlands around Point Sal and low, rocky shores at Purisima Point.

Human modification of this coast has been sporadic in both time and space, with moderate-to-intense development being restricted to the shores of San Luis Obispo Bay. Here, a considerable variety of land uses is found within a limited area, ranging from single- to multiple-family dwellings, hotels, recreational facilities, and commercial fisheries. The coast south of Pismo Beach has either remained undeveloped or been modified to a limited degree owing to substantial landholdings or leases maintained by the petroleum industry, the state's park system (Pismo State Beach), the U.S. Department of Defense (Vandenberg Air Force Base), and private ranches devoted to agricultural and livestock businesses. Thus development of this portion of the California coast is more limited than in other selected areas within the state, for example, the Ventura and Los Angeles county coastlines. Nevertheless, a broad spectrum of existing and potential hazards exists. This is particularly evident around the shores of San Luis Obispo Bay where recreational and residential development has frequently ignored the hazard potential of the coast.

Physical setting

This section of the coast is characterized by a variety of physical features including seacliffs that rise to some 750 feet near Point San Luis and lower, active seacliffs along the San Luis Obispo Bay segment; rocky promontories such as Point Sal and Mussel Rock (fig. 14.2), Purisima Point, Point Pedernales, and Point Arguello, which punctuate long, sandy stretches of beach; and an extensive dune complex along the Santa Maria coast (fig. 14.3).

Materials for the sandy beaches are derived from 3 principal

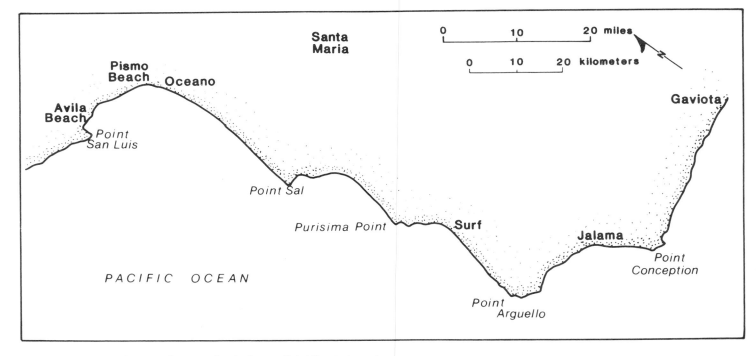

Figure 14.1. Location map for coastline between Point San Luis and Gaviota.

Figure 14.2. Active (nonvegetated) sand dunes and paleodunes (vegetated) at Mussel Rock, November 1982. Photo by Kevin Mulligan.

Figure 14.3. Mouth of Arroyo Grande Creek, Pismo State Beach, and Pismo Dunes complex, March 1983. Photo by Kevin Mulligan.

sources: neighboring coastal watersheds, seacliff erosion, and dune sand. The largest contributing drainages are Arroyo Grande, the Santa Maria River, and the Santa Ynez River. Smaller contributing catchments are San Luis Obispo Creek, San Antonio Creek, and Jalama Creek.

Between Point San Luis and Point Conception, some 30–50 percent of swells approach the coast from the northwest, the remainder mostly from west-northwest or west. Thus with the exception of short sections sheltered in the lee of Point San Luis, Point Sal, and Point Arguello, most of this coast is fully exposed to the predominant swells and their attendant winds. This is

reflected in the direction of the prevailing longshore current and littoral drift, from north to south, and of sand transport across the coastal dunes. Nevertheless, southerly and southwesterly swells associated with late summer storms off western Mexico, with southern hemisphere storms, and with the passage of winter storms across southern California, can cause serious erosional problems on the limited stretches of south-facing coast as well as promoting limited longshore current reversal. Close to the headlands cited above, currents from south to north may be associated with northwesterly swells. Mean tidal range for this section of coast is around 3.6 feet, while the range between mean lower low water (MLLW) and mean higher high water (MHHW) is 5.2 feet. The extreme range at spring tides is 8.8 feet.

Development background and planning issues

As a portion of the Manila-Acapulco trade route, the coastline between Point San Luis and Point Conception was reasonably well known to Spanish navigators from the sixteenth century —this being reflected in the names of many major headlands. When Europeans first saw this coast, its early inhabitants were the Chumash people. With the arrival of the Franciscan clergy in the eighteenth and nineteenth centuries and the establishment of the mission system along the California coast, the Chumash were eventually concentrated within the confines of mission lands. Whereas no mission existed directly on the coast between Point San Luis and Point Conception, two centuries ago the missions at San Luis Obispo (1772) and Purisima Conception (1787) incorporated considerable rangeland, which stretched to the coast along a broad front. However, following the independence of Mexico in 1821 and the secularization of the missions, these lands were granted to individuals. Much of the land along the terraces between Avila Beach and Pismo Beach remained agricultural until the mid-twentieth century (fig. 14.4). Since the 1950s and especially during the past 10 years, much of this same land was converted to commercial and residential developments, which proceeded with little regard for potential dangers of landslides and seacliff retreat. Today, pressure to develop the coast more fully is exemplified by continued attempts to construct a major

Figure 14.4.   Shell Beach between Fossil Point and Pismo Beach, 1931. Photo by Fairchild Air Photos, UCLA.

marina at Port San Luis and additional dwellings, as well as motel and tourist facilities, at Avila, Shell, and Pismo beaches. Despite the recommendations of local coastal plans, the reliable data needed to effectively manage this coast are still scarce. Farther south, the Jalama coastal stretch has remained relatively undeveloped to this time. However, with the recent leasing of offshore properties near Point Arguello and Point Conception to the petroleum industry, the construction of drilling platforms and pipelines will add a new dimension to the management of the coast.

San Luis Obispo Bay (fig. 14.5)

This 9-mile stretch of coast is the most developed and presents the most problems of the entire coast between Point San Luis and Point Conception. It contains the settlements of Port San Luis, Avila Beach, Shell Beach, and Pismo Beach.

Port San Luis, a commercial fishing port and recreational boat harbor, is located within the lee of Point San Luis, which offers excellent protection against the predominant northwesterly swells but little shelter from southerly storms. The commercial fishing fleet based here presently comprises about 50 full-time and 100 part-time vessels. Their total catch in 1982 was valued at $1.1 million. Some 248 moorings for recreational vessels are also available, but plans to develop more extensive marina facilities have so far been unsuccessful. The port area is backed by 100-foot-high cliffs descending eastward into 30-foot cliffs. The local harbor district, created in 1954, maintains a rubble mound breakwater 2,300 feet long, a wharf, a floating fuel dock, and a service area for vessels. The harbor district wants to construct a boat repair and storage yard on a landfill to be built with materials taken from the adjacent hillside. The base of the hill would be leveled for future construction. The principal limitations on further development are the unstable nature of the cliffed hillside behind the port and the possibility of contamination (through increased turbidity and marine fuel) of the heavily used bathing beach at nearby Avila Beach.

Farther east, the more exposed residential and recreational community of Avila Beach extends eastward across a 300-foot-wide barrier beach at the mouth of San Luis Obispo Creek onto the cliffed slopes leading to Fossil Point. The vulnerability of this portion of the coast is illustrated by recent events. The San Luis Obispo Harbor District formerly maintained a recreational pier at Avila Beach (fig. 14.6) but this was severely damaged during the winter of 1983 (fig. 14.7). Additional damage was incurred east of the primary bathing beach along the seacliffs adjacent to the Union Oil storage area. Owing to undermining, seawall collapse, saturation of cliff material, and failure within and behind the cliff top, several houses experienced foundation damage and local roads were cracked and breached. Furthermore, behind the barrier beach a former lagoon has been developed for mobile homes. Although these are protected by a road dike, they are subject to flooding during stormy periods associated with increased water level along the San Luis Obispo Creek.

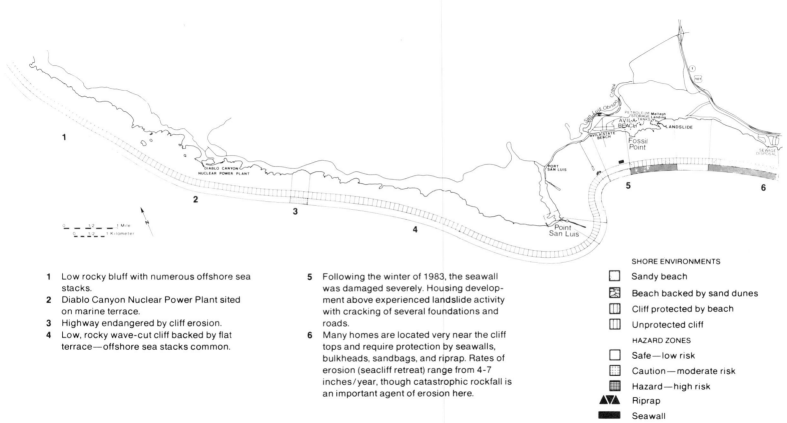

1 Low rocky bluff with numerous offshore sea stacks.
2 Diablo Canyon Nuclear Power Plant sited on marine terrace.
3 Highway endangered by cliff erosion.
4 Low, rocky wave-cut cliff backed by flat terrace—offshore sea stacks common.

5 Following the winter of 1983, the seawall was damaged severely. Housing development above experienced landslide activity with cracking of several foundations and roads.
6 Many homes are located very near the cliff tops and require protection by seawalls, bulkheads, sandbags, and riprap. Rates of erosion (seacliff retreat) range from 4-7 inches/year, though catastrophic rockfall is an important agent of erosion here.

SHORE ENVIRONMENTS

☐ Sandy beach
▨ Beach backed by sand dunes
▥ Cliff protected by beach
▥ Unprotected cliff

HAZARD ZONES

☐ Safe—low risk
▦ Caution—moderate risk
▦ Hazard—high risk
▲▲ Riprap
▬ Seawall

Figure 14.5. Site analysis: Diablo Canyon through Oceano.

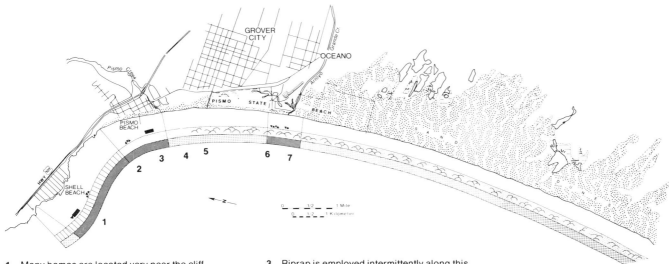

1 Many homes are located very near the cliff tops and require protection by seawalls, bulkheads, sandbags, and riprap. Rates of erosion (seacliff retreat) range from 4-7 inches/year, though catastrophic rockfall is an important agent of erosion here.

2 Bluffs 50-65 feet and dune sand subject to collapse, partially protected by a gunnite wall.

3 Riprap is employed intermittently along this stretch of coast to the south side of Pismo Creek.

4 Foredunes south of Pismo Creek are subject to erosion during HHW with a storm surge. Protection of the dunes is hindered by continuous use of vehicles on the beach and among the dunes.

5 Ramp providing beach access destroyed, winter, 1983.

6 Ramp providing beach access destroyed, winter, 1983.

7 Riprap utilized at the mouth of Arroyo Grande Creek to protect dwellings. Though the homes here are fronted by dunes, strong storms can damage the property, as evidenced in winter, 1983.

Figure 14.6. Avila Beach Pier and Fossil Point, 1931. Photo by Fairchild Air Photos, UCLA.

Figure 14.7. Avila Beach Pier and Port San Luis, July 1983. Photo by Amalie Jo Brown.

East of Avila Beach the coast is characterized by active sea-cliffs, shore platforms exposed at low tide, and pocket beaches covered by higher high tide. From Fossil Point to Mallagh landing, vertical 50–100-feet-high cliffs are exposed. The cliffs at Fossil Point reach 200 feet in height and are prone to rockfall with debris common at the cliff base. East of Mallagh Landing a marine terrace 100–200 feet high reaches the coast in cliffs. A massive landslide, cut in the cliff at the west end of the embayment, provides sediment for the pocket beach fronting the cliff. Farther east, the terrace widens to 2,300 feet and is the site of single-family dwellings, some of which are clustered dangerously close to the cliff top. There is considerable evidence of recent cliff collapse, damaging or destroying the beach access stairs to the homes in several places.

Although active seacliff erosion continues southeast toward Spyglass Inn, where the terrace narrows, this has not deterred further cliff-top development in the form of single- and multiple-family dwellings. The northwestern section of Shell Beach also displays active cliff retreat that local residents have sought to counter with sandbags, wood bulkheads, and concrete seawalls (fig. 14.8). Controversy has recently been stirred by the proposed development of a cliff-top motel and recreation complex along the eastern section of Shell Beach where the cliff has retreated laterally up to 15 feet since 1965. From aerial photographs, average rates of cliff retreat ranging from 1 to 2.4 inches per year have been estimated for selected areas of Shell Beach. Such estimates, projected for a 100-year bluff setback required by the

Figure 14.8. Housing development at Shell Beach, July 1983. Photo by Amalie Jo Brown.

Local Coastal Plan, would cause portions of the seacliff to retreat onto U.S. Highway 101. Furthermore, a monitoring program begun in the fall of 1978 has yielded initial rates of cliff retreat on the order of 4 to 7 inches per year with catastrophic rockfalls at some locations, causing retreat to within 160 feet of U.S. 101. Ironically, in 1973 the eastern portion of Shell Beach was designated by the City of Pismo Beach as a "scenic and sensitive area" where "cliff retreat is a major problem," and was classified as "proposed open space."

The problem of cliff retreat persists to the southeast at Pismo Beach bluffs, 50–60 feet high. The bluffs have been gunnited (unsuccessfully) for protection. Damage from the winter storms of 1983 is evident near the Pismo Beach pier where riprap, rubble, and a concrete seawall have failed to provide adequate protection for the commercial and residential structures behind the sandy beach, 60 to 200 feet wide. At the present time a seawall is under construction to help protect a city parking lot that serves the beach community. Pismo Beach and its neighbor Grover City have become the focal points for relatively intense recreational use, owing to their proximity to Pismo State Beach. The state beach offers the opportunity for street-legal vehicle use on the beach, beach camping, and off-road vehicle (ORV) activity within a designated area of the Pismo dune complex. Access to the beach is provided by a ramp that is subject to damage during intense storm activity (fig. 14.9) and requires continued repair. At the present time the San Luis Obispo County Board of Supervisors has recommended closure of the entire dune area to ORVs. The foredunes of the Pismo complex are subject to erosion during storm surges at high high water (HHW). Additional management considerations have been imposed on this section of beach by the continued vehicular activity that has both destroyed vegetation critical to the maintenance of the dunes and interfered with archeological sites within the dunes.

The dune system here is characterized by 3 phases: (1) foredunes comprising low, active dunes adjacent to the beach; (2) an active middle transverse dune complex that rises to some 100 feet above sea level; and (3) an inner, old dune complex now

Figure 14.9. Beach access ramp at Pismo State Beach, March 1983. Photo by Kevin Mulligan.

largely vegetated with coastal scrub and some trees. Restoration of heavily damaged areas near Oso Flaco Lake has begun with the construction of enclosures and sand fences, while planners from the California departments of Fish and Game and of Parks and Recreation, from San Luis Obispo County, and from the California Coastal Commission are seeking to provide a clearly designated, limited-access area for camping, day-use parking, and dune use. It is clear that this section of coast is subject to moderate hazards, which intense human and vehicular abuse have amplified in areas to high hazards.

Santa Maria coast (figs. 14.5, 14.10, 14.11, 14.12, and 14.13)

The area between Pismo Beach and Point Pedernales is largely undeveloped owing to substantial land holdings by the state (Pismo State Beach) and the U.S. military (Vandenberg Air Force Base and the Point Arguello Naval Missile Facility). Characterized by long stretches of sandy beach with intervening rocky headlands, the area is generally designated as low hazard. High-hazard segments exist, for example at the mouth of the Santa Maria River where a sand barrier and dunes are breached during flood periods, and at Ocean Beach County Park near Surf where the Santa Ynez River debouches at the coast. Additionally, active seacliff erosion is found at Point Sal and the adjacent state park. The cliffs here are composed of the Franciscan melange and are subject to landsliding, providing rates of retreat on the order of 5–12 inches per year. Here and farther south near Lion's Head, landslides and debris flows have undermined or totally removed park access roads and hiking trails. Near Point Pedernales, the main line of the Southern Pacific Railroad between Los Angeles and San Francisco lies within 30 feet of the cliff top where it is protected by a concrete seawall and rock revetment.

Jalama coast (fig. 14.13)

The stretch of coast between Point Pedernales and Point Conception is a cliffed coastline with small pocket beaches. This area is also largely undeveloped, lying partially within the Point Arguello Naval Missile Facility and partially in private ownership.

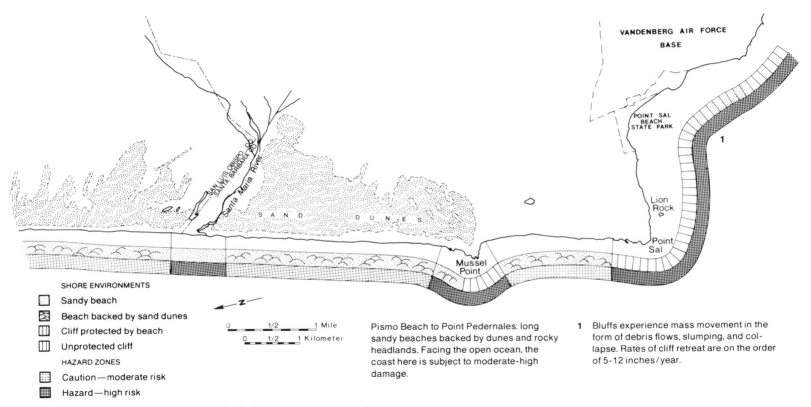

SHORE ENVIRONMENTS

- Sandy beach
- Beach backed by sand dunes
- Cliff protected by beach
- Unprotected cliff

HAZARD ZONES

- Caution—moderate risk
- Hazard—high risk

Pismo Beach to Point Pedernales: long sandy beaches backed by dunes and rocky headlands. Facing the open ocean, the coast here is subject to moderate-high damage.

1 Bluffs experience mass movement in the form of debris flows, slumping, and collapse. Rates of cliff retreat are on the order of 5-12 inches/year.

Figure 14.10. Site analysis: Santa Maria River through Vandenberg Air Force Base.

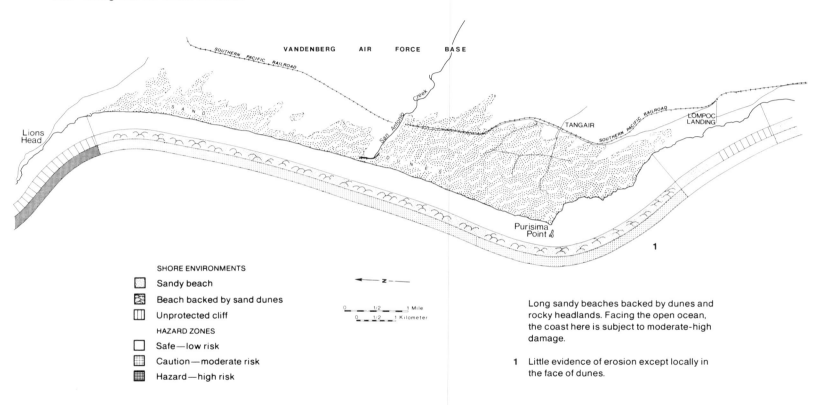

Figure 14.11. Site analysis: Lions Head through Lompoc Landing.

SHORE ENVIRONMENTS

- Sandy beach
- Beach backed by sand dunes
- Unprotected cliff

HAZARD ZONES

- Safe—low risk
- Caution—moderate risk
- Hazard—high risk

Long sandy beaches backed by dunes and rocky headlands. Facing the open ocean, the coast here is subject to moderate-high damage.

1 Little evidence of erosion except locally in the face of dunes.

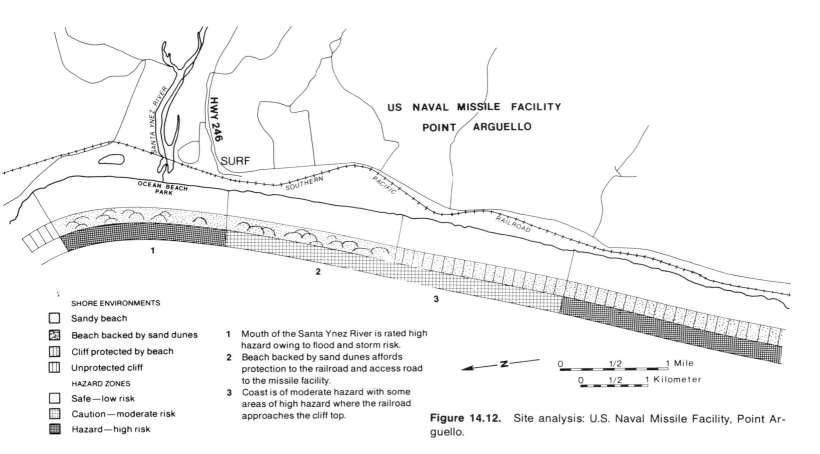

Figure 14.12. Site analysis: U.S. Naval Missile Facility, Point Arguello.

SHORE ENVIRONMENTS

Sandy beach

Beach backed by sand dunes

Cliff protected by beach

Unprotected cliff

HAZARD ZONES

Safe—low risk

Caution—moderate risk

Hazard—high risk

1 Mouth of the Santa Ynez River is rated high hazard owing to flood and storm risk.
2 Beach backed by sand dunes affords protection to the railroad and access road to the missile facility.
3 Coast is of moderate hazard with some areas of high hazard where the railroad approaches the cliff top.

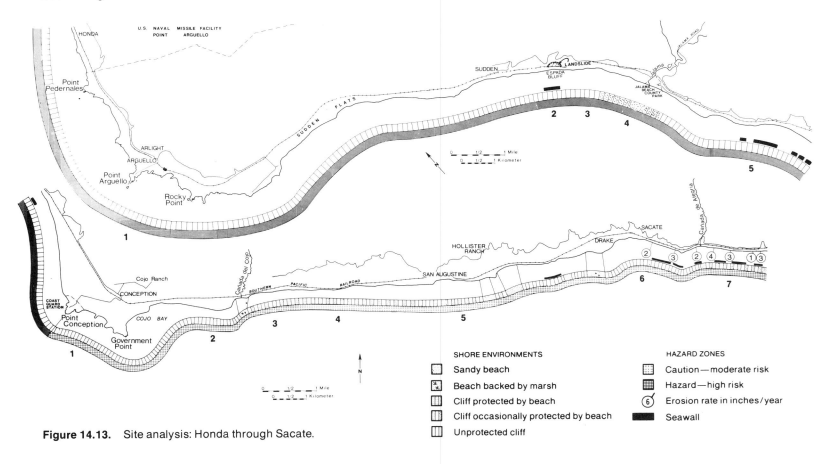

Figure 14.13.   Site analysis: Honda through Sacate.

Point Pedernales to Point Conception: active seacliffs rising up to 230 feet in height and broad marine terraces backed by higher bluffs of the Santa Ynez Mountains.

A

1 Frequent landslides along cliff face.
2 Espada Bluff is a landslide approximately 200 feet in height which is supported along its base by a seawall.
3 Espada Bluff to Jalama Creek: cliff is subject to rockfall.
4 At the mouth of Jalama Creek the county park was formerly protected by low dunes which, during winter, 1983, suffered severe erosion. Immediately south of the park, an access road lies dangerously close to the cliff top.
5 Seawalls afford partial protection for the railroad line along this section of the coast.

B

1 Seacliffs composed of Monterey Shale. Rocky promontories are separated by narrow sandy beaches. Coast is uninhabited, and any future development should be set back from coast.
2 Weak siltstone and shale. Erosion rates not determined due to lack of development, but probably similar to those to east with similar rock type. Beach sand commonly absent during winter.
3 Proposed site, liquified natural gas facility. Weak south-dipping siltstone and shale. Erosion rates are probably similar to those to east (2-3 inches/year). Sand commonly absent from abrasion platform.
4 Uninhabited coastline.
5 Sparse habitation on ~100 acres ranchos. Erosion rates are probably low to moderate (2-3 inches/year) based on geomorphic evidence (near vertical cliffs, small hanging drainages at seacliff).
6 Weak shale with average annual erosion rates of 2-3 inches/year. Due to lack of coastal development, railroad tracks are only structures with potential problems.
7 Monterey shale dips seaward. Gradual undercutting results in failure along bedding surfaces. Historic erosion rates vary between 1 and 4 inches/year. Sandy beaches present during only some years.

Figure 14.14. Sudden Flats area with the Southern Pacific Railroad line, Jalama Coast, 1931. Photo by Spence Air Photos, UCLA.

At Point Arguello and to the immediate south-southeast at Rocky Point, the U.S. Coast Guard lighthouse, radio tower, and access road are in danger as the bluffs composed of shales and siltstones suffer undermining, sliding, and rockfall. Rubble breakwaters, revetments, and concrete seawalls provide some protection for the Southern Pacific Railroad line near Sudden Flats, though arroyo incision (fig. 14.14) and mass movement are widespread. Although the cliffs are prone to rockfall (fig. 14.15), rates of seacliff retreat near Jalama Beach on the order of 12–20 inches per year have been recorded. Again the access roads and rail line are dangerously close to the cliff top and are partially protected by

Figure 14.15. Cliff collapse, Jalama Coast, July 1983. Photo by Vatche Tchakerian.

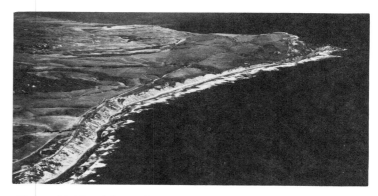

Figure 14.16. Jalama Coast, Southern Pacific Railroad line along cliff top in foreground and Point Conception in background, 1931. Photo by Spence Air Photos, UCLA.

Point Conception to Gaviota State Beach (figs. 14.13 and 15.2)

Historical development. Most of the beach land along the coastline from Point Conception to Gaviota State Beach is part of either the Hollister or Bixby ranches, which originated as Spanish land grants during the early nineteenth century. Hollister Ranch was sold in the early 1970s and approximately 2 square miles were separated as a proposed nuclear power plant site (never realized and subsequently proposed as a liquefied natural gas facility). The remainder of this large ranch was subdivided into small ranches of about 100 acres each. Present development is

crumbling seawalls 2.5 miles north of Point Conception. A narrow, sandy beach with exposed shore platforms backed by a mantle of dune sand form cliffs 50 to 80 feet high that terminate at Point Conception (fig. 14.16) where frequent erosion along the cliff base and face warrant a high-hazard rating.

Figure 14.17. Railroad tracks follow the seacliff on Hollister Ranch, about 2.5 miles west of Gaviota State Beach. Note absence of sand on intertidal abrasion platform. Photo by Tom Rockwell.

Figure 14.18. Dated seawall protecting railroad tracks on Hollister Ranch. Photo by Tom Rockwell.

light with most structures set considerably back from the coastline. Most of the land is presently used for grazing cattle and horses. Bixby Ranch to the west, which extends from Canada del Cojo to Point Conception, is virtually undeveloped except for dry farming and cattle grazing. There are no structures associated with the ranch that are endangered by seacliff retreat.

The Southern Pacific Railroad tracks nearly parallel the coast through this entire 17-mile distance (fig. 14.17). The tracks were laid at the turn of the century with the first trains running by 1901. Coastal erosion was apparently recognized fairly early on because the first seawalls designed to protect the tracks in this

area were constructed in 1909 (the dates for seawall construction are engraved into the ends of the concrete structures, see fig. 14.18). Because the individual seawalls were built and repaired and/or extended at different times, and because each wall is dated, the walls preserve a good record of the amounts of erosion over different time periods. Unfortunately, the seawalls were apparently only constructed where tracks were specifically endangered. Thus, control is excellent in the eastern part of the study area but virtually lacking to the west.

Geology and geomorphology. The stretch of coast between Government Point and Gaviota Beach State Park is relatively straight, with narrow, sand-deficient beaches and low to moderately high (15–60 feet on average) seacliffs fronting a marine terrace that is usually quite wide (fig. 14.19). The sand volume on the beach fluctuates from year to year and season to season. During the spring and summer of 1980, long stretches (up to several miles) of this coast were completely denuded of sand cover leaving only a bare rock platform (figs. 14.1–14.3). The presence of marine sand deposits on the raised Pleistocene marine terrace in this area suggests that this sand-deficient condition is probably not the norm. Much of the sand formerly supplied to the beaches by coastal streams in this area has now been trapped in reservoirs behind dams. About 66 percent of the drainage basin of the Santa Maria River, for example, is now blocked by dams. The effects of this type of sand loss on the extent and width of southern California beaches has been well documented and was discussed earlier in this book (see chapter 2).

A sand deficiency or decreased beach width along the coast directly increases coastal erosion or seacliff retreat. Sand beaches are the principal buffer zones to wave attack and protect seacliffs against abrasive erosion. With the removal (or nonsupply) of this protection, erosion rates increase. This historic erosion record from Point Conception to Gaviota is sparse and incomplete, and thus it is difficult to correlate directly periods of dam construction with acceleration in coastal erosion.

The seacliff and the uplifted marine terrace that form the coast-

Figure 14.19. Typical coastal morphology north of Gaviota State Beach. A wide, elevated terrace is fronted by a seacliff exposing shale beds dipping about 45 degrees seaward. Photo by Tom Rockwell.

line throughout this area consist of sedimentary rocks, dominantly shales and mudstones of the Sisquoc and Monterey formations. The rocks are distinctly layered or well bedded for the most part and dip or are tilted moderately steeply seaward (fig. 14.19). These factors facilitate cliff erosion as individual layers are undercut at the beach and the overlying material fails as this support is removed.

cating this retreat is an addition to the original structure. The geometry of the original seawall in figure 14.20, for example, does not allow for a precise determination of the amount of erosion in the time span between the installation of the 2 wall segments. High erosion rates over short time periods, interspersed with longer periods characterized by low rates, indicate that erosion is an episodic process here and elsewhere along the California coast.

At the end of one seawall west of Alegria Canyon on Hollister Ranch, 6.5 feet of erosion occurred between 1909 and 1915, when the original wall and an addition were built, respectively. Subsequently, only 5.5 feet of erosion has occurred (1915–80). The average long-term rate since 1909 is only 2 inches per year, the 1909 to 1915 rate is 13 inches per year, and the 1915–80 rate is only 1 inch per year. Similarly at a seawall west of Agua Caliente Canyon about 1 mile east of Alegria Canyon, the average retreat rate is 3 inches per year since 1928, but an addition in 1937 indicates that 6.2 feet of the total 12.8 feet occurred between 1928 and 1937 yielding a short-term rate of 8 inches per year. The post-1937 rate is only about 2 inches per year. Similar erosion patterns are indicated by data from other seawalls.

From the above, it appears that a large part of the erosion along this section occurred over short time periods and perhaps during single seasons due to abnormally large storms or severe winters. The data were collected during the summer of 1980, and it would be useful to document subsequent erosion in light of the severity of the winter of 1983.

Figure 14.20.   A seawall built in 1909 with an addition dating from 1915, showing seacliff retreat. Photo by Tom Rockwell.

Seacliff erosion. The amount of seacliff retreat was determined by directly measuring the distance from the end of the seawall to the present seacliff (fig. 14.20). This method is based on the assumption that the wingwalls of each seawall were initially attached to the seacliff, substantiated by bedrock fragments cemented into the end of the seawall where the wall turns toward the seacliff. In some cases, however, the part of the seawall indi-

15. Southern Santa Barbara County, Gaviota Beach to Rincon Point

Robert M. Norris

Southern Santa Barbara County (figs. 15.1, 15.2, 15.3, 15.4, and 15.5) lies in the western Transverse Range province and as a result has an east-west trending coast in marked contrast to much of the west coast of North America. This coast is dominated by the steep-fronted Santa Ynez Range for its entire length, and the coastal plain is narrow throughout, no more than 2.5 miles wide at any point. West of Dos Pueblos Canyon it is almost absent. Locally, the coastal plain is interrupted on its shoreward side by a few short, elongated hills, some with elevations of as much as 600 feet. All rocks and geologic structures in this area have a strong east-west trend.

Much of the coast is cliffed, and the average height of the cliffs is about 50 feet, but some cliffs are more than 200 feet high. All cliffs are cut in sedimentary rocks that are extensively deformed. Because none of the rocks forming the cliffs are particularly resistant to erosion, all cliffs are undergoing fairly rapid retreat.

Three segments of this coast, however, lack cliffs and are marked instead by low sandbars enclosing salt marshes or low-lying areas. The longest of these is the relatively undisturbed salt marsh at Carpinteria which is due in part to the presence of an offshore rocky reef exposed only at the very lowest tide. The wave shadow of this reef has produced a sandy, cuspate headland, Sand Point (fig. 15.6). This sandy headland encloses Carpinteria Salt Marsh.

The second segment of low-lying coast is at Montecito, mainly to the west of Fernald Point. This part of the coast lies at the foot of a gently sloping plain formed from flood deposits left by the various short and intermittent streams that drain the south face of the Santa Ynez Range in the vicinity of Montecito. Cliffs, where present, are only a few feet high.

Prior to urbanization, which began about 150 years ago, the coast at the city of Santa Barbara was also low and marshy. It remains low, of course, but little remains of the former marsh apart from the Andree Clark Bird Refuge. This stretch of shoreline has been greatly modified by urban development and by the construction of Santa Barbara harbor and breakwater in the late 1920s, about which more will be said.

About 10 miles west of Santa Barbara is Goleta Slough, once a rather large salt marsh. This feature is mostly cut off from the sea by coastal hills and by a short sand spit blocking most of the entrance. Only a small amount of the original salt marsh remains. Much of it was filled during the development of Santa Barbara Airport in World War II days. Goleta Slough, like Carpinteria Salt Marsh, has been reduced in area by periodic severe floods that have brought large amounts of sediment into the marsh from the nearby mountains. The large areas of open water that charac-

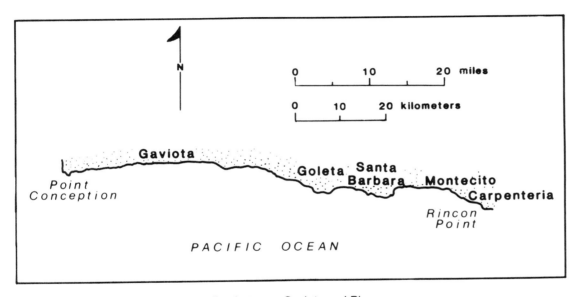

Figure 15.1. Location map for coastline between Gaviota and Rincon.

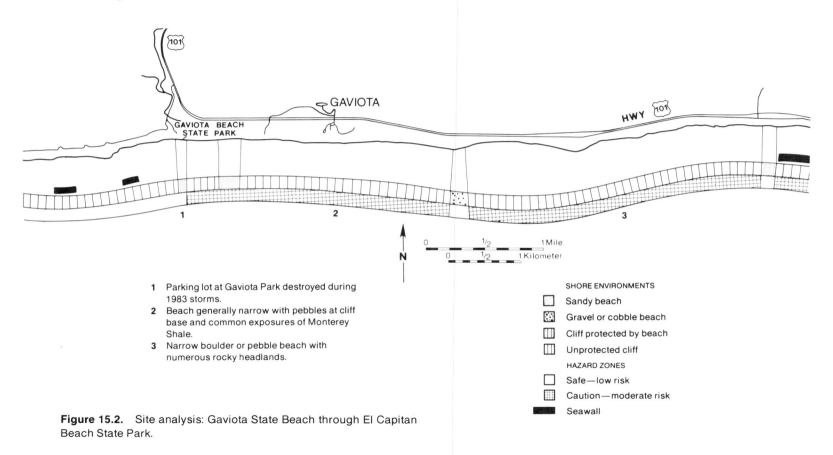

1 Parking lot at Gaviota Park destroyed during 1983 storms.
2 Beach generally narrow with pebbles at cliff base and common exposures of Monterey Shale.
3 Narrow boulder or pebble beach with numerous rocky headlands.

SHORE ENVIRONMENTS

☐ Sandy beach
▦ Gravel or cobble beach
▥ Cliff protected by beach
▥ Unprotected cliff

HAZARD ZONES

☐ Safe—low risk
▦ Caution—moderate risk
■ Seawall

Figure 15.2. Site analysis: Gaviota State Beach through El Capitan Beach State Park.

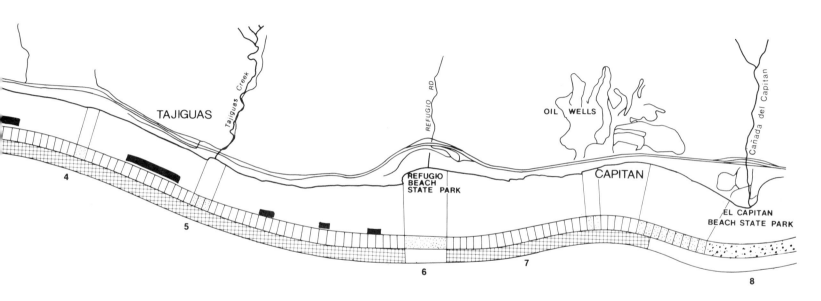

4 Moderately protected shoreline with two 3000-foot long concrete seawalls.
5 Beaches generally narrow with pebbles or boulders at cliff base. Few sandy pocket beaches.
6 Park damaged by storm debris during winter, 1983.
7 Storm damage to the bike path during winter of 1982–83.
8 Sandy beach and boulder delta at El Capitan Beach State Park.

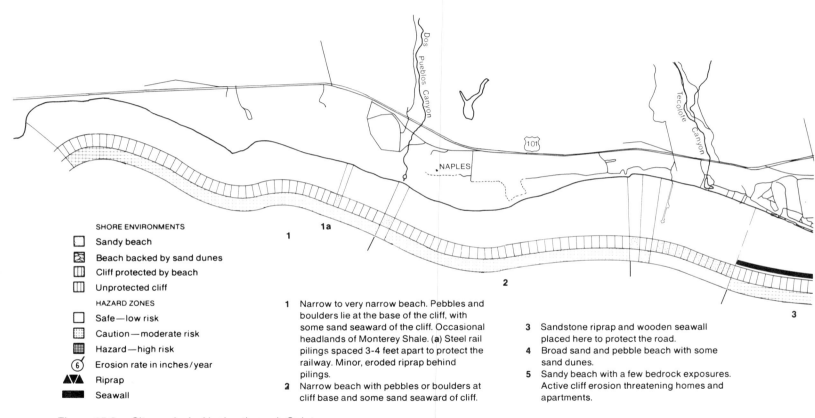

SHORE ENVIRONMENTS

☐ Sandy beach

▨ Beach backed by sand dunes

⊞ Cliff protected by beach

⊞ Unprotected cliff

HAZARD ZONES

☐ Safe—low risk

▨ Caution—moderate risk

▨ Hazard—high risk

⑥ Erosion rate in inches/year

▲▼▲ Riprap

▬▬ Seawall

1 Narrow to very narrow beach. Pebbles and boulders lie at the base of the cliff, with some sand seaward of the cliff. Occasional headlands of Monterey Shale. **(a)** Steel rail pilings spaced 3-4 feet apart to protect the railway. Minor, eroded riprap behind pilings.

2 Narrow beach with pebbles or boulders at cliff base and some sand seaward of cliff.

3 Sandstone riprap and wooden seawall placed here to protect the road.

4 Broad sand and pebble beach with some sand dunes.

5 Sandy beach with a few bedrock exposures. Active cliff erosion threatening homes and apartments.

Figure 15.3. Site analysis: Naples through Goleta.

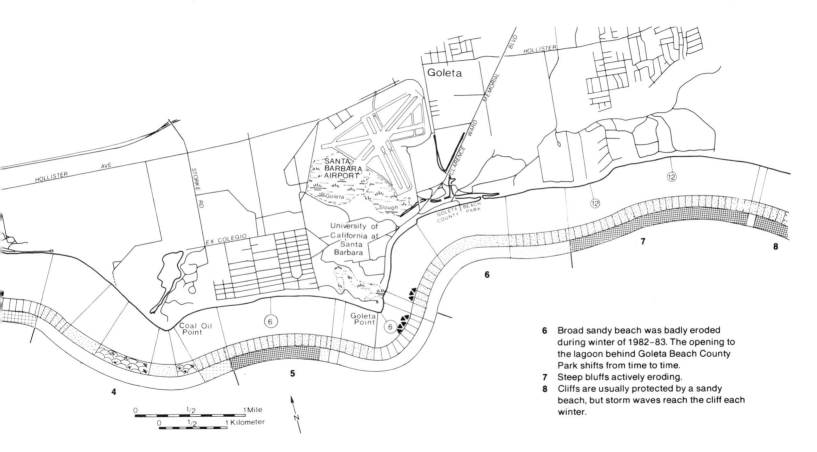

6 Broad sandy beach was badly eroded during winter of 1982–83. The opening to the lagoon behind Goleta Beach County Park shifts from time to time.

7 Steep bluffs actively eroding.

8 Cliffs are usually protected by a sandy beach, but storm waves reach the cliff each winter.

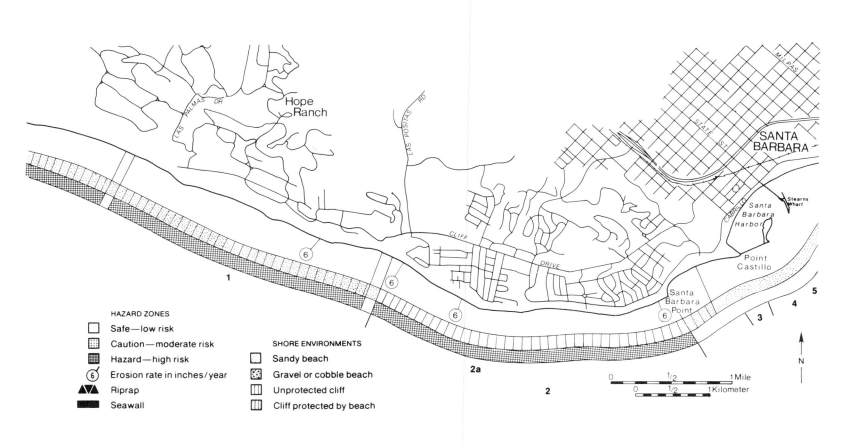

Hope
Ranch

SANTA
BARBARA

Stearns
Wharf

Santa
Barbara
Harbor

Point
Castillo

Santa
Barbara
Point

1

6

6

6

6

2a

2

3

4

5

N

HAZARD ZONES

☐ Safe—low risk

▨ Caution—moderate risk

▣ Hazard—high risk

⑥ Erosion rate in inches/year

▲▼▲ Riprap

■■ Seawall

SHORE ENVIRONMENTS

☐ Sandy beach

▦ Gravel or cobble beach

▥ Unprotected cliff

▥ Cliff protected by beach

0   1/2   1 Mile

0   1/2   1 Kilometer

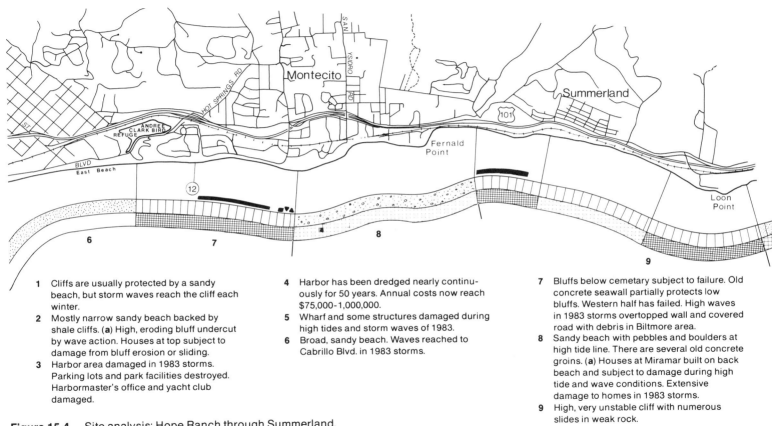

1 Cliffs are usually protected by a sandy beach, but storm waves reach the cliff each winter.

2 Mostly narrow sandy beach backed by shale cliffs. (**a**) High, eroding bluff undercut by wave action. Houses at top subject to damage from bluff erosion or sliding.

3 Harbor area damaged in 1983 storms. Parking lots and park facilities destroyed. Harbormaster's office and yacht club damaged.

4 Harbor has been dredged nearly continuously for 50 years. Annual costs now reach $75,000-1,000,000.

5 Wharf and some structures damaged during high tides and storm waves of 1983.

6 Broad, sandy beach. Waves reached to Cabrillo Blvd. in 1983 storms.

7 Bluffs below cemetary subject to failure. Old concrete seawall partially protects low bluffs. Western half has failed. High waves in 1983 storms overtopped wall and covered road with debris in Biltmore area.

8 Sandy beach with pebbles and boulders at high tide line. There are several old concrete groins. (**a**) Houses at Miramar built on back beach and subject to damage during high tide and wave conditions. Extensive damage to homes in 1983 storms.

9 High, very unstable cliff with numerous slides in weak rock.

Figure 15.4. Site analysis: Hope Ranch through Summerland.

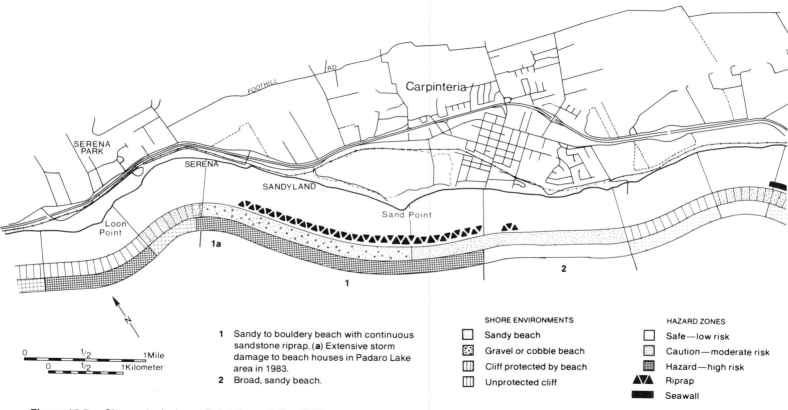

SERENA
PARK

SERENA

SANDYLAND

Loon
Point

Sand Point

Carpinteria

FOOTHILL RD

1a

1

2

N

| 0 | 1/2 | 1Mile |
| 0 | 1/2 | 1Kilometer |

1 Sandy to bouldery beach with continuous
sandstone riprap. (**a**) Extensive storm
damage to beach houses in Padaro Lake
area in 1983.
2 Broad, sandy beach.

SHORE ENVIRONMENTS

- Sandy beach
- Gravel or cobble beach
- Cliff protected by beach
- Unprotected cliff

HAZARD ZONES

- Safe—low risk
- Caution—moderate risk
- Hazard—high risk
- Riprap
- Seawall

Figure 15.5. Site analysis: Loon Point through Sea Cliff.

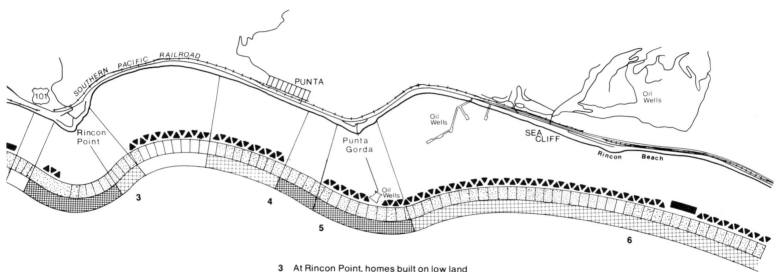

Rincon Point to Ventura River mouth: low, erodible bluffs now largely protected by riprap, revetments, and short seawalls. Developments within this shorezone are subject to periodic erosion as in winter, 1983 when some houses collapsed and road destruction occurred.

3 At Rincon Point, homes built on low land were extensively damaged during 1983 storms.

4 The revetment protecting the freeway ends at the south end of Punta, and over a distance of ½ mile southward low bluffs up to 20 feet high are subject to periodic erosion which, during winter 1983, undermined the freeway.

5 Mussel Shoals development at Punta Gorda is protected by poorly designed riprap, and the headland is prone to erosion.

6 From Sea Cliff southwards, the Rincon Parkway, offering access to the shore, is protected by a poor rubble revetment and a short seawall.

terized these marshes in the eighteenth and nineteenth centuries have been substantially reduced by both natural and man-made causes.

A very small salt marsh, known locally as Devereux Lagoon, lies just west of Goleta Slough at Coal Oil Point. This feature is now flooded by high tides for short periods chiefly during the winter and early spring months, but it is dry most of the year and supports very little salt marsh vegetation.

Coastal headlands

The coastline is fairly straight, particularly from Point Conception eastward to El Capitan Beach, but beginning there and continuing beyond Rincon Point the coastline is interrupted by a number of small headlands. At El Capitan Creek a small, cuspate boulder delta has been formed by the deposition from Cañada del Capitan Creek (fig. 15.7). A similar feature, but much smaller, occurs at the mouth of Gato Canyon. More prominent headlands occur near Goleta and at Santa Barbara, as does the broadest coastal plain in the Santa Barbara region. At Rincon Point, on the Ventura–Santa Barbara County line, another prominent deltaic headland has been produced by Rincon Creek. This feature is similar to the one at El Capitan. Several other smaller deltaic features occur along the Santa Barbara coast, but because the streams that supply them are short and intermittent, the deltas are quite small and have only minor effects on the general trend of the coast. As a rule, where boulder deltas have developed,

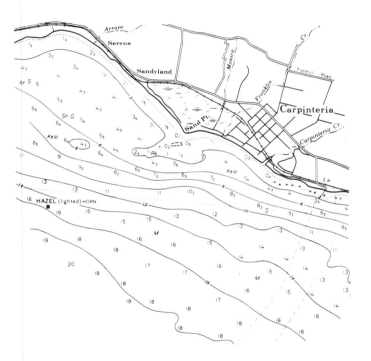

Figure 15.6.   A portion of the 1972 U.S. Coast and Geodetic Survey coastal chart showing location of offshore reef at Sand Point.

the tributary streams are longer, have relatively larger drainage basins that extend up into the high ridges, and have either permanent or nearly permanent flows.

A prominent but low headland occurs at the mouth of Carpinteria Salt Marsh, although the natural shoreline is formed of loose, sandy beach deposits. The cause of this feature is a prominent rocky reef lying just offshore from the salt marsh inlet (fig. 15.6). This reef is exposed only during the very lowest tides, perhaps 2 or 3 times a year, but the reef is nonetheless enough of an impediment to approaching waves to refract and reduce the inshore wave height much as a breakwater would. As a result, both the longshore current and beach drift falter and together deposit a greater amount of sand there. It is a tombolo (long sandbar) of sorts.

Figure 15.8. Steeply dipping, thin-bedded Monterey Formation. Photo by Robert Norris.

Figure 15.7. Small boulder delta at the mouth of Cañada del Capitan. Photo by Robert Norris.

Rocks of the seacliffs

Apart from the narrow canyon mouths the entire coastline westward from Goleta to Point Conception is cliffed, and the bedrock exposed in the cliff face belongs either to the Monterey or Sisquoc formations (fig. 15.8). Both are weak, extensively fractured, and easily eroded rocks. Numerous small rockslides occur as marine erosion undercuts the base of the cliff (fig. 15.9).

Figure 15.9.   Rockslide in Monterey Formation about 2 miles west of Santa Barbara Point. Photo by Robert Norris.

Figure 15.10.   Wave-cut platform, just west of Santa Barbara Point. Photo by Robert Norris.

As a consequence of rock weakness, a prominent wave-cut platform has developed in front of the cliffed portions of the Santa Barbara coast despite the fact that the offshore islands provide a barrier to waves from the open Pacific. The wave-cut platform extends seaward perhaps as much as 1 mile in some places, but generally a shorter distance. In winter, particularly after severe storms when beaches have been eroded, the inshore edge of this wave-cut platform is well exposed (figs. 15.10, 15.11).

East of Goleta, cliffs expose other rocks in addition to the Monterey and Sisquoc formations. At More Mesa, for example, prominent cliffs about 100 feet high have been cut in massive siltstones (fig. 15.12). East of More Mesa to Hope Ranch and as far as Santa Barbara Point, the cliffs are again composed of Monterey Formation. From Santa Barbara harbor to the eastern side of the Carpinteria lowland, the lower seacliffs are composed mainly of boulders at Montecito, or the sandy Casitas Formation at Summerland where the seacliff is nearly 200 feet high. From Carpinteria eastward to Rincon Creek, the seacliffs are once again made of the Monterey Formation.

Although the largest part of the seacliff is cut from the formations just mentioned, the uppermost part often consists of nearly horizontal terrace and dune sands, typically 50 to 10 feet thick.

As a rule, the basal part of these nearly horizontal deposits contain many fossil mollusks, including numerous rock-boring clams that have drilled into the underlying bedrock (fig. 15.22C).

Beaches

For the most part beaches along the Santa Barbara coast are thin and narrow and are subject to marked degradation during stormy winters. This is usually followed by erratic recovery during the ensuing summer and autumn. This seasonal pattern is certainly not unique to the Santa Barbara area, but is particularly prominent there because of the normally small sand supply that can be removed quickly in winter, exposing or partially exposing the rocky platform that fronts the seacliffs.

These seasonal beach changes vary in duration and magnitude from year to year. Some years show little change, but other years, such as the abnormally stormy winter of 1982–83, produced the most severe stripping of the beach seen in many decades (fig. 15.11).

The seasonal cycles are further modulated by other long-term changes such as those in total sand supply. The Santa Barbara coast from Point Conception as far eastward as Point Hueneme in Ventura County is considered a littoral cell, or beach compartment. Little or no sand is transported eastward from the west-facing beaches north of Point Arguello around that headland and Point Conception to south coast beaches. As a result, south coast beaches are entirely dependent upon sand sources on

Figure 15.11. Shore platform cut in Sisquoc Formation at Coal Oil Point. This platform is normally covered by sand but was stripped by winter storm waves. Photo by Don Norris.

the south side of the Santa Ynez Range, delivered by the intermittent, short, steep streams draining that mountain face. One minor exception occurs at Gaviota Creek, which has breached the axis of the range and drains a small area on the north side of the range. The total amount of sand in transit along the coast increases eastward and has been measured at Santa Barbara harbor as a result of annual dredging. Volume approximates 300,000 to 350,000 cubic yards per year.

Fortunately for south coast beaches, very few of the tributary creeks are dammed or channelized so that sand supply to beaches is in a nearly natural state to date. Rainfall, in this region of Med-

Figure 15.12.   Cliffs at More Mesa, west of Santa Barbara, cut in the Santa Barbara Formation. Massive, inactive tar seeps in foreground and in front of the stairway. Photo by Robert Norris.

iterranean climate, produces a highly erratic pattern of streamflow.

Higher than normal rainfall, that is, in excess of the usual annual 18 inches at the coast and 35 inches at the mountain crest, is no assurance that streamflow will be adequate to transport appreciable new sand to the beaches; the rains must be closely spaced and of sufficient magnitude to produce a sustained and vigorous runoff. The chances are that the wetter years will produce more runoff and therefore more sand supply than the drier years, but this is not always true. The U.S. Army Corps of Engineers studies showed a tendency for beach widening to correlate best with rainfall totals for 2-year periods. Because sand delivered to stream mouths takes time to reach beaches downcoast, the Corps found a 3.5-year time lag in response at Santa Barbara harbor. For example, if the 1970–71 and 1971–72 rainfall totals exceeded the norm, an increased influx of sand would be expected at Santa Barbara harbor during the calendar year 1975.

No significant barriers to the eastward longshore movement of sand exist between Point Conception and Santa Barbara, though several small headlands are present. At Santa Barbara, however, an L-shaped breakwater, open to the east, was constructed in the late 1920s and early 1930s. This structure has seriously interfered with the sand supply to beaches lying east of the city at least as far as Rincon Point. For various reasons discussed more fully later in this chapter, the beaches between Santa Barbara and Rincon Creek were denied their normal sand supplies during most of the 1930s. The only sand reaching these beaches during that period came from the immediate offshore area as a result of normal seasonal processes, or from the streams reaching the coast east of Santa Barbara. As a result, serious sand losses occurred, and the beaches did not recover until the 1940s. Further damage has been prevented by more-or-less regular sand bypassing at the harbor since then.

To the observant reader, it will now be evident that sand transport along the Santa Barbara coast is dominantly from west to east. This comes about because most waves enter the Santa Barbara channel from the west and are refracted so as to approach the shore obliquely.

Because of the nearly continuous chain of islands on the south side of Santa Barbara Channel very few waves approaching from the south ever reach the Santa Barbara coast. The 10-mile-wide gap between Anacapa Island, at the eastern end of the island chain, and the mainland at Port Hueneme does allow some waves to enter the channel from the southeast. Such waves are produced, typically, by tropical storms off Mexico and Central America and are most apt to occur during the late summer and early fall. In addition, some winter seasons are characterized by "southeasters." Both these situations may reverse the normal direction of sand movement alongshore. The number of days on which beach sand is moved toward the west varies widely from year to year, but would very seldom exceed 30 days in any single year. Despite such occasional reversals, beach and harbor maintenance is largely dependent upon the west-to-east pattern, and corrective measures can safely ignore the infrequent reversals.

Immediate offshore environment

As was noted earlier, the Santa Barbara coast is dominantly a cliffed one fronted by a wave-cut platform cut in the rocks that compose the cliffs. The width of the bedrock platform is, of course, variable, but averages perhaps .5–1 mile in most places. The outer margin of this rock platform is marked in many places by a band of kelp attached, for the most part, to bare rock. Between the kelp band and the beach, the bottom is generally sandy, due in part no doubt to the effectiveness of the kelp in damping out short, steep winter waves approaching the beach from open water. Long, low summer waves are scarcely affected by the kelp and tend to sweep some of the sand shoreward onto the beach.

A man-made submerged ridge off East Beach in Santa Barbara is of special interest because it represents an early, unfortunate attempt to bring about sand bypassing at Santa Barbara Harbor. In 1933 sand first began to pass around the outside of the breakwater and accumulate in the harbor entrance. Dredging was initiated in 1935. Because a hopper dredge was used on that occasion, some clearance below the hull was required in order to discharge the load of sand. A disposal site was chosen off East Beach in about 20 feet of water where 202,000 cubic yards were dumped. It was expected that this ridge of sand would be moved inshore and supply the eastflowing current with new supplies of sand to nourish the depleted beaches farther east.

Unfortunately, this did not occur to any appreciable degree. The ridge of sand, about 2,000 feet long, 1,000 feet offshore, has remained nearly unaltered since 1935 and it can be seen plainly on the local navigation chart (fig. 15.13). This result also shows that wave activity in the Santa Barbara area is generally too weak to move sand shoreward if it is in water more than about 15

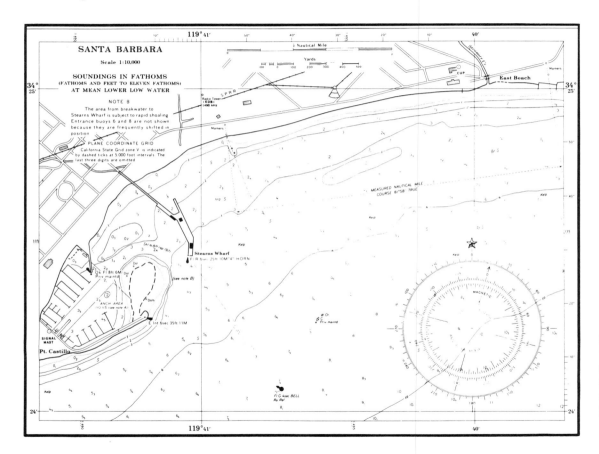

Figure 15.13. A portion of the 1972 edition of the U.S. Coast and Geodetic Survey chart 5120 showing the dredge spoil deposited by the hopper dredge in 1935.

feet deep. Subsequent dredgings, in recognition of this problem, have utilized suction dredges, and the spoil has been pumped onto the face of East Beach near the foot of Stearns Wharf. This has stabilized if not fully restored the beaches lying to the east.

History of Santa Barbara Harbor

Until the late 1920s there was no harbor at Santa Barbara, only an open roadstead and a wharf. As a result, there was no safe anchorage for fishing, commercial, or recreational craft, and all vessels calling at Santa Barbara were at the mercy of the weather. Local interests 5 times submitted requests for federal advice about the feasibility of building a harbor at Santa Barbara. The first of these requests was made in 1873 and the last in 1921. The Army Corps of Engineers in every case recommended against any new construction. Nevertheless, local interests and desires prevailed, and a $750,000 L-shaped breakwater parallel to the beach was constructed in 1928 (fig. 15.14). Initially, the breakwater had a gap between the beach and the short arm of the L, ostensibly to permit the longshore currents to carry sand through the harbor to the beaches to the east. What was then not sufficiently appreciated was that the current and the associated beach drifting can be maintained only by the very waves the breakwater is designed to block. Unfortunately, this same lesson has had to be relearned and relearned at other places, such as at Santa Monica near Los Angeles, where the same mistake was made only 5 years later.

Figure 15.14.   Aerial view of Santa Barbara breakwater under construction in 1928. Photo by Fairchild Aerial Surveys.

The open, L-shaped breakwater was completed in early 1929, and by the fall of that year shoaling inside the harbor had become so serious that the short end of the L was extended to the beach, a distance of about 600 feet. This was followed by a rapid deposition of sand on the west side of the breakwater. By 1933, a triangular area west of the breakwater had been completely filled, the shoreline realigned, and sand began moving along the long arm of the breakwater into the harbor entrance (figs. 15.15, 15.16, 15.17).

Figure 15.15. Santa Barbara breakwater, attached to Point Castillo, with about 1 year of sand accumulation to the west in October 1930. Photo by Spence Air Photos, UCLA.

While beaches immediately west of the breakwater built up and widened from 1928 to 1933, beaches as far as Carpinteria, 13 miles to the east, suffered varying degrees of erosion. Some previously sandy beaches were stripped of most of their sand, leaving only a narrow band of pebbles and boulders. It is curious that the full effects of impounding the sand upcurrent of the breakwater were not anticipated because the general west-to-east drift of sand had long been known in the area. A load of boulders with a mean diameter of 2 feet dumped on the beach just west of Summerland in 1917, for example, had been moved east as much as 2,500 feet by 1938. They traveled mainly in the band between low and high water and were observed to travel more rapidly after 1938 when sand depletion began to affect that beach. Yet another example occurred in 1931 when a 20 horsepower boiler broke through a wharf and fell onto Summerland beach. It was deliberately filled with loose rock to prevent its movement, but despite this, it had moved 900 feet to the east by 1938.

Sand moving along the outer edge of the breakwater and into the harbor entrance starting in 1933 had, by 1935, almost entirely blocked the harbor entrance. This was to happen numerous times in later years, most recently in 1983, when the harbor was blocked for about a month after an abnormally stormy winter.

The 1935 hopper dredging, of course, did nothing to check the erosion of beaches east of Santa Barbara because the spoil was not put on the feeder beach. Moreover, because the breakwater had caused accretion on its west side and because no significant amount of sand was bypassed to the feeder beach until 1938, the eastern beaches were subjected to sand starvation for roughly 10 years, resulting in severe erosion. By 1942, however, the beaches for about 3 miles east of Santa Barbara had been restored almost to their pre-1928 condition, but such was not the case farther east. At Sand Point in January 1940 the beach eroded landward 245 feet and caused about $2 million in property damage. By 1941, however, with continued dredging and sand bypassing, sand had begun to reappear on the rocky beaches as far east as Fernald and Loon points.

Because dredging has continued more or less regularly to the

Figure 15.16. Sand deposition west of Santa Barbara breakwater, December 1934. In 1933 sand had begun to move along the outer arm of the breakwater and into the harbor. Photo by Spence Air Photos, UCLA.

Figure 15.17. Sand beginning to develop a bar or spit at the end of the breakwater. Santa Barbara Point in foreground. Photo by Spence Air Photos, UCLA.

present time, beaches in southern Santa Barbara County have remained generally stable. However, owing to the vagaries of rainfall and winter storms, erosion is noticeably more severe in some winters than in others. The winter of 1982–83 is a case in point, and beaches were severely eroded along the entire Santa Barbara shore (fig. 15.11). Although some substantial recovery was evident by early summer of 1983, it remains to be seen whether or not full recovery will occur.

Some method of continuous sand bypassing at Santa Barbara Harbor has been discussed since the early 1940s, but no permanent pumping plant has ever been installed. Although such a plant would have many advantages over the periodic and irregu-

lar dredging, it is very probable that any plant, no matter how well designed, would miss some of the incoming sand, perhaps even a large amount. This sand would then pass around the intake and be deposited in the harbor entrance, requiring occasional dredging even though the sand bypassing plant was doing most of the job. In recent years annual dredging costs have ranged from $350,000 to over $1.7 million to move from 125,000 to over 600,000 cubic yards of sand.

Over the years it has been amusing to read in the local newspapers imaginative suggestions for bypassing sand without dredging. Many of these, of course, have been tried at other places and found wanting. A recurring suggestion involves some variation of a detached breakwater that is supposed to allow the sand to be transported along the foreshore through the harbor. All such suggestions overlook the fact that sand transport is generated by waves, and breakwaters are designed to reduce waves (and therefore reduce the shore current as well), not to mention the fact that this very method was tried and failed at Santa Barbara in 1928 and 1929.

Other suggestions offered lie more in the realm of magic and legerdemain. Some years ago a local inventor claimed to have designed a machine that could be housed in a small building at the end of the breakwater. This machine, the nature of which was secret, allegedly would bypass sand across the harbor entrance to East Beach where nature would take care of it. This particular inventor thoughtfully recommended that the small building housing his machinery be given a Spanish tile roof in keeping with the city's dominant architectural theme.

Nature, of course, has her own way of taking care of the problem and if allowed to work her will would restore the beaches to their pre-1928 condition. This would come about (as it has nearly done several times) by deposition of a curving sandbar from the tip of the breakwater to East Beach near the foot of Stearns Wharf. Once this bar or new beach was in place, sand would move uninterruptedly along the outside face of the breakwater, across the harbor entrance, and eastward down the coast. The obvious trouble with this solution is that it would eliminate Santa Barbara Harbor as a useful boat facility and convert it into a shallow lagoon.

Seacliff erosion

While the most dramatic examples of coastal erosion in the Santa Barbara area involve retreat of low, sandy beaches, such as the 245 feet of beach erosion that occurred at Sand Point near Carpinteria in 1940, the sea cliffs are also eroding steadily if unspectacularly. Rates of cliff retreat in the Santa Barbara–Goleta area are well documented and are probably typical for the rest of the south coast because the rocks are similar throughout, the coastline has about the same trend, and exposure to wave activity is very similar from Gaviota to Rincon Point.

Over the last 60 years particularly, a number of bench marks and triangulation station monuments have been placed along the top of the seacliff in the Santa Barbara area by such agencies as the U.S. Coast and Geodetic Survey (U.S.C.&G.S.) and the U.S. Geological Survey (U.S.G.S.). Because the original descriptions of these monuments often include rather precise measurements to the upper edge of the seacliff, they can be very useful for determining rates of retreat. Other dated or dateable structures such as storm drains, roadways, houses, and the like may also provide useful data. Aerial photographs likewise can be very useful, though they may not show small changes over short periods as clearly as do bench marks and triangulation stations.

Numerous other features demonstrate the reality of cliff retreat, but they do not generally give very accurate rates. Mature trees at the cliff edge obviously were not planted in these precarious positions, and where some of their root systems are exposed, erosion is clearly demonstrated (fig. 15.18).

Figure 15.18. Mature eucalyptus tree at Santa Barbara Point show-ing erosion of the root system in 1968. Photo by David Doerner.

In the Santa Barbara area, as doubtless is the case elsewhere, it is often assumed that cliff retreat is largely or entirely due to wave activity, and protective measures are then designed to cor-rect only this part of the problem. There is plenty of evidence to show that nonmarine processes such as runoff, landsliding, and groundwater seepage are important and may account for half or even more of the retreat. Not infrequently, of course, marine processes set the stage for nonmarine processes to act. Wave

erosion at the cliff base often removes support, and heavy rains may trigger landslides in the upper parts of cliffs well above the reach of waves (figs. 15.9, 15.19).

Numerous examples of landsliding occur along the Santa Bar-bara coast, including some long-inactive examples as well as currently active ones. One of the more dramatic slides of recent date occurred in the city of Santa Barbara in February 1978, following heavy winter rains in the preceding weeks. This slide destroyed or damaged several houses, garages, and properties in a few hours' time, beginning in the evening (fig. 15.20). By noon the next day most of the motion had stopped, but when the sea trims off the toe of this slide, some renewal of movement may occur. This particular slide caused a new, nearly vertical cliff to be established about 150 feet inshore from the previous cliff. The slide was a typical rotational slump on a curved failure surface that was nearly vertical at the cliff but flattened seaward.

One rather unusual type of landslide affects a number of sites along the Santa Barbara coast as well as the coastline east of Rincon Point in Ventura County and the north coast of Santa Cruz Island to the south, across the Santa Barbara Channel. This type of slide is induced by burning tar seeps. The Monterey Formation particularly contains fractures and voids filled with asphaltum. For various reasons these deposits occasionally get ignited and burn or smoulder for long periods, often producing smoke and steam that bakes or even fuses the enclosing rock. In some cases the heating probably induces fracturing, but in others the asphaltum has been acting as a binder or cement for per-

Figure 15.19. Debris slide in weakly consolidated older alluvium, about one-quarter mile east of Loon Point. Photo by Robert Norris.

Figure 15.20. Landslide of 14–15 February 1978 at Santa Barbara. Most of this slide occurred during a single 24-hour period following a 2-month period in which about 16 to 18 inches of rain had fallen. Photo by D. W. Weaver.

vasively fractured rock. Once the asphaltum is burned, the rock crumbles readily and produces rockfalls or small slides.

Examples occur just east of Rincon Point in Ventura County where the Southern Pacific has had to erect electric warning fences at the cliff base to warn trains of slides blocking the tracks. These slides have been ignited several times. Another example of a burning tar seep creating problems for cliff-top development occurred west of Coal Oil Point at the time the Sandpiper Golf Course was developed in the early 1960s. This seep was smouldering and was extinguished by bulldozing a pit on the top of the seacliff into which seawater was pumped. When the pit was dug, the smouldering abruptly changed to open flames from which the bulldozer operator narrowly escaped. The water eventually put out the burning and produced a number of temporary steam vents in the golf course area as water reached the smouldering asphaltum.

In some cases, tar seeps have protected the seacliff from erosion by producing a tough conglomerate rock. An area of currently inactive tar seeps just east of the mouth of Goleta Slough is an example (15.21). At the east side of Carpinteria Beach State Park there are low cliffs composed of tar-cemented rubble, and at one point a small, active seep has formed a fan-shaped mass of asphaltum on the beach.

Figure 15.21. Recently active tar seeps flowing onto the beach at More Mesa, just east of Goleta Slough. Photo by David Doerner.

Hazards

Shoreline hazards in the Santa Barbara area, as elsewhere, are chiefly the result of some sort of construction within an actively changing environmental setting. Where the shoreline is undeveloped or in parkland, there is little problem with geologic change. But where railways, roads, and houses have been built too close to the shoreline, the costs of protection, maintenance, and repair after storms, high tides, or even normal erosion may ultimately be so great that money would have been saved for governments and private owners alike had the shorefront been initially acquired by some public entity and maintained as open space. Some sort

of in-depth cost-benefit analysis certainly should accompany every new plan to develop land within 100 to 200 yards of the high-tide line in the Santa Barbara area.

Along the Santa Barbara shore, cliff retreat averages between a few inches and a foot or more per year. For this reason alone it is foolish to permit any permanent construction closer than 100 feet from the cliff top, with the possible exception of roads. Even where it will take 40 or 50 years for cliff retreat to reach a building, it is naive to suppose that the building owners will then be prepared to write off his or her investment on the ground that the useful life or cost of the building is fully amortized. Buildings change hands, and few owners will be able to move the building to a safer site or be prepared to write off the cost, so pressure is inevitably brought on governmental agencies either to permit or to build protective seawalls and the like.

Furthermore, the progress of cliff erosion is notoriously erratic. Some years pass with little or no visible erosion, but stormy years are characterized by sudden slab failures at many places, and several feet of retreat may occur during a single storm. The average of 3–12 inches per year is based on 40 or 50 years of experience. If the past 50 years has been an unusually quiet period, as some suggest, the hundred-year average could be appreciably larger than the currently established rate.

In the Santa Barbara area all seacliffs are cut from weak rocks, and all can be expected to erode landward well into the future, especially if sea level remains stable or rises. The rate of erosion is not high as worldwide rates go, but it is enough so that a substantial setback should be required for anything but the most temporary structures, such as fences, walks, and possibly public roads. Numerous examples of inadequate setback occur in the Santa Barbara area, as well as numerous examples of very expensive protective structures (figs. 15.22A–E, 15.23A–C). It is difficult to generalize about an appropriate setback because the rate of cliff retreat varies depending upon rock structure and stability, height of cliff, susceptibility to landsliding, and so on. Further, the setback should be based on some specific time

period. Is 100 years appropriate, or is 50 years better? Anything less than 50 years is certain to create problems because it is close to the useful life of many buildings. Long-term public interests as opposed to short-term private interests would seem to call for a minimum setback of 100 feet or so. Where a public road parallels the cliff within perhaps 100 yards, public interests (and ultimate costs) clearly would be best served over the long term if such land was acquired at the time the road was built for open space or park purposes. This would eliminate the need for expensive seawalls, groins, or other protective structures.

Regrettably, there are many houses and other buildings on cliff-top properties in the Santa Barbara area that are within 100 feet of the edge, and even a larger number within 25 feet or less of the brink (fig. 15.23C). As long as these buildings remain, demands for shore protection will be heard and at least some public money will inevitably be required to protect private investment.

Because developers of coastal properties often quickly sell their handiwork to others, pressures for less than adequate setback will continue unabated. Local governments do not have a very encouraging track record when it comes to resisting such pressures and protecting the long-term public interest, including that of future cliff-top property owners. Anyone contemplating the purchase of a cliff-top building or building site should look long and hard at the potential hazards before investing in any property that allows less than 100 feet of setback. Were we living in an ideal world, state or local government would acquire all bluff-top land lying within the 100-year danger zone, thus saving tax-

Figure 15.22 (A-E) Concrete-faced seacliff at Isla Vista about 1 mile west of Goleta Point. Photos by Robert Norris.

A. March 1971.

B. April 1974.

C. February 1978.

D. February 1980.

E. February 1981.

A. March 1971.

B. Projecting porch added and front of house redesigned, April 1974.

C. House and porch removed, leaving porch-supporting beams, February 1980.

Figure 15.23. (A–C) House at cliff edge, Isla Vista, about 1 mile west of Goleta Point.

payers and property owners alike the heavy cost of the inevitable protective measures.

Although much of the Santa Barbara coast is cliffed, there are several stretches of low, sandy shore from Santa Barbara east to Carpinteria and near Goleta. Much of this low-lying shoreline has had to be protected by riprap or seawalls to protect homes, industrial activities, and railways (figs. 15.24 and 15.25). Prior to installation of these protective works, for example, storm erosion cut back the beach from 150 to more than 200 feet at some places in the Carpinteria area following construction of the Santa Barbara breakwater. It is true that these structures have, for the most part, protected the houses, but they are unsightly, often make it quite difficult to reach the beach, and occupy part of the beach itself. During severe storms, such as occurred in 1982–83, many of these protective structures were damaged and required extensive costly repairs.

The low sandbars or spits that separate the sea from the low-lying salt marshes at Goleta and Carpinteria are particularly unstable environments. The differences in the history of these 2 areas are instructive. Goleta Beach is in public ownership with little in the way of construction apart from a public pier. On the other hand, the sandbars that enclose Carpinteria salt marsh have long been used for exclusive private homesites. Unfortunately, this is the very area that suffered the most severe beach erosion following construction of the Santa Barbara breakwater. This extensive erosion dramatically reduced the margin of safety once provided for buildings by the wide, sandy beach, and in

Figure 15.24.   Sandstone riprap seawall constructed about 1960, just west of Sand Point near Carpinteria. Photo by Robert Norris.

Figure 15.25. Undercut concrete seawall at Biltmore Hotel, just east of Santa Barbara, July 1983. Photo by Robert Norris.

about 1960 led to the installation of the present continuous rip-rap seawall, except at the salt marsh entrance. Although this seawall has provided considerable protection during most winters, it as well as the nearby houses have suffered some damage during particularly severe storms. Costs of repair are substantial. Most people would agree that the seawall is an unattractive structure and that it makes beach access from their homes a major undertaking.

Although Goleta Beach, being upcurrent from Santa Barbara Harbor, has never experienced the sort of severe beach sand starvation and landward shift of the high-tide line seen at Carpinteria, it is equally subject to winter season erosion; had it been developed for waterfront housing, instead of parkland, it is very likely indeed that it too would now possess an unsightly seawall of some sort. Instead, it is a broad, sandy beach most of the year and a major local recreational resource. In short, the price of putting housing on low, sandy shorelines is high in terms of money as well as aesthetics.

16. Rincon Point to Santa Monica

Antony Orme and Linda O'Hirok

This portion of the California coast embraces the entire 42-mile shoreline of Ventura County between Rincon Point and Sequit Point and a farther 33 miles of Los Angeles County's shoreline from Sequit Point to Marina del Rey (fig. 16.1). The coast extends obliquely across the seaward edge of the Ventura Basin, then along the south front of the Santa Monica Mountains, and finally turns south across the western edge of the Los Angeles Basin.

The coast embraces 4 natural units: (1) the 12.5-mile-long Rincon coast between Rincon Point and Ventura River; (2) the 22-mile-long Oxnard coast between the Ventura River and Point Mugu; (3) the 32-mile-long Malibu coast from Point Mugu to Santa Ynez Canyon where Sunset Boulevard reaches the shore; and (4) the 8-mile-long Santa Monica coast from Santa Ynez Coast to Marina del Rey. Along the Rincon and Malibu coasts, mountains have restricted urban growth to ribbon development close to the shore where congestion is heightened by the inherent physical instability of the coastal zone. In the Oxnard and Santa Monica sections, flatland behind and above the coast has favored primarily agricultural use on the former and essentially urban use of the latter. Each unit presents its own individual management problems, but transcending these is the overwhelming pressure for development and, conversely, the urgent need for coastal conservation created by the vast Los Angeles metropolitan area.

Physical setting

The Ventura Basin and the Santa Monica Mountains are part of the several complex east-west structures that comprise the Transverse Ranges of southern California. Among California's physical provinces the Transverse Ranges are unusual because their orientation departs from the normal north-northwest–south-southeast structural grain of the state's other ranges. These mountains extend westward for 350 miles from the Eagle Mountains north of the Salton Trough to Point Arguello. The Rincon Point–Santa Monica coastal segment thus lies across the west-central part of this province. Rising abruptly from the coast, the Transverse Ranges also serve as a partial climatic barrier, trapping moist Pacific air against their seaward slopes while sheltering the coast from the interior's summer heat and winter cold. Under natural conditions, therefore, these seaward slopes carry a dense vegetation of coastal sage and chaparral, which inhibits erosion but is susceptible to fire during drought, while the canyons and wider valleys support vegetation dominated by oak and sycamore.

Climate also imposes a distinct geomorphic personality on the region, both along the coast and immediately inland. In summer, prevailing winds and predominant swells reach most of California from the northwest. In winter, the eastward progression of

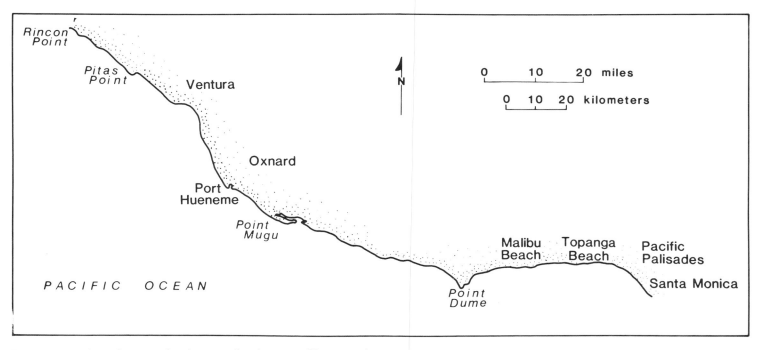

Figure 16.1.   Location map for the coastline between Rincon and Santa Monica.

mainly cold fronts across the coast maintains westerly swells for much of the time. Thus, north of Point Conception, 30–50 percent of swells approach from the northwest, and most others from west-northwest or west. South of Point Conception, however, the changing coastal orientation, strong refraction of waves, and the presence of offshore islands cause 70 percent of all swells to pass up the Santa Barbara Channel toward the Ventura coast from due west, while 80 percent of all swells along northern Santa Monica Bay approach from west-southwest. Here also southerly swells set up by late summer hurricanes off western Mexico, by southern hemisphere winter storms, and by local winter depressions passing inland along more southerly tracks may all cause erosion on south-facing beaches, especially during periods of high spring tides. Wave heights along this coastline average 3 feet but range from 1 foot to 24 feet. With westerly swells, the predominant littoral drift is from northwest to southeast along the Oxnard coast, moving about 1 million cubic yards per year of sand downcoast. Much of this littoral drift is siphoned off into Hueneme and Mugu submarine canyons, such that a substantially reduced volume of sediment passes east around Point Mugu. Nevertheless, a strong longshore current moves such sediment as does exist eastward along the Malibu coast, and then southward along the Santa Monica coast, as clearly revealed by sediment plumes and accumulation of sand against engineering structures. Mean tidal range along this coast is around 4 feet, and tides are mixed. From mean low low water to mean high

high water the range is around 5.2 feet. The extreme range at spring tides is 10.5 feet.

Climate also imposes a distinct if erratic personality on the area immediately inland from the coast where the processes of erosion and landslides are important to an understanding of coastal zone behavior. There are usually 2 distinct seasons: a long, hot, dry period from May through October, and a mild, moist period from November through April. Long-term annual precipitation ranges from 12 inches at the coast to 48 inches in the high San Gabriel Mountains where the Santa Clara and Los Angeles rivers have some of their headwaters. But averages are misleading. A few intense storms usually produce most of the precipitation; and exceptionally wet years, such as 1916, 1938, 1969, 1978, 1980, and 1983, may be separated by rainfall-deficient periods. Intense storm events, such as that of 16 February 1980, when 8 inches of rain fell in one day in the central Santa Monica Mountains amid a sequence of storms lasting 10 days, have a predictable effect on slope stability and flooding. Indeed, most of the annual work of landscape sculpture occurs in a few stormy days, even hours, in winter. For the rest of the year the land lies mainly dormant, recovering from storms and healing its scars. Along coastal slopes, however, fogs and drizzle mitigate the summer drought and promote a denser vegetation cover, which, if left alone, inhibits erosion and mass movement.

Fractured bedrock, steeply sloping hills, and periodic drenching rains are essential prerequisites for the deep weathering and devastating landslides that periodically plague the Malibu coast, and to a lesser extent the Rincon coast. Additionally, fire has a damaging effect on slope stability by removing the vegetation that protects the hillsides from raindrop impact and runoff. Fires, caused by lightning or human activity, are fairly frequent in the Santa Monica Mountains in the tinder-dry months of late summer and fall when hot, dry, dessicating Santa Ana winds from the interior promote severe vegetal drought.

Hillside failures along this coastline vary widely in the amount of material involved, the depth of failure, and the rate of movement. Where weak materials lie unsupported above seacliffs or highway cuts along the south side of the Santa Monica Mountains, massive landslides or slumps are common. On other slopes where bedrock is solid, the overlying soils may fail during high-intensity rainstorms, producing soil slips or debris flows. These flows may have a tragic impact on life and property downslope, and too little attention has been given previously to their disaster potential in developments along the Malibu coast and elsewhere. Like flooding, mass movement does not usually begin with the first winter rains because the earth is relatively dry after the long summer drought. Indeed, some large landslides or deep failures may not occur until several weeks, even months, have passed, for such time is often needed for rainwater to infiltrate the ground to some potential failure plane deep beneath the surface. Alternatively, if rainfall intensity quickly exceeds the infiltration capacity of the soil, as sometimes occurs during midwinter storms, shallow failures, surface erosion, and gullying may soon occur.

Furthermore, landscape watering, septic tanks, and increased surface runoff in tract developments have created wetter slope environments, causing old landslides to reactivate and new mass movements to occur. Such problems are a perennial hazard along the Malibu coast.

Stream sediment discharge, so important for nourishing local beaches, also reflects the precipitation and runoff regime, as well as the availability of sediment and the extent to which stream transport is blocked by dams. With most annual precipitation falling in a few intense winter storms, large infrequent floods account for most stream erosion and sand transport. Tributary channels are scoured, main channels become debris chutes, and transported debris may be discharged violently onto lower land near the coast. Whether or not such debris reaches the coast depends on the magnitude of individual and successive stormflows. For example, most of the 10 million tons of sand delivered to the coast by the Santa Clara River between 1933 and 1938 arrived in 6 days of flood in 1938. Further, sediment discharged by this river in the 1969 floods was 52.4 million tons, compared with 1 or 2 million tons in relatively dry years. The coastal sediment budget thus sees a few years of plenty (most recently 1969, 1978, 1980, and 1983) and long years of famine (the 30 years preceding the 1969 floods).

Over the past 50 years the addition of river sand to beaches has been restricted by dams in the contributing watersheds and by mining of river floodplain gravels. About 42 percent of the Ventura River watershed is blocked by dams (Matalija built in

1948, and Casitas in 1959). About 37 percent of the Santa Clara River watershed is blocked by dams (Bouquet, 1934; Piru, 1955; Pyramid, 1971; Castaic, 1972). The reduced volume of sand reaching the coast because of these dams seems incapable of restoring eroding beaches, and during a succession of relatively dry years serious beach starvation occurs. Malibu Creek, which cuts across the Santa Monica Mountains, is the most important contributor of stream sediment to the Malibu coast, but here again sediment delivery is reduced by a number of dams and artificial lakes in developed areas north of the mountains. Other local canyons may also supply sand during intense storms.

Seacliff erosion as a source of beach sediment is now much reduced as a result of the extensive protection afforded the Santa Monica, Malibu, and Rincon coasts by riprap, revetments, and seawalls. Along the Malibu coast in particular, cliff erosion must once have been a very important sediment source, but construction of the Pacific Coast Highway during the 1920s and 1930s effectively curtailed this supply. Much of this sediment would have been delivered to the shore in the form of mass movement. Rockfalls, slumps, landslides, and soil slips persist today, but the debris generally falls first on the coastal highways and is then sidecast into the sea.

Development history and planning issues

Development between Rincon Point and Santa Monica has been strongly influenced by the physical characteristics of the coast—the mountains restricted growth to ribbon development close to the shore along the Rincon and Malibu coasts, while extensive alluvial flatland favored more open growth along the Oxnard coast. As will be noted, despite some earlier events, most development of significance to contemporary shoreline management has occurred during the twentieth century, most intensively between 1950 and 1970. With various federal, state, and local government agencies now involved in coastal management, many current problems and issues trace their origins to past irrational or unwise developments. Nevertheless, the proximity of the vast Los Angeles metropolitan area and the recent discovery of large petroleum resources offshore continue to generate development pressures along the coast and, with these pressures, the need for very careful evaluation of all proposals.

When the Portuguese navigator Juan Rodriguez Cabrillo steered 3 Spanish ships toward this area in 1542, the coast was settled by Hokan-speaking Chumash peoples who maintained fishing camps at intervals along the shore, most significantly near Point Mugu and Ventura. Their wicker-framed hemispherical huts were roofed with reeds from nearby marshes, while their pine-plank canoes were caulked with tar from local seepages. The Spanish navigator, Sebastian Vizcaino, described similar features in 1602, and little appears to have changed when Gaspar de Portola began the formal Spanish occupation of California in 1769. A diet of fish, shellfish, local fruits, and porridge made of acorn meal allowed these early peoples to maintain a population density of over 130 persons per 100 square miles in Ventura and

Los Angeles counties, and at the immediate coast this figure was almost certainly much higher. East of Malibu Creek, the Chumash people gave way to Uto-Aztekan–speaking groups from the interior (the most important local group were later called the Gagrieleno people). Traces of their encampments can still be found along the coast in the form of extensive kitchen middens and occasional burial grounds, but the people themselves have long since disappeared. A few Chumash placenames survive: for example, *muwu* as Point Mugu.

The initial impetus to development was provided by the joint religious-military occupation of coastal California beginning in 1769 with the establishment of an army presidio and Franciscan mission in San Diego. As one in a chain of missions extending at intervals northward to San Francisco Bay, Mission San Buenaventura was established locally in 1782 and for the next 50 years was the dominant force in the area. This mission was situated where the important north-south coastal route, El Camino Real, reached the coast from the inland missions of San Gabriel Arcangel (1771) and San Fernando (1797) en route to the presidio and mission at Santa Barbara (1786). Aided by the military, Mission San Buenaventura resettled and concentrated the local Chumash population and introduced agriculture and basic crafts. As elsewhere, however, the mission population fell prey to disease, and when Mexican independence in 1821 removed the protecting central authority, political power fell to local civilian interests. With the Chumash population already waning, Mission San Buenaventura was secularized in 1836 and its lands distributed as private ranches over the next decade. Vast areas of coastal and valley land were granted by the Mexican governor to soldiers and other fortunate applicants, among whom 8 families formed a colony immediately south of the Santa Clara River mouth in the vicinity of modern Oxnard. The Chumash wandered away from the decaying mission to work on the large ranches, and for many years cattle raising became the essential economic base of the coastal plains and terraces below the sheep pastures on nearby hillsides. During this time much of the native vegetation was converted to grassland, while a small cow town developed at San Buenaventura, much as a large cow town was growing around the Los Angeles pueblo farther east. Along the Malibu coast a land grant of Topanga-Malibu-Sequit had been made during the Spanish period. Land around Santa Monica was subject to grants very early in the Mexican period, including recognition of the Ballona wetlands as important cattle saltings. Owing to its inaccessible and rough terrain, the coast between Sequit and Point Mugu was not specifically allocated at this time.

After California's transfer to the United States in 1848, the first legislature divided the state into 27 counties in February 1850. Three of these—San Diego, Los Angeles, and Santa Barbara—embraced the entire coast from the new Mexican border to the Santa Maria River. Existing land grants were confirmed, but depressed cattle prices followed by severe drought in the early 1860s wrought havoc with the cattle business. Ranches were divided and divided again, and, as new settlers arrived, the coastal land began to lose its pastoral character. The small cow town of San Buenaventura, a collection of some 70 to 80 houses and a hotel in 1856, acquired a post office in 1861 and was incorporated as a city in 1866, the same year the first formal schoolhouse was built. A regular stagecoach line was established in 1868 as coaches from Los Angeles made their way into Ventura County through the San Fernando, Simi, Conejo, and Santa Clara valleys before proceeding along the beach to Santa Barbara. In 1873 Ventura County was carved from the southeast corner of Santa Barbara County and given the bleak, by then depopulated offshore islands of Anacapa and San Nicolas to administer. As the new county seat, San Buenaventura soon acquired a courthouse, library, bank, hotels, newspaper, churches, fire department, and other trappings of developing townhood. A wharf was built, while farther downcoast Hueneme (the Chumash word *huenumu* means halfway place) was settled by squatters. The arrival of the Southern Pacific Railroad in 1887–88 promoted a real estate boom along its route and represented the first major disturbance of the narrow coastal terrace west of San Buenaventura, now abbreviated for timetable purposes to Ventura. To the south, Oxnard was founded in 1898 and incorporated in 1903, but Port Hueneme was not incorporated until 1948. Farther east, Santa Monica was incorporated in 1886 and witnessed considerable residential and recreational development thereafter, while to the south an enterprising attempt to transform part of the Ballona wetlands into a "Venice of the West" was initiated with a modest canal system in the early 1900s.

Meanwhile, the long-recognized tar seepages of the Ventura area had prompted the search for and development of petroleum resources as early as the 1850s. In 1857 a small refinery was constructed near Ventura to produce marketable oil from the seepages. In 1865 the first deep well was sunk near Sulphur Mountain on the Ojai Road, at a place contiguous to tar seepages, and further wells followed including some along the coast west of Ventura. It was not until the present century, however, that the development of deep drilling methods and the demand for petroleum products led to a major boom in the oil industry. The important Ventura Avenue oil field reaches the coast west of Ventura, most notably between Pitas Point and Punta Gorda, before passing westward down the Santa Barbara Channel. The presence of oil production, storage, and pipeline facilities adds a further component to an already congested coastal corridor.

Today the city of San Buenaventura is an important shopping center with a 1978 population of 68,200, but whose further growth is limited by the physical constraints of its site. Oxnard (96,400 persons in 1978), on the broad alluvial plain created by the Santa Clara River, is not so constrained, although its rapid growth in recent years has not proceeded along the most rational or orderly lines. Oxnard had its origins near a sugar beet factory established in 1897 with the backing of the Oxnard family and is today the center of a rich agricultural area yielding sugar beets, lima beans, lemons, avocados, strawberries, and many vegetables. More recently it has also developed an important industrial and service base to its economy, avenues that seem likely to increase population pressures on the area and its coast. Immediately to the south, the old Port Hueneme, with its picturesque early homes, lighthouse, and small wharf, has since 1942 been dominated by the U.S. Navy, specifically by the vast Construction Battalion Center ("Seabees") operating an expanded naval harbor, docks, warehouses, training facilities, and various government buildings. A short distance farther southeast, Point Mugu Naval Air Station has operated the Pacific Missile Range since 1946. This facility, although excluding the public from about 16 miles of shoreline along the southern edge of the Oxnard Plain, has also protected perhaps the last surviving seminatural wetland in southern California, namely Mugu Lagoon, from the undesirable depredations of irrational development. The coast and nearby mountains immediately to the east of this base now form part of Point Mugu State Park.

Rincon coast (fig. 16.2)

This 12.5-mile-long coast, which stretches in 3 shallow curves from Rincon Point southeastward to the Ventura River mouth, is a congested corridor of activity developed upon one or more low terraces never more than 440 feet wide. Behind the terraces, steep unstable cliffs rise 300 to 650 feet above sea level, while seaward lies a line of low, erodible bluffs now largely, but not effectively, protected by revetments and seawalls. The corridor contains 6 small nodes of linear development—from northwest to southeast, Rincon Point, Punta, Punta Gorda, Seacliff, Pitas

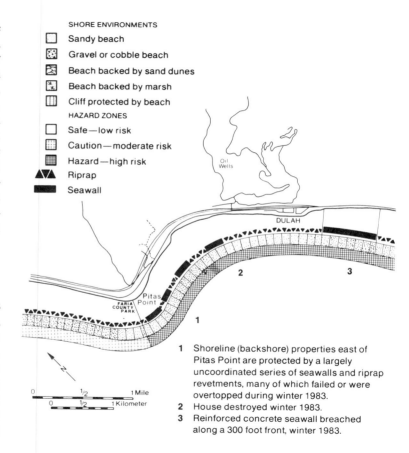

SHORE ENVIRONMENTS

☐ Sandy beach

▨ Gravel or cobble beach

▧ Beach backed by sand dunes

▦ Beach backed by marsh

▥ Cliff protected by beach

HAZARD ZONES

☐ Safe—low risk

▦ Caution—moderate risk

▦ Hazard—high risk

◣◥◣ Riprap

▬ Seawall

1 Shoreline (backshore) properties east of Pitas Point are protected by a largely uncoordinated series of seawalls and riprap revetments, many of which failed or were overtopped during winter 1983.
2 House destroyed winter 1983.
3 Reinforced concrete seawall breached along a 300 foot front, winter 1983.

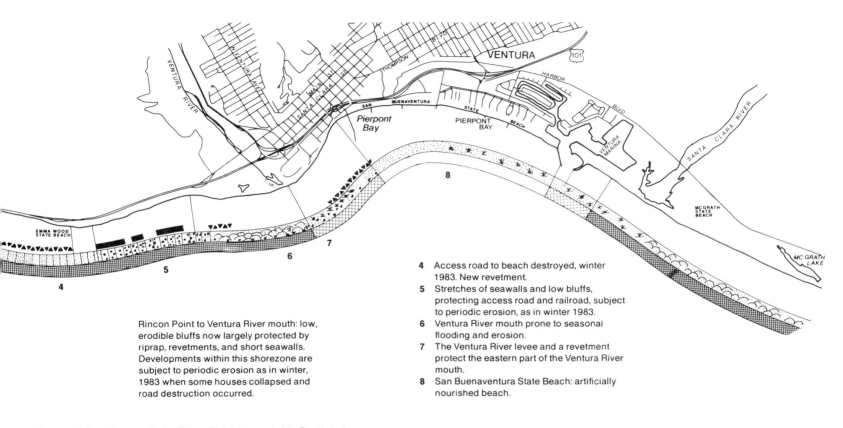

Rincon Point to Ventura River mouth: low, erodible bluffs now largely protected by riprap, revetments, and short seawalls. Developments within this shorezone are subject to periodic erosion as in winter, 1983 when some houses collapsed and road destruction occurred.

4 Access road to beach destroyed, winter 1983. New revetment.
5 Stretches of seawalls and low bluffs, protecting access road and railroad, subject to periodic erosion, as in winter 1983.
6 Ventura River mouth prone to seasonal flooding and erosion.
7 The Ventura River levee and a revetment protect the eastern part of the Ventura River mouth.
8 San Buenaventura State Beach: artificially nourished beach.

Figure 16.2. Site analysis: Pitas Point through McGrath Lake.

Point, and Dulah—together with a railroad first built in 1887–88, old coast highways now reduced to parkways and access roads, a major freeway (U.S. 101), and various onshore and offshore petroleum facilities. Limited park development has occurred at Rincon Point County Park (Santa Barbara County), Hobson County Park at Seacliff, Faria County Park at Pitas Point, Emma Wood State Beach, and at the Ventura River mouth, Seaside Wilderness Area and Surfers' Point Park.

Two major problems characterize this coast: the unstable cliffs and the irrational proximity of development to the shore. First, the coastal zone is composed mainly of shales, mudstones, sandstones, and conglomerates. The cliffs formed in these rocks are prone to frequent mass movement and gully erosion, most notably between Pitas Point and Emma Wood State Beach. While the lower parts of these old cliffs are thus mantled with slide and slump debris, the upper parts are prone to rockfall. The potential instability of the old cliffs poses a continuing threat to developments at their base.

The second problem is essentially attributable to human intervention. The shore faces generally southwest along an east-west part of the California coast. Direct wave energy associated with westerly swells in the Santa Barbara Channel is thus somewhat softened by refraction and sheltering headlands, so that most damage potential is associated with less frequent swells from the southwest, south, and rarely the southeast. Before development this coast was characterized by low, erodible bluffs 15–30 feet high. Under natural conditions these bluffs would have continued

Figure 16.3. Rincon Point development, Ventura–Santa Barbara county line. Photo by Linda O'Hirok.

to retreat slowly, but eroded debris would have contributed to the beaches. However, residential developments have been placed on these deposits, as at Rincon Point (fig. 16.3), and beneath low bluffs, while either the railroad or highway has been built immediately above the bluffs. All these developments have required protective riprap, revetments, or seawalls, which in turn have inhibited erosion of sand from the cliffs and therefore beach replenishment. As a result, strong swells superimposed on high tides cause serious damage from time to time, as during the win-

Figure 16.4. Collapse of seawall along Rincon Parkway, Winter 1983. Photo by Amalie Jo Brown.

ter of 1983 when several properties were lost to storm damage, many more were damaged, and long sections of highway and railroad were undermined (fig. 16.4).

Oxnard coast (figs. 16.5 and 16.6)

The 22-mile-long Oxnard coast extends in a broad arc, convex seaward, across the edge of the Oxnard Plain between the mouth of the Ventura River and Point Mugu. The shore on the Oxnard Plain originally comprised a narrow, sandy barrier beach, with low dunes backed by lagoons, marshes, and sparsely vegetated alluvial flats. To this coast flowed the Ventura and Santa Clara rivers and Calleguas Creek, but only the last-mentioned appears to have changed its course much over the past 200 years. This natural shore zone has been much modified by human interference, creating a variety of management problems. Interference has taken 2 main forms: restrictions on the amount of sand reaching the shore as a result of upstream dams and downstream levees, discussed earlier; and construction within the shore zone (fig. 16.7).

Levee construction and reduced sand yields from the Ventura River during the relatively dry years between 1948 and 1959 were largely responsible for encouraging the shoreline between the Ventura and Santa Clara rivers to erode 300 feet landward during that period. To offset this, an erosion control project was completed between 1962 and 1967 at San Buenaventura State Beach comprising 7 rock groins and deposition of sand in a beach 130–260 feet wide and over 2 miles long. The beach is presently maintained with additional sand dredged from Ventura Marina. This present situation is in significant contrast to the evidence from old maps of the area, which suggests that between 1855 and 1933 the shoreline was built seaward at an average annual rate of 6.5 feet.

Diminished beach sand supplies and more frequent episodes of coastal erosion are also responsible in part for the problems of beach undernourishment at Oxnard Shores, downdrift from the

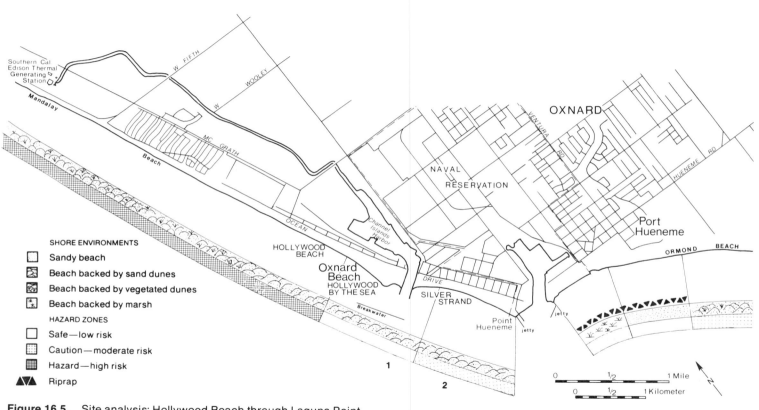

SHORE ENVIRONMENTS

- Sandy beach
- Beach backed by sand dunes
- Beach backed by vegetated dunes
- Beach backed by marsh

HAZARD ZONES

- Safe—low risk
- Caution—moderate risk
- Hazard—high risk
- Riprap

Figure 16.5.   Site analysis: Hollywood Beach through Laguna Point.

1   Channel Islands Harbor: the presence of
    massive rubble mound jetties and offshore
    breakwater at the harbor entrance reduce
    the hazard rating in the immediate vicinity to
    one of low risk.

2   Silver Strand is essentially an artificial
    pocket beach contained within flanking
    jetties. Occasional sand losses are
    replenished by sand bypassing operations.

3   Laguna Point eroding.

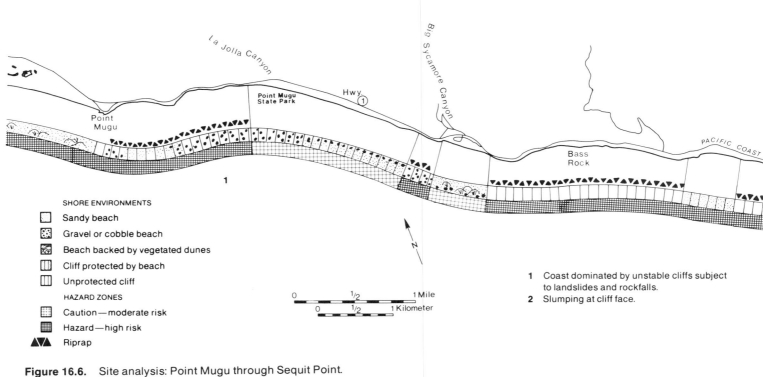

SHORE ENVIRONMENTS

☐ Sandy beach

☐ Gravel or cobble beach

☐ Beach backed by vegetated dunes

☐ Cliff protected by beach

☐ Unprotected cliff

HAZARD ZONES

☐ Caution—moderate risk

☐ Hazard—high risk

▲▼▲ Riprap

0   1/2   1 Mile
0   1/2   1 Kilometer

1 Coast dominated by unstable cliffs subject to landslides and rockfalls.
2 Slumping at cliff face.

Figure 16.6.   Site analysis: Point Mugu through Sequit Point.

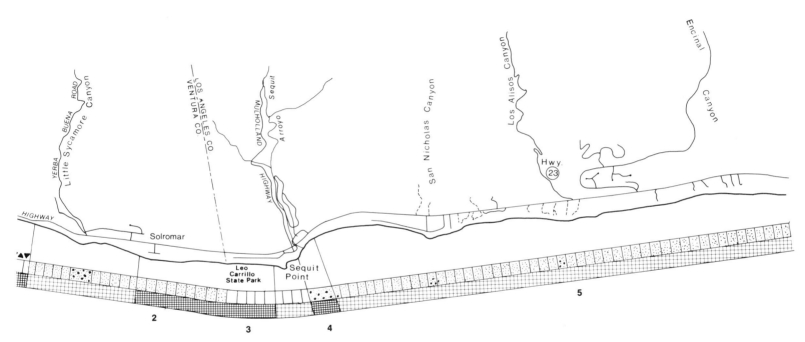

3 Major beach erosion in winter 1983.
4 Arroyo Sequit: flood risk. Beach is formed of boulders, cobbles, and sand.
5 Some houses on beach or bluff face damaged in 1983 storms.

Figure 16.7.   Coastal barrier and artificial levee at mouth of Ventura River, November 1982. Photo by Linda O'Hirok.

Figure 16.8.   House at northern end of Oxnard Shores development, February 1978. Photo by Antony Orme.

Santa Clara River mouth. Here, during the 1960s, residential development placed homes along the backshore of an exposed and little-understood shoreline. Over the past fifteen years periodic storm waves have removed much of this beach and destroyed many houses, leaving the remainder stranded out on the foreshore (fig. 16.8). Further human intervention along this shoreline certainly should not be undertaken without a careful evaluation of long-term beach behavior. Erosion attributable in part to reduced beach sand supplies is also a significant problem along the seaward edge of the Point Mugu Naval Air Station, but here the pattern of damaging waves and currents is more complex. In any case, under natural conditions frequent shoreline changes were to be anticipated along the Oxnard coast because of its narrow barrier beach and occasionally flooding rivers. Problems have arisen because human beings have sought to freeze a moderately unstable shoreline with residential and recreational developments that have paid little attention to the vagaries of the natural system.

The other form of interference with the local shore zone has been the construction of major engineering works for military and recreational needs. The new harbor at Port Hueneme was completed by the Oxnard Harbor District as a civil project in 1938–40, but was converted for use by the U.S. Navy in 1942. The harbor comprises an entrance channel protected by 2 jetties. After jetty construction, accretion or sand buildup occurred along the upcoast or northwest shore, while by 1948 erosion extended 7 miles downcoast beyond the south jetty. Fill dredged from upcoast and a 3,000-foot rock revetment were then placed downcoast, but with minimal effect as erosion continued beyond the end of the revetment and dunes that were formerly more extensive were eroded and breached. Shoreline surveys between 1856 and 1938 show this downcoast area to have been relatively stable over that 82-year period. Clearly a bolder plan was needed to stabilize the beaches, and this was linked to the construction of Channel Islands Harbor in 1961. There, 2 jetties and a detached breakwater provide a sand trap from which great quantities of sand are dredged every 1.5 to 2 years to nourish the shoreline downcoast of Port Hueneme, 1 mile to the southeast. This has had some beneficial impact, but reduced erosion problems continue to confront Point Mugu Naval Air Station. Channel Islands Harbor is a major marina still under development as an excavation among the dunes and wetlands of the former barrier-lagoon complex behind Hollywood Beach.

Immediately to the north of the Santa Clara River mouth, the public Ventura Marina and private Ventura Keys are 2 further developments in former wetlands behind the coastal barrier. Ventura Marina was badly designed in a poor location (fig. 16.9). It originally comprised an entrance channel protected by 2 rock and tribar jetties, a middle groin, turning basin, and 3 basins with 520 berths 9–16 feet deep. Opened in 1963, rapid shoaling in the entrance channel generated breaking waves, averaging 8 feet high at low water, thereby effectively closing the marina for 66 days every year. The need to bypass sand downcoast every 2–3 years had been anticipated, but shoaling had not been predicted, nor was it alleviated by dredging 190,000 cubic yards of sand annually from the entrance. Before remedial measures could be taken, the marina's problems were compounded by the 1969 winter floods of the Santa Clara River, which breached its north bank and discharged directly through the marina. All but 2 docks were destroyed, 490 berths and 88 boats demolished, 5 20,000-gallon gasoline tanks washed into the harbor, and, with main trunks to a nearby sewage treatment plant destroyed, raw sewage flowed into the marina at 4 million gallons per day for 2 weeks. The marina silted throughout. After this sediment was removed, a 1,500-foot-long detached rock breakwater was completed in 1971, offset to the north to trap 785,000 cubic yards of sand upcoast of the north jetty. Despite maintenance dredging every year or two, shoaling remains a problem and the marina remains as testimony to improvident coastal development.

With the extreme northern and southern ends of the Oxnard coast located within San Buenaventura State Beach and Point Mugu State Park, respectively, and the shore south of the Santa Clara River lying within McGrath State Beach while much of the rest falls within existing marinas or military installations, the

Figure 16.9.   Mouth of Santa Clara River and Ventura Marina, March 1983. Photo by Linda O'Hirok.

opportunities for further development along this shore are confined to 2 sections—Mandalay Beach (including Oxnard Shores) between McGrath State Beach and Hollywood Beach, and Ormond Beach between Port Hueneme and Point Mugu Naval Air Station. Both sections are sites for operating thermal power stations that do little to grace the shoreline, and both sections are also battlefields for pending management decisions. It is to be hoped that experiences gained as a result of previous developments will be evaluated before further growth is permitted.

Malibu coast

The 32-mile-long Malibu coast extends from Point Mugu more or less due east to Santa Ynez Canyon. Of this distance, approximately 12.5 miles are absorbed into state and county beaches and parks, while much of the remainder has been exposed to unplanned or poorly conceived development during the present century. Management problems along this coast are somewhat akin to those described for the Rincon coast, namely unstable cliffs and an irrational proximity of development to the shore, but of a far greater intensity. Furthermore, the nearness of the vast Los Angeles metropolitan area, of which Malibu developments are certainly an extension, creates continuing pressures on this coastal zone.

The dominating geological feature along this coast is the Malibu Coast Fault. This fault passes out to sea near Temescal Canyon but remains close to shore and comes ashore again between Carbon Canyon and Arroyo Sequit. This fault is seismically active, as emphasized during the February 1971 San Fernando earthquake (Richter magnitude 6.5) when hot water emerged off Malibu pier, and further indicated by a series of subsequent earthquakes off Point Mugu. No development along this shore should ignore the seismic potential of this dangerous geological structure.

The shales, sandstones, and conglomerates that are exposed in bluffs and canyon walls immediately east and west of Topanga Canyon are all extensively fractured and deformed, reflecting proximity to the Malibu Coast Fault. These rocks tend to crumble and collapse onto the Pacific Coast Highway with monotonous regularity, as is to be expected along the line of old sea cliffs that have been undercut to make room for the highway and housing development.

Over a distance of approximately 19 miles between Carbon Canyon and Little Sycamore Canyon, flat, well-developed terraces occur. Two of these are particularly visible in the vicinity of Point Dume where their gently rolling topography has permitted the widest extent of residential development to be met along this coast. Farther west, notably west of Little Sycamore Canyon, unstable cliffs again present a serious threat of landslides to the Pacific Coast Highway.

As we have seen, therefore, mass movement is a perennial problem along this coast and has undoubtedly been accentuated by construction of the Pacific Coast Highway along the base of the cliffs. Rockfalls are common throughout the year. More mas-

Figure 16.10. Big Rock landslide, Malibu coast, November 1982. Photo by Kevin Mulligan.

Figure 16.11. Coast between Point Mugu and Little Sycamore Canyon, August 1983. Photo by Linda O'Hirok.

sive failures also occur fairly frequently, commonly some time after winter rains have drenched the coastal slopes. The 4-mile stretch of coast between Carbon Canyon and Topanga Canyon is particularly notorious for landslides and related phenomena. One of the largest of these, at Big Rock, has necessitated the wholesale regrading of the mountain front (fig. 16.10). Elsewhere, notably between Little Sycamore Canyon and Point Mugu, mass movement combines with marine erosion to undermine or bury the Pacific Coast Highway at intervals during most winters (fig. 16.11).

In the days of the Spanish land grant of the Topanga-Malibu-Sequit cattle range, access to this section of coast was by way of Malibu Canyon or along the beach west of Santa Monica, the coast around Point Mugu being quite inaccessible. As late as the 1920s there was no through route along the south side of the Santa Monica Mountains west of Malibu, although local access roads led over the narrowing terraces and down local canyons. During the 1930s, however, a broad coastal road—formerly called the Roosevelt Highway but now termed the Pacific Coast Highway—was completed between Santa Monica and Oxnard, the section west of Little Sycamore Canyon being cut with difficulty into crumbling and erodible coastal bluffs. This highway has since caused nothing but trouble, being rerouted in several places and requiring frequent repair or debris removal elsewhere (fig. 16.12). The highway also enhanced access to the Malibu coast in general, an accomplishment of dubious value in that extensive and often unwise development followed in its wake.

Residential development along the Malibu coast began in earnest following completion of the Roosevelt highway in the

Figure 16.12. Point Mugu, showing rerouting of coastal highway behind headland, with Mugu lagoon and Oxnard Plain in background, March 1983. Photo by Kevin Mulligan.

1930s. The first major development was that of the "Malibu Colony" on the barrier spit at the mouth of Malibu Creek, a suite of expensive homes linked in part to the then-burgeoning Hollywood film industry (fig. 16.13). Despite the somewhat precarious location of this colony, near the mouth of a major creek subject to dangerous winter floods and on a barrier created by strong wave action, the houses have proved remarkably persistent over the years, although often threatened by damaging storm seas and high tides. As the amount of land along the immediate shoreline was consumed by subsequent housing, however, more and more structures were built on pilings in potentially dangerous locations at the base of crumbling bluffs (fig. 16.14). Furthermore, these houses and their various protective structures in turn hindered the longshore movement of beach sand, leading to downdrift starvation of beaches to the east. Over the past 60 years, therefore, the pattern of beach erosion has grown in significance until many houses formerly built at the rear of broad backshores now find themselves stranded high above eroding foreshores, the waves periodically pummelling the underlying bluffs that connect the houses to the highway. The management problems facing this coast can only increase with time, as society as a whole has to pay the penalty for unwise, uncoordinated, and irrational developments of the past.

Figures 16.15 and 16.16 show the Malibu coast earlier in this century, just prior to the initiation of development. Figure 16.15 shows Point Dume and the broad marine terraces lying seaward of the Malibu Coast Fault as they were in 1924, and probably

Figure 16.13.   Mouth of Malibu Creek with "Malibu Colony" in foreground and pier and commercial district in background, March 1983. Photo by Linda O'Hirok.

Figure 16.14.   Housing on beach below Big Rock landslide, April 1979. Photo by Amalie Jo Brown.

much as they were during the years of the Spanish land grant. The area is now moderately developed, affording as it does the broadest expanse of flat land along this coast, and further development is in progress or imminent. Figure 16.16 shows the coast between Tuna Canyon and Big Rock just after completion of the Roosevelt Highway in 1934. Toward the headland in the middle distance, building lots constructed of debris sidecast from undercut coastal bluffs can be seen seaward of the highway. This was to become the site of the Big Rock landslide seen in figure 16.10.

Santa Monica coast (figs. 16.6, 16.17, and 16.18)

The 8-mile-long Santa Monica coast between Santa Ynez Canyon and Marina del Rey is really a transitional unit between the foothills of the Santa Monica Mountains and the extensive Los Angeles Basin, at whose western edge it lies. This coast is readily divided into 2 parts: a 5-mile stretch of unstable cliffs behind a broad, nourished beach between Santa Ynez Canyon and Santa Monica Pier, and a further 3-mile stretch along the barrier beach forming the western margin of the former Ballona wetlands. This

Figure 16.15. Point Dume, marine terraces, and Malibu Ranch, 1924. Photo by Spence Air Photos, UCLA.

Figure 16.16. Malibu coast between Tuna Canyon and Big Rock, 1934. Photo by Spence Air Photos, UCLA.

area is thoroughly developed, and its problems are not ones of further growth but of managing existing residential areas and recreational facilities for the benefit of both the local people and the larger metropolitan population.

In the north the cliffs of Pacific Palisades at first appear in deformed and fractured shales, which give rise to steep, irregular slopes prone to mass movement. Near Temescal Canyon, south of Malibu Coast Fault, these rocks are replaced by sandstones and conglomerates. Over thousands of years wave action has

been able to cut back significantly into these deposits, thereby creating a vertical cliff face that descends from a 230-foot height at its northern end to disappear south of Santa Monica Pier. When the highway and then residential and recreational facilities were developed along the cliff base, this cliff was protected from wave action but continued to experience mass movements. Rockfalls are common, and massive slumps are both spectacular and disastrous. One of the latter occurred at the seaward end of Via de la Paz immediately south of Temescal Canyon in April 1958,

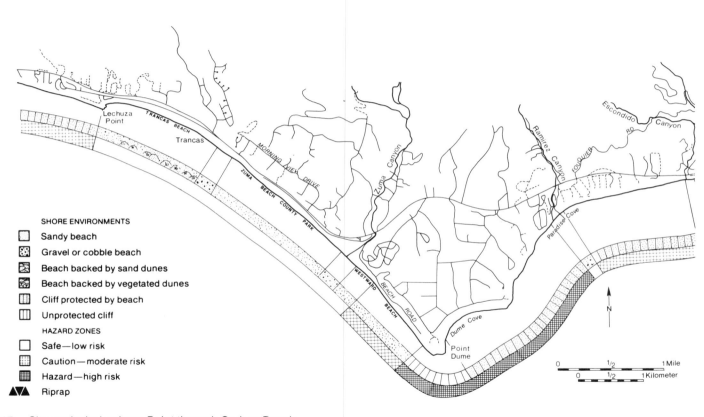

SHORE ENVIRONMENTS

- ☐ Sandy beach
- ▣ Gravel or cobble beach
- ▣ Beach backed by sand dunes
- ▣ Beach backed by vegetated dunes
- ▥ Cliff protected by beach
- ▥ Unprotected cliff

HAZARD ZONES

- ☐ Safe—low risk
- ▦ Caution—moderate risk
- ▦ Hazard—high risk
- ▲▼▲ Riprap

Figure 16.17. Site analysis: Lechuza Point through Carbon Beach.

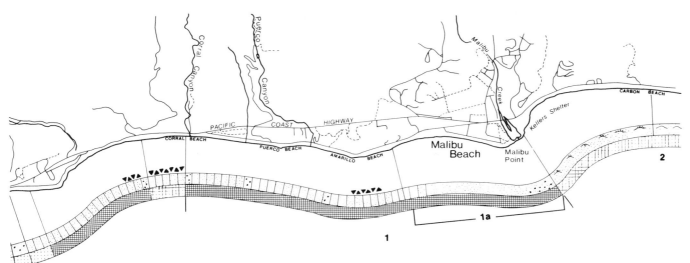

1 Low bluffs fronting the terrace are prone to
erosion and many houses are constructed
on pilings seaward of high water mark. (**a**)
The high hazard nature of this area was
demonstrated during the damaging storms
in winter 1983.

2 Originally cliffs prone to minor gullying and
slumping, fronted by a 325-490 foot wide
sandy beach with low dunes. Development
over the past sixty years has largely obliterated the beach and may have caused
increased landsliding behind Carbon Beach
and LaCosta Beach. Dwellings built on piles
seaward of the highway are often subject to
wave battering. Protection offered by Malibu
Point and the more modest scale of landsliding justify a moderate hazard rating.

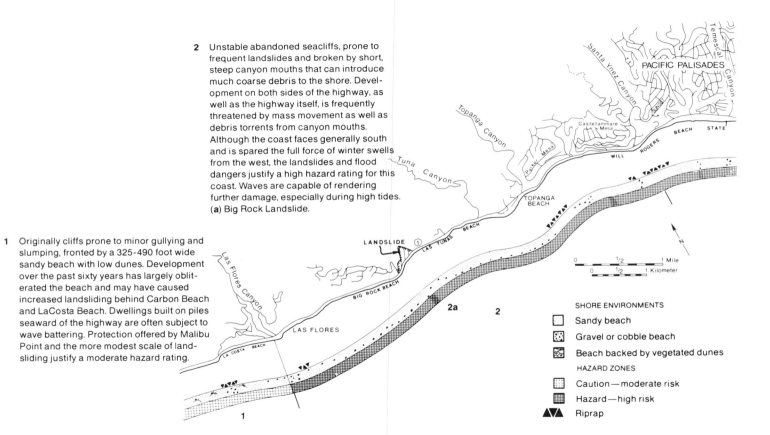

2 Unstable abandoned seacliffs, prone to frequent landslides and broken by short, steep canyon mouths that can introduce much coarse debris to the shore. Development on both sides of the highway, as well as the highway itself, is frequently threatened by mass movement as well as debris torrents from canyon mouths. Although the coast faces generally south and is spared the full force of winter swells from the west, the landslides and flood dangers justify a high hazard rating for this coast. Waves are capable of rendering further damage, especially during high tides. (**a**) Big Rock Landslide.

1 Originally cliffs prone to minor gullying and slumping, fronted by a 325-490 foot wide sandy beach with low dunes. Development over the past sixty years has largely obliterated the beach and may have caused increased landsliding behind Carbon Beach and LaCosta Beach. Dwellings built on piles seaward of the highway are often subject to wave battering. Protection offered by Malibu Point and the more modest scale of landsliding justify a moderate hazard rating.

PACIFIC PALISADES

Santa Ynez Canyon

Temescal Canyon

Topanga Canyon

Castellammare Mesa

Palal Mesa

WILL ROGERS BEACH STATE

Tuna Canyon

TOPANGA BEACH

LANDSLIDE ①

LAS TUNAS BEACH

Las Flores Canyon

BIG ROCK BEACH

2a

2

LAS FLORES

LA COSTA BEACH

N

0   ½   1 Mile

0   ½   1 Kilometer

1

SHORE ENVIRONMENTS

☐ Sandy beach

▨ Gravel or cobble beach

▨ Beach backed by vegetated dunes

HAZARD ZONES

▦ Caution—moderate risk

▦ Hazard—high risk

▲▼▲ Riprap

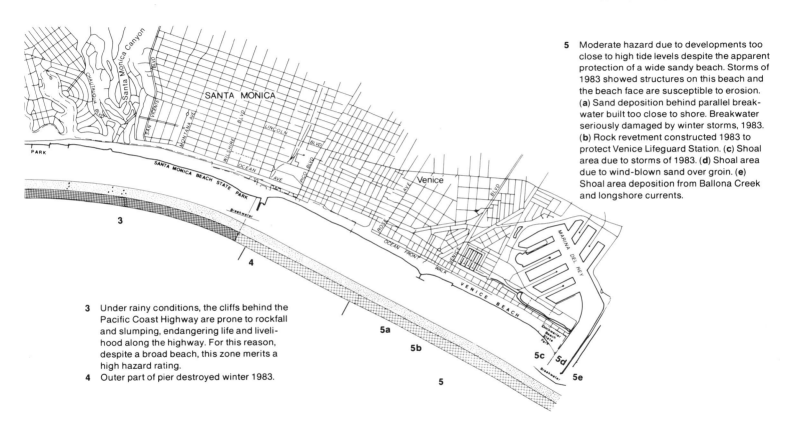

5 Moderate hazard due to developments too
close to high tide levels despite the apparent
protection of a wide sandy beach. Storms of
1983 showed structures on this beach and
the beach face are susceptible to erosion.
(**a**) Sand deposition behind parallel breakwater built too close to shore. Breakwater
seriously damaged by winter storms, 1983.
(**b**) Rock revetment constructed 1983 to
protect Venice Lifeguard Station. (**c**) Shoal
area due to storms of 1983. (**d**) Shoal area
due to wind-blown sand over groin. (**e**)
Shoal area deposition from Ballona Creek
and longshore currents.

3 Under rainy conditions, the cliffs behind the
Pacific Coast Highway are prone to rockfall
and slumping, endangering life and liveli-
hood along the highway. For this reason,
despite a broad beach, this zone merits a
high hazard rating.

4 Outer part of pier destroyed winter 1983.

Figure 16.18. Site analysis: La Costa Beach through Marina del Rey.

burying the Pacific Coast Highway which has since been diverted around the toe of the slide (fig. 16.19). Farther south, the cliffs stand vertical, and rockfalls are frequent. The beach along this stretch of coast falls within the Will Rogers Beach State Park, where a groin system helps to maintain a partly nourished beach, and within Santa Monica State Beach, where an offshore rubble breakwater, built in 1933 but presently much damaged, has helped to promote a broader recreational beach and an anchorage for small boats. During the winter of 1983 this stretch of coast was seriously eroded by storm waves approaching from the southwest. The outer part of Santa Monica Pier was entirely lost, the offshore breakwater was severely modified, and portions of the beach and various facilities farther south were seriously damaged.

From Rose Avenue southward to Playa del Rey, urban and recreational development during the present century conceals what was formerly a broadly triangular barrier beach lagoon system extending 4 miles inland from Playa del Rey. This lagoon was developed in an estuary abandoned by the Los Angeles River in the early nineteenth century, and the barrier developed as a result of buildup across the estuary. During the nineteenth century the lagoon was gradually transformed into a wetland rich in marsh plants, animal and bird life, and small lagoons. During the early years of the present century the northwest portion of the lagoon was dredged to form the Venice canal system, and in the next few decades various activities occupied the wetlands until in the 1950s the Marina del Rey development was initiated.

Figure 16.19. Large slump southeast of Temescal Canyon, Pacific Palisades, April 1958. Photo by Spence Air Photos, UCLA.

The few surviving areas of wetland, mainly south of the canalized Ballona Creek, remain a bone of contention gnawed upon by potential developers, while various private groups and some government agencies seek to preserve the area. The struggle over the last wetlands is but one illustration of the conflicting human players at work in the continuing drama of coastal management, a drama that in April 1984 was adjudicated by the United States Supreme Court.

17. Santa Monica to Dana Point

Bernard Pipkin

When Juan Rodriguez Cabrillo sailed into the lee of the Palos Verdes Peninsula he saw a pall of smoke from signal fires of the Gabrielino Indians and named the refuge "Bahia de los Humos." He could hardly realize that the "Bay of Smokes," renamed San Pedro Bay by Sebastian Vizcaino, would be part of a supercity of over 10 million people, and that El Pueblo de Nuestra Senora La Reina de Los Angeles de Porciuncula would extend from the mountains to the sea.

Southern California has experienced phenomenal growth, and we find here the most heavily used recreational beaches in the state. They include Manhattan, Redondo, Huntington, and Newport beaches (fig. 17.1) of Beach Boys' fame as well as the lyrically reknowned Santa Catalina Island "26 miles across the sea." Along this beautiful stretch of coastline we find all kinds of beaches. There are broad, sandy ones backed by low coastal plains, intimate pocket beaches nestled between rugged points of land, and high-cliffed shores such as those at Palos Verdes and Laguna Beach. "Going to the beach" is a major recreation for people in the high-density environments of Los Angeles and Orange counties. The beaches are an important factor in the quality of life in southern California and few would argue against the aesthetic and emotional value of seashore to its inhabitants. There is also an unseen land, the one beneath the sea on the mainland shelf and adjacent borderland. It contains spectacular canyons, deep basins, and minerals of great economic importance, all of which have an impact on living in the coastal zone in southern California.

Population pressure in the southern part of the state has placed a great need for expanded recreational and pleasure-boating facilities along the coast. To this end, wetlands were transformed into marinas and canal living became a major life-style for coastal dwellers. The contrast between what was and what is can be seen at Huntington Harbour, a marina constructed by dredging and filling a part of Bolsa Chica Lagoon in Orange County. The photograph in figure 17.2 was taken about 1952 and shows Bolsa Chica and Bolsa Chica Beach, then known as "Tin Can Beach," as they were. The beach had a reputation as a hangout for squatters, transients, and victims of the Great Depression who simply disposed of empty containers by dropping them on the spot. Figure 17.3 is the same lagoon 25 years later, photographed in about the same direction but slightly to the north. The marina is a luxurious and densely populated development, but the trade-off is loss of ecologically important and equally beautiful tidal wetlands.

Los Angeles and Orange counties are the fastest-growing areas in the state. Rapid growth has resulted in restrictive legislation, such as the Coastal Zone Initiative, to control this expansion. The power of the coastal commission has been abused in some

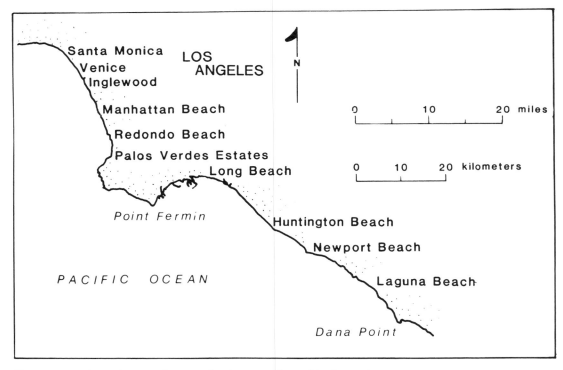

Figure 17.1. Location map for coastline between Santa Monica and Dana Point.

Figure 17.2. "Tin Can Beach" and Bolsa Chica Lagoon looking northeast as they were in 1952. Photo by Robert Stevenson, Office of Naval Research.

Figure 17.3. Seal Beach, Huntington Harbour, and Bolsa Chica Lagoon as they appear in 1983. Looking east into an area slightly north of view in figure 17.2. Photo by Bernard Pipkin.

cases, but, on balance, it has caused local government to assess the direction of development within its domain. This is important in these 2 counties because of the high percentage of land in private ownership (48 percent in Los Angeles County, and 38 percent in Orange).

Even if all the shoreline were public, some areas would still not be available for recreation. These are the high cliffs and also the wetlands so necessary to migrating birds and spawning marine life. Between Santa Monica and Dana Point there are 71 miles of shoreline of which 43 percent is high cliffs with only limited access to a few small pocket beaches. The remaining 40 miles of broad, sandy beaches backed by the low plain of the Los Angeles Basin bears the brunt of the recreational needs of 10 million people. This works out to about one-quarter inch per person! Thus this chapter is really about the "care, feeding, and nourishment" of this scarce and marvelous resource.

The coastline

Geologists love to study beaches because they are constantly in motion. A beach 150 feet wide may disappear in a single storm and mysteriously reappear over several tidal cycles with more gentle waves. This trait that makes them so interesting geologically also makes them ephemeral and worthy of close scrutiny. How many times have you been carried along the shore by the littoral current that is the conveyor belt for our dynamic beach? This same current moves sand grains along the shore, and when the supply of sand to our favorite beach is equal to or more than the amount carried away, then we know the beach will be there for our next visit. However, along most southern California beaches the amount of sand delivered each year is less than that being removed by currents. As a result, the beaches are eroding. Not all this erosion is critical, and we can live with it until it begins to threaten the existence of the beach and beach-front structures. Curiously, there are a few places where sand supplied is greater than that removed by wave action. Where this occurs the beach will widen until wind action takes over and coastal sand dunes are formed. Such dunes are found at El Segundo where huge volumes of sand were deposited by the Los Angeles River until it changed to its present course in 1884. Spectacular sand dunes, some over 100 feet high and extending several miles inland (fig. 17.4), formed where we now find Los Angeles International Airport and Manhattan and Redondo beaches. These dunes were the original movie locations for *The Sheik* and other silent films involving desert settings.

Figure 17.4. El Segundo Dunes in 1947. Ballona Creek is at bottom, and the cities of El Segundo and Manhattan Beach are at top. The flat area left-center is now the extended runways of Los Angeles International Airport. Photo by Pacific Air Industries, courtesy of Pat Merriam.

Where do beaches come from?

There is a child's poem that goes something like this: "Little drops of water / Little grains of sand / Run away together / And destroy the land." Beach materials, usually sand or gravel, are transported to the beach by rivers. These sediments are stockpiled at the mouths of large rivers, and waves subsequently distribute them to nearby beaches. Human activities in the sediment source area far inland from the beaches may have a profound effect on beach nourishment. Every rainy season we read about destructive mudflows and flooding in the foothills of the San Gabriel and Santa Monica mountains. The U.S. Army Corps of Engineers and county flood control agencies try to mitigate this flooding and mud damage by building debris basins and flood control dams and by channeling rivers. These structures are built principally to control and store floodwaters, but they trap sediment as well. The trade-off for trying to protect our inland citizens against flooding is greatly reduced contributions of sand to the coastal zone. Figure 17.5 shows the percentage of the drainage areas of major coastal rivers in Los Angeles and Orange counties behind flood control structures. As a rule of thumb, you could say that it also represents the percentage of sand being lost to the beach. Oh, that life were so simple! Additional sand has been cut off by urban expansion, which covers the sediment sources in the coastal watersheds. At San Juan Capistrano, for example, San Juan Creek is the sole source of sand for Doheny and Capistrano beaches at its mouth. There are excellent data to indicate that the sand supplied by this stream will be reduced 30

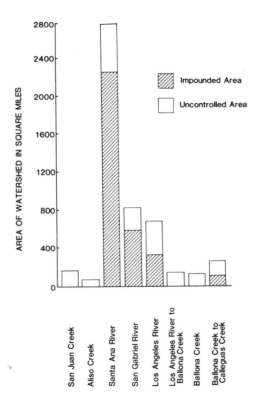

Figure 17.5. Watershed areas of major rivers behind dams in Los Angeles and Orange counties.

percent by the year 2000 because of the need for housing, streets, and infrastructure to support new residents. On top of all this, the construction industry in southern California annually mines over 20 million tons of sand and gravel from beaches, dunes, and riverbeds.

A final note on getting our little grains of sand onto the beach. Waves never stop, but rivers do! Rainfall is seasonal here and, contrary to our senses and logic, southern California has been in a period of protracted drought. During the 30 years from 1946 to 1976, only the 1969–70 winter had floods of consequence, and it is the large floods that transport great volumes of sand downstream to the oceans. The 8 million cubic yards of sand deposited at the mouth of the Santa Clara River during the floods of 1969, for example, nourish downcurrent beaches even to this day. During the remainder of the years, however, far less sand than normal was available to the beaches.

Where are the beaches going?

In general, longshore currents in southern California have a net movement toward the south or southeast. This is because most of our weather comes from the north Pacific and the resulting waves move out in a southerly direction. During short periods of the year, usually summer, swell from the south strikes our shore and produces some of the biggest waves we experience. This is the legendary "south swell" of the surfer and is due to southern hemisphere storms (during their winter) of great duration and strength. Because of the prevailing north swell, however, you would think a grain of sand starting at Santa Monica would travel southward and eventually pass all our beaches until it reached Mexico. Not so! The shoreline is divided into smaller depositional packets or units known as *littoral cells* (fig. 17.6). A sand grain set in motion by the longshore current at the beginning of a cell moves in a southerly or downcoast direction until it reaches a submarine canyon or headland that marks the end of a cell. These sand grains may be trapped in the head of the canyon to eventually flow into the deep water offshore, be blown inland on sand dunes, or trapped on the beach at the end of the cell to move on- and offshore with the seasons. A good estimate is that over 1 million cubic yards of sand per year are lost into submarine canyons between Oxnard and Newport Beach. This represents about 20 acres of beach removed from the littoral system forever; this amount of sand would half fill the Los Angeles Coliseum, or if replaced on the beach by trucks, would cost on the order of $3 million!

Man's obstructions

Man has repeatedly built shoreline structures in an attempt to control nature and reduce the impact of waves. Every coastal structure will have an effect on the coastal zone, and it is only the magnitude of the effect that is not predictable. Many coastal engineering structures that have been built between Santa Monica and Dana Point (table 17.1) illustrate the problems faced by

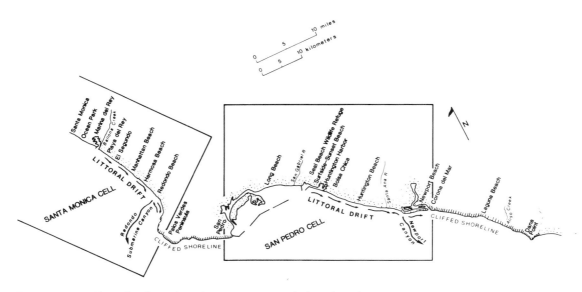

Figure 17.6. Littoral cells or beach compartments in Los Angeles and Orange counties.

Table 17.1 Artificial barriers between Santa Monica and Dana Point

Locality	Structure type	Purpose	Effect
Santa Monica	Parallel breakwater	Harbor	Downcoast erosion Noncritical
Marina del Rey	Jetties/ breakwater	Harbor inlet	Downcoast erosion Noncritical
Ballona Creek	Jetties	River outlet	Downcoast erosion Noncritical
Redondo Beach	Attached breakwater	Harbor	Loss to canyon Critical
Los Angeles– Long Beach harbors	Break- water complex	Harbor	L.A. River sediment cut off
Anaheim Bay	Jetties	Harbor inlet	Downcoast erosion Critical
San Gabriel R.	Jetties	Harbor outlet	Downcoast erosion
Newport Harbor/ Santa Ana River	Jetties	Harbor inlet	Downcoast erosion Critical
Dana Point	Attached breakwater	Harbor	Harbor siltation Noncritical

shoreline engineers. Groins, jetties, and breakwaters, depending on their orientation and placement, all tend to produce deposition upcurrent and erosion downcurrent. One way to minimize these effects in some cases is to build permeable groins that allow some sand to pass through the structures.

Such a successful groin was built at Topaz Street in Redondo Beach (fig. 17.7). The groin was designed and placed at the north end of the beach to prevent sand from moving northward into the Redondo Submarine Canyon. Because sand moves both north and south along the beach during the year, the engineers did not want the groin to totally block sand movement and create an erosion problem adjacent to the structure. The groin was designed with a solid concrete wall below the desired sand level of the beach and with connected openings above this level. Thus, in theory at any rate, sand can move slowly through the structure with the prevailing longshore current. Observations of the movement of fluorescent-dyed sand indicate that this design does work as intended.

Jetties are long structures built in pairs to improve and maintain a river outlet or harbor entrance. If jetties were not built, sand moving in both directions on the littoral conveyor belt could be deposited in the opening and eventually block the harbor entrance. It would be extremely difficult to enter Newport Harbor, for instance, if the long jetties that mark the entrance did not keep it free from sandbars and shoals. A fringe benefit of these particular jetties is an accumulation of sand on the upcurrent side of the north jetty known as "The Wedge." During

Figure 17.7. Seven-hundred-foot-long groin at Topaz Street in Redondo Beach. Note buildup of sand on south side of groin, indicating a northward-moving littoral drift. View looking east. Photo by Bernard Pipkin.

most of the year it is a tranquil beach, but when big swells from distant storms arrive here the resulting surf is legendary. Because of the orientation of the beach and the jetty, waves from a particular direction reflect off the jetty and combine with the incoming unreflected waves to form a wedge-shaped wall of water. Body surfing under these conditions is for "professionals only" and becomes a spectator sport even for strong swimmers. "The Wedge" has taken several lives and caused many severe injuries.

Construction of entrance jetties to natural harbors in the Los Angeles–Orange County area has resulted in a variety of problems; some are relatively minor, but they are problems nevertheless. For example, the jetties built at Marina del Rey were to keep littoral drift and sediment from Ballona Creek from shoaling the harbor entrance. However, it was not anticipated that accumulation of sand on Venice Beach would be blown by the wind over the top of the north jetty causing shoaling in the entrance channel that requires periodic dredging.

Rapid beach erosion may occur on downcurrent sides of structures that impede or stop longshore transport. An appreciation of the magnitude of the problem can be gained by determining the amount of sand that accumulates on the upcurrent side of groins or jetties. The amount accreted represents the capability of the waves to move sand and thus the amount of erosion to anticipate if supply is totally cut off (table 17.1).

Los Angeles County

Marina del Rey (fig. 16.18)

Marina del Rey is the largest man-made pleasure-boating and residential marina in the world (fig. 17.8B). It provides moorings for 6,000 boats, and storage and launching facilities for thousands more. Because natural protected waters are rare in southern California, unlike the east coast of the United States, harbors are often created by building breakwaters and by dredging lagoons and estuaries. Marina del Rey, for example, was dredged from the swampy flood plain of Ballona Creek, an abandoned outlet of the Los Angeles River, and is buffered from the sea by the sandy spit of Venice Beach (fig. 17.8).

The U.S. Army Corps of Engineers and its contractors, prior to construction of the marina in the early 1960s, studied every aspect of ocean engineering relating to marina construction. Included in the study were wave energy, longshore currents and littoral drift, impact for the drainage from Ballona Creek, and general foundation conditions at the site. In order to determine the highest probable waves to strike the entrance channel jetties, oceanographers "hindcasted" wave conditions from wind and swell records for many past years. The jetties were designed accordingly with the expectation that wave energy would be dissipated outside the marina proper and not create problems within the artificial bay. Oh, how man's best-laid plans can go awry when dealing with natural force in the coastal zone! Occasional large waves of the proper dimensions and approach direction set up constructive interference patterns in the entrance channel that were carried into the marina to create surge and standing waves. Simply translated, this means that there were strong horizontal currents and an up-down sloshing of the water much like that

A

B

Figure 17.8. Marina del Rey before and after. Photo by Los Angeles County Department of Beaches. (A) Ballona Creek and lagoon just prior to dredging but after construction of entrance jetties. Venice Beach is left center. (B) Marina del Rey in 1981 (note the narrow beaches).

observed in a pan of water that is tilted and set down abruptly. The height of the waves was as much as 6 feet, resulting in tremendous damage to boats and slips at certain points in the marina. Many boat owners fled, and concessionaires sued Los Angeles County to make restitution for lost business.

Temporary walls or baffles were built across the main channel to cut down surf, and a large-scale model of the marina was constructed in a wave tank to assess the problem more accurately. The study showed that a detached breakwater 1,200 feet long across the mouth of the entrance jetties would dissipate wave energy in the harbor. At last, you might think, yacht owners had a safe refuge. Not so quick. The new breakwater decreased wave action but also decreased the sediment-transporting capacity of the waves. Runoff from Ballona Creek is controlled by a channel parallel to the south jetty of the marina (fig. 17.8B). The sediment from the creek had formerly been transported by long-shore currents to Playa del Rey beaches downcoast. Now, instead, it forms a bar across the southern entrance channel that makes the channel unnavigable without periodic and expensive dredging. To add insult to injury, so much sand accumulates at the north jetty that the north side of the entrance channel is being blocked by sand that percolates through the jetty or is blown over the top by strong prevailing winds.

The Marina del Rey one sees today is a fine facility providing space for water sports activities for thousands of Angelenos. Few people are aware of the design problems and the continuing problems and maintenance costs of siltation around the entrance channel (fig. 17.9). It was economically and politically expedient to build the marina without first building a scale model to test it under various wave conditions. The result was an engineering disaster but a financial success. Construction bonds were paid off 19 years early in 1980, and annual receipts to the county exceed $10 million exclusive of taxes.

Figure 17.9. Marina del Rey north jetty shoal area. This deposit formed as a result of longshore sand transport from the 1983 winter storms. The shoal is charted as a hazard to boating. Photo by Marina del Rey *Reporter*, April 1983.

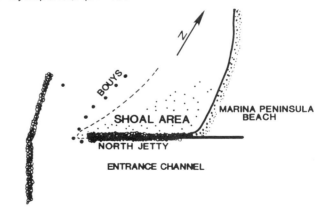

Playa del Rey to Redondo Beach (fig. 17.10)

These beaches are wide and mostly stable or only slightly eroding. This is unusual for southern California but is due in large part to presettlement supplies of sand from the old Los Angeles River outlet, now Ballona Creek. In addition, where large engineering projects have been built along the coast, the excess sand has been placed on the beach to become part of the littoral system. This was true of the Hyperion Sewage Disposal plant and the Scattergood Steam Generating Station of the City of Los Angeles.

In 1913 Standard Oil established a refinery and tanker-loading facility at "El Segundo," so called because it was the second oil refinery in California. The refinery and tank farm that evolved have given the incorporated city of El Segundo, christened by city fathers "El Segundo a nada," second to none, the lowest taxes in southern California. However, the facility was constructed with an impermeable groin system and an erosion problem developed directly downcoast. To protect the waterfront facility and restore the beach, a 900-foot-long permeable groin is to be constructed at the south end of the facility and backfilled with .5 million cubic yards of sand. The permeable groin allows some sand to pass through the structure so that serious erosion does not occur downcoast. At the same time enough sand is held in the restored area to protect the refinery during winter storms and to allow it to act as a feeder or sand reservoir for downcurrent beaches. Periodic sand replenishment will be required in the feeder-beach

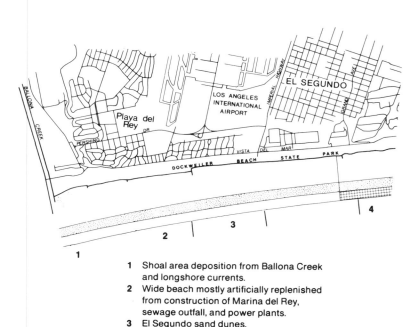

1   Shoal area deposition from Ballona Creek and longshore currents.
2   Wide beach mostly artificially replenished from construction of Marina del Rey, sewage outfall, and power plants.
3   El Segundo sand dunes.
4   Severe erosion during winter storms of 1983.

Figure 17.10.   Site analysis: Ballona Creek through Palos Verdes Estates.

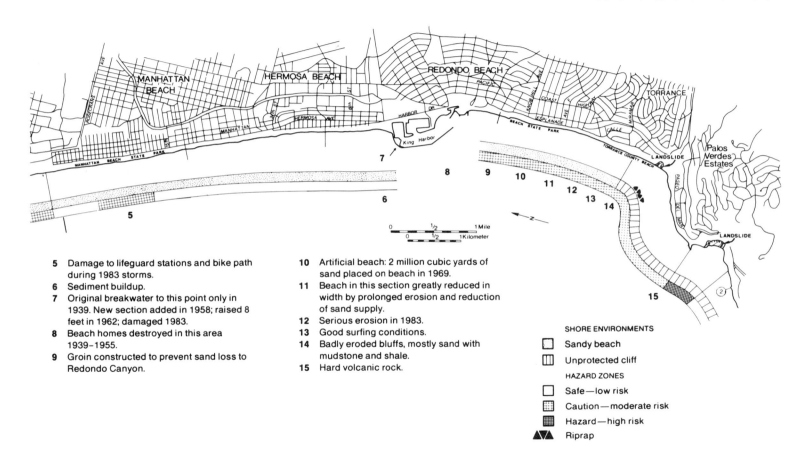

5 Damage to lifeguard stations and bike path during 1983 storms.

6 Sediment buildup.

7 Original breakwater to this point only in 1939. New section added in 1958; raised 8 feet in 1962; damaged 1983.

8 Beach homes destroyed in this area 1939–1955.

9 Groin constructed to prevent sand loss to Redondo Canyon.

10 Artificial beach: 2 million cubic yards of sand placed on beach in 1969.

11 Beach in this section greatly reduced in width by prolonged erosion and reduction of sand supply.

12 Serious erosion in 1983.

13 Good surfing conditions.

14 Badly eroded bluffs, mostly sand with mudstone and shale.

15 Hard volcanic rock.

SHORE ENVIRONMENTS

▢ Sandy beach

▥ Unprotected cliff

HAZARD ZONES

☐ Safe—low risk

▦ Caution—moderate risk

▦ Hazard—high risk

▲▲ Riprap

area, however, as longshore currents move sand through and around the groin.

Manhattan and Hermosa beaches are almost unique in southern California because they have grown demonstrably wider over the last 50 years. This growth is most likely at the expense of beaches immediately to the north and also from the addition of beach material to the littoral zone from construction at El Segundo and Marina del Rey. Although there has been considerable residential development along this stretch of coast, there has been no disruption of the littoral stream and therefore some stability has been attained.

Redondo Beach is at the end of the Santa Monica littoral cell and has suffered spectacular wave erosion for a variety of reasons. The severe storms of January and February 1983 were no exception. It all started in 1890 when a state engineer discovered a deep canyon close to shore just off what is now the Rainbow Pier. This allowed large freighters of the time to dock at Redondo, and thus, quite by accident, a seaport was born. It was not by accident, however, that the Santa Fe Railroad pushed for the port because they had just completed laying track to Redondo in 1889. Cargo could be unloaded from the ships directly into freight cars for the short trip to Los Angeles (fig. 17.11). The railroad, being enterprising, then established passenger service to Redondo Beach and built a 225-room hotel to attract tourists and paying passengers. Los Angeles needed a port, and the thought of turning Redondo into the main harbor for the city had been proposed for some time. Collis Huntington, who con-

Figure 17.11. Passengers carrying luggage on long walk to steamer on Wharf 3 at Redondo Beach in 1910. Photo courtesy *Daily Breeze*.

trolled the Southern Pacific Railroad, was promoting San Pedro and Santa Monica as ports because his track serviced those areas. San Pedro had a protective breakwater. Because of this, it became the Port of Los Angeles and eventually the third largest port in the U.S., while Redondo returned to tourism as its number one attraction. The Huntington family owned most of Redondo Beach. Henry Huntington, a nephew of Collis Huntington, headed a syndicate in 1905 that brought a young Hawaiian to Redondo to entertain tourists. George Feth rode Redondo's waves on a long wooden board for 5 years in what surely must have been the beginning of this sport in the United States.

Santa Barbara, Santa Monica, and Venice all had breakwaters protecting small craft harbors, and all had erosion problems. Redondo Beach, unfortunately, learned nothing from these earlier efforts. In 1938 a long breakwater perpendicular to the beach was approved by Franklin D. Roosevelt at a cost of $50,000. A year later in 1939 it was completed. On 2 February 1940 the *Redondo Breeze* reported "It was men against the sea at dawn today. . . . A mass of enormous billowing ground swells rolled onto the sand and pounded against the underpinning of several homes. The water took huge hunks of sand from the foot of Second Place and Third Street. Damage became less as one neared the breakwater" (fig. 17.12A).

The same submarine canyon that made Redondo famous as an early port also caused the demise of beaches north and south of the canyon head. Not only does the canyon head catch and remove sand so that beaches to the south are impoverished, it also bends or refracts wave crests causing wave energy to concentrate just north of the canyon head, destroying the beach and adjacent homes. The canyon has always been there, but construction of the breakwater interrupted the normal flow of sand along the beach and diverted it from the shore directly into the canyon where it eventually disappears into the 3,000-foot-deep San Pedro Basin.

Repeated storms throughout the 1940s and 1950s destroyed several rows of houses along the Redondo waterfront (figs. 17.12B and 17.12C), and a new plan emerged. The city bought the properties damaged by waves and formed a redevelopment district that included harbor improvement. The breakwater was extended in a dog-leg fashion to intercept waves refracted over the canyon, and an additional segment was built to form an entrance to the harbor and protect a turning basin and shopping mall. The end result is a beautiful harbor named after Congressman Cecil R. King. However, even the redesign and rededication to a congressman failed to halt the wave energy focused in the area. The breakwater was repeatedly overtopped by storm waves, and it was necessary to raise the northwestern section of the rock wall by 8 feet in 1962. In the winter storms of 1983 both the raised section and the older section to the south were overtopped, causing millions of dollars of damage to facilities inside the harbor (fig. 17.13). The Portofino Inn, a major landmark in the harbor, was inundated by standing waves. This is an up-down sloshing of the water in the harbor caused by storm-wave energy that passes through and over the top of the breakwater (fig. 17.14). Farther south, a beach that was artificially restored by importing sand in 1957 was destroyed along with volleyball courts and portions of the bike path and bluff (fig. 17.15).

It is said that nature abhors a vacuum, and the epilogue to this story is that the beaches came back with astounding rapidity almost to their former levels. Los Angeles County lifeguards reported that with normal summer sand rebuilding and work crews redistributing and leveling sand, some beaches between Manhattan and Redondo were as good as ever by July of 1983. However, missing sections of the bicycle path and eroded bluffs still testify to what happened only 6 months earlier.

A

B

C

Figure 17.12. (A) Crowd gathers to watch high tides at Redondo Beach in 1914. Photo courtesy *Daily Breeze*. (B) High tides and giant surf destroy homes at Redondo Beach in the early 1950s. Photo courtesy *Daily Breeze*. (C) The problem of storm damage at Redondo Beach continues into the late 1950s. Photo courtesy *Daily Breeze*.

A

B

Figure 17.13. (A,B) Damage and overtopping of King Harbor break-water during winter storms of 1983. Photo by Bernard Pipkin. (C) Giant breaker overtopping Galveston wall at King Harbor Yacht Club. Note collapsed section of wall. This wall takes repeated beatings as it is oriented toward the prevailing storm-wave approach. Photo by Jim Bailey, Director Public Services, City of Redondo Beach.

C

Figure 17.14. High water entering Portofino Inn in King Harbor Marina, 1983. Long-period storm waves penetrate permeable break-waters and generate waves within the harbor. Photo by Suzanne E. Butler, Emergency Services Coordinator, City of Redondo Beach.

Figure 17.15. Redondo Beach looking northward toward King Harbor, February 1983. Note gentle shore slope and lack of any well-defined beach berm. Bicycle path is covered by sand at base of bluff. Somehow the lifeguard stations survived. Photo by Bernard Pipkin.

Palos Verdes Hills (figs. 17.10 and 17.16)

Palos Verdes is a peninsular upland 25 miles south of Los Angeles. The hills have the distinction of being part of the first private land concession in California, granted in 1784 to Juan José Dominguez for faithful service. He named his 75,000 acres Rancho San Pedro. It included all the coastline from Redondo Beach to Long Beach and included what are now the cities of Torrance,

Gardena, Lomita, and Harbor City. The peninsula is also noted for its cliffed shoreline and terraced hillsides that make it a subdivider's dream. The terraces are ancient beaches formed by wave action and subsequently uplifted to their present elevation. There are some 13 of these flat, wave-eroded beaches between present sea level and San Pedro Hill, 1,480 feet above sea level (fig. 17.17). Although raised or uplifted marine terraces are common features along much of California's geologically active coastline, nowhere is a more complete sequence preserved than at Palos Verdes.

The beauty of this area did not escape early inhabitants of California. Archeological excavations suggest significant pre-Columbian cultures at 4 distinct levels. Also, no fewer than 150,000 Indians are estimated to have lived in the area, perhaps more than in any other part of the United States. With plenty of food from the sea, chert and opal for ornaments and knives, earth for paint, the yucca bush for shampoo and fibers, and asphaltum to seal serving dishes, this must have been an idyllic setting for the Gabrielino Indians who settled here about A.D. 1200.

Active faults border the Palos Verdes hills on their north and south sides, and the hills have been uplifted along these faults. Impressive seacliffs, up to 200 feet high, and areas of land instability are products of the rapid uplift. Landslides are common and constitute 53 percent of the shoreline between Abalone Cove and Point Fermin. Some form of downslope mass movement is to be found along 90 percent of the peninsula's shoreline, mostly soil creep and rockfalls. Almost all the slides are predictable and occur in areas where shale layers are inclined seaward out of the cliffs. Such slides are found at Portuguese Bend, South Shores, and at Point Fermin (see fig. 17.16).

Portuguese Bend saw the largest of these slides, and recent movement began here in early 1956. What initially appeared as minor cracks in a few houses ended in total destruction of nearly 150 homes on 300 acres of land 3 years later. The slide moves down a seaward-tilted layer of slippery clay, known as the Por-tuguese Tuff, which can be seen emerging from the slide on the beach at Abalone Cove. By 1959 all structures below the highway were destroyed, and only a few remain "livable" today in the main slide mass (fig. 17.18). Ironically, prior to development, this area had been mapped by geologists of the U.S. Geological Survey as an ancient slide mass with an area of about 1,000 acres. The well-known report was ignored by developers and homebuyers alike with tragic consequences. Reactivation of this ancient slide mass was inevitable with growth and development of the hills. Ultimately, the people of the County of Los Angeles had to foot the bill, in excess of $7 million.

In 1976 slight cracking appeared in the road above Abalone Cove west of the active landslide but within the mapped historic slide area. The new slide encompasses about 80 acres of the ancient slide area and extends 1,200 feet along the beach and 2,000 feet inland. Involved in the slide are some 25 houses, a county beach facility, and the famous Wayfarer's Chapel, designed by the late Frank Lloyd Wright, Jr. The cause is an increase in the amount of underground water due to heavy rains, septic tank and cesspool effluent, and irrigation water. This resulted in a weakening, and ultimate failure, of weak clays at depth. Six wells have been drilled to remove this excess water at considerable expense to the homeowners. The movement has been arrested but not stopped, but this has come a little late for owners of the many homes so badly damaged that they had to be abandoned.

The South Shores landslide is not now active, but it is at least

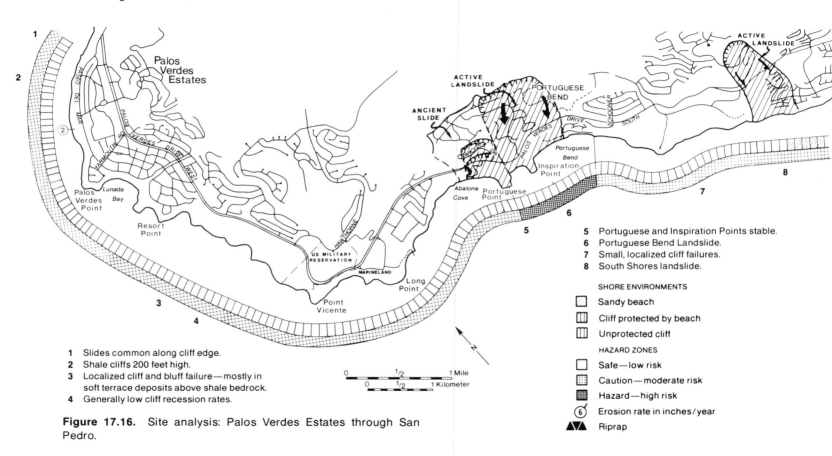

Figure 17.16. Site analysis: Palos Verdes Estates through San Pedro.

1 Slides common along cliff edge.
2 Shale cliffs 200 feet high.
3 Localized cliff and bluff failure—mostly in soft terrace deposits above shale bedrock.
4 Generally low cliff recession rates.

5 Portuguese and Inspiration Points stable.
6 Portuguese Bend Landslide.
7 Small, localized cliff failures.
8 South Shores landslide.

SHORE ENVIRONMENTS

Sandy beach
Cliff protected by beach
Unprotected cliff

HAZARD ZONES

Safe—low risk
Caution—moderate risk
Hazard—high risk
⑥ Erosion rate in inches/year
Riprap

0   1/2   1 Mile
0   1/2   1 Kilometer

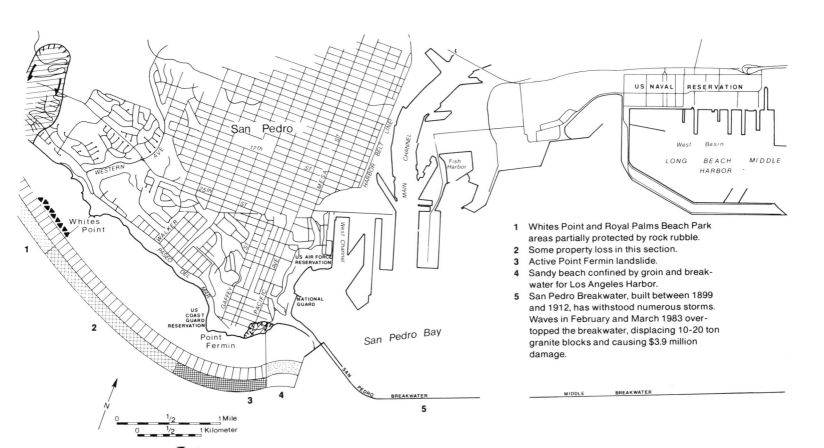

1 Whites Point and Royal Palms Beach Park areas partially protected by rock rubble.
2 Some property loss in this section.
3 Active Point Fermin landslide.
4 Sandy beach confined by groin and breakwater for Los Angeles Harbor.
5 San Pedro Breakwater, built between 1899 and 1912, has withstood numerous storms. Waves in February and March 1983 overtopped the breakwater, displacing 10-20 ton granite blocks and causing $3.9 million damage.

16,000 years old. The slide mass is 3,000 feet wide along the beach and extends inland for 4,000 feet. It is structurally similar to Portuguese Bend and has been partially developed, after extensive geologic study, into a mobile home park. This appears to be a reasonable use for an ancient landslide mass inasmuch as the trailers, presumably, can be moved should the slide reactivate. In addition, no permanent dwellings are permitted to be built on the slide.

The first historic landslide at Point Fermin occurred in 1929. Many homes and a quarter-mile section of the road were destroyed. Failure was caused by underground water weakening the shale layers, which sloped seaward out of the cliff face just like those at Portuguese Bend. Total movement has been about 200 feet, and recurrent movement takes place during heavy rains. It has become something of a tourist attraction in the area (fig. 17.19).

This litany of failure-prone cliff areas should be warning enough against building too close to the cliff edge where weak geologic materials are found. In addition, wave erosion at the foot of the seacliffs undermines the cliff face and contributes to failure. Extensive geologic investigations should precede any building along the top of the cliff and, even then, may not provide 100 percent insurance against ground movement. Average cliff erosion rates as high as 9 inches per year have been determined for cliffs composed of soft sedimentary deposits, whereas 0.2 inches per year has been calculated for hard volcanic rocks. Two inches a year is a reasonable average erosion rate over a

Figure 17.17. Palos Verdes Hills looking north showing outlines of Portuguese Bend landslide. Approximate limit of an ancient slide area is shown by the fine dashed line. The 1956 slide area and the smaller, more recent slide areas are shown by heavier dashed lines. Stepped surfaces or terraces represent ancient beaches uplifted to their present elevation. Photo by Los Angeles County Engineers Office.

front property is sold by the pound and glamour by the ton. Two of the largest rivers in both counties reach the sea here and were responsible for the expanses of sand from Seal Beach to Newport Beach. However, too much man and too little nature has resulted in reduced sand supplies where they are wanted most. The Los Angeles, San Gabriel, and Santa Ana rivers are now either dammed or paved or both. This combined with a long period of drought, imprudent jetty and groin construction, and a sewage outfall or two has produced trouble in paradise.

Figure 17.18. The Portuguese Bend landslide in November 1983. Pier in figure 17.17 has been overrun by landslide mass and no longer exists. Photo by Bernard Pipkin.

long period of time for rocks exposed in seacliffs at Palos Verdes. Thus, if a home is built at the top of the cliffs with a life expectancy of 50 years, a minimum setback from the edge can be determined. However, each site is different and an expert should be consulted.

Orange County (figs. 17.20 and 17.21)

This is probably the most beautiful and surely the most expensive stretch of coastal real estate in southern California. Beach-

Figure 17.19. Point Fermin and landslide mass that failed in 1929. Slide is still unstable and poses a minor threat to homes but a major threat to unwary hikers. Photo by Bernard Pipkin.

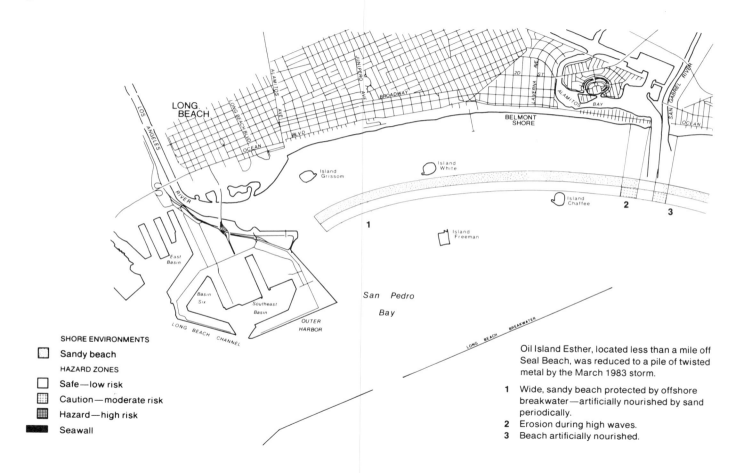

SHORE ENVIRONMENTS

Sandy beach

HAZARD ZONES

Safe—low risk

Caution—moderate risk

Hazard—high risk

Seawall

Oil Island Esther, located less than a mile off Seal Beach, was reduced to a pile of twisted metal by the March 1983 storm.

1 Wide, sandy beach protected by offshore breakwater—artificially nourished by sand periodically.
2 Erosion during high waves.
3 Beach artificially nourished.

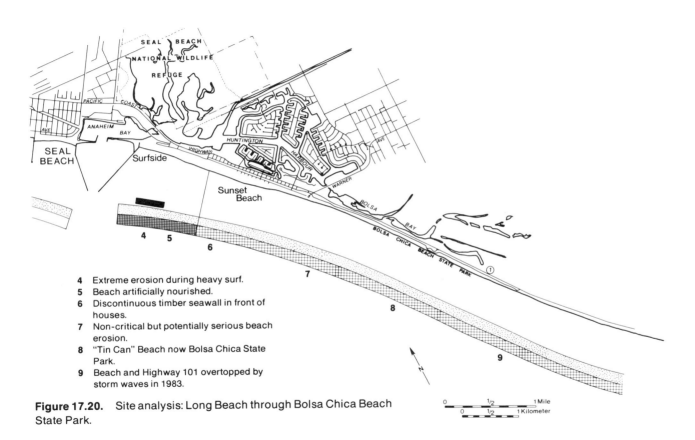

4   Extreme erosion during heavy surf.
5   Beach artificially nourished.
6   Discontinuous timber seawall in front of houses.
7   Non-critical but potentially serious beach erosion.
8   "Tin Can" Beach now Bolsa Chica State Park.
9   Beach and Highway 101 overtopped by storm waves in 1983.

Figure 17.20.   Site analysis: Long Beach through Bolsa Chica Beach State Park.

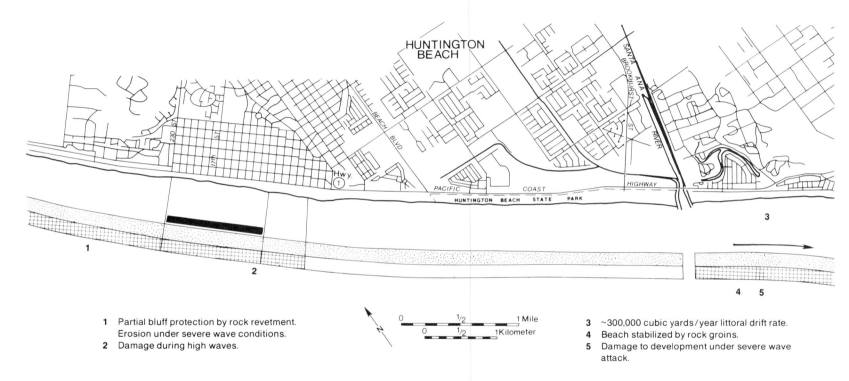

1 Partial bluff protection by rock revetment. Erosion under severe wave conditions.
2 Damage during high waves.

3 ~300,000 cubic yards / year littoral drift rate.
4 Beach stabilized by rock groins.
5 Damage to development under severe wave attack.

Figure 17.21.   Site analysis: Huntington Beach through Corona del Mar.

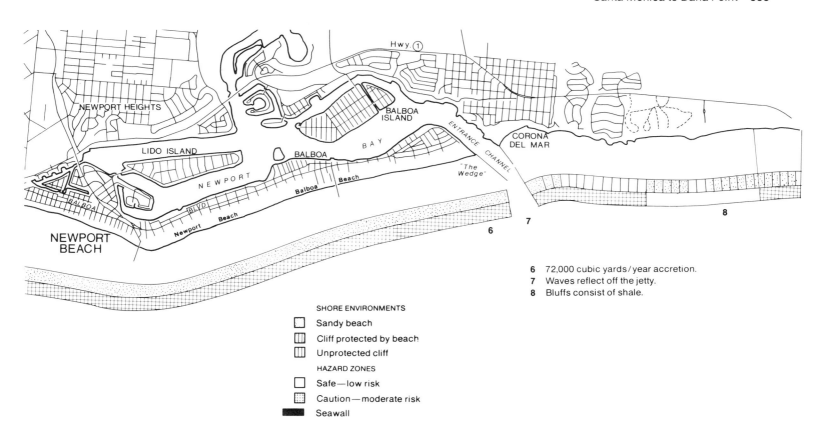

6 72,000 cubic yards / year accretion.
7 Waves reflect off the jetty.
8 Bluffs consist of shale.

SHORE ENVIRONMENTS

☐ Sandy beach

▥ Cliff protected by beach

▥ Unprotected cliff

HAZARD ZONES

☐ Safe—low risk

▦ Caution—moderate risk

▬ Seawall

Seal Beach (fig. 17.20)

Since the Anaheim Bay East Jetty was constructed in 1944, Surfside–Sunset Beach, just south of Seal Beach, has been totally removed by storm waves more times than memory can recall. The U.S. Navy created the harbor as a facility for antisubmarine nettending equipment, and Sunset Beach, which was nourished by sand from the San Gabriel River, was cut off and greatly reduced in width (fig. 17.22). Large northwest winter swells pound the beach just south of Anaheim Bay and the same beach-front houses are imperiled with every storm. Periodic sand replenishment by the Corps of Engineers has maintained the beach for recreation and preserves it as a major sand-feeder for Huntington Beach to the south. Homeowners have been lucky to this point, but there comes a time when luck changes and funding for replenishment projects runs out. Beach restoration treats the symptoms but not the problem. To correct the problem the responsible agencies would have to remove the breakwater or provide a permanent sand bypass system, and neither is likely to occur.

Huntington Beach (fig. 17.21)

Bolsa Chica Lagoon is a coastal marsh of over 1,600 acres that serves as home refuge for migrating wildfowl, spawning marine animals, and indigenous animal and plant populations (fig. 17.23). In 1960 it was almost twice as long as it is today, extending parallel to the coast from Seal Beach to Huntington Beach. In the early 1960s Huntington Harbour was developed, and 600

Figure 17.22. Anaheim Harbor showing south jetty and breakwater. Critical erosion has taken place many times at Surfside Beach just south (right) of jetty. Photo by Robert Stevenson, Office of Naval Research.

acres near the center of the marsh were eventually converted to luxurious marina residences. As seen today, Huntington Harbour separates the Seal Beach National Wildlife Refuge to the north from Bolsa Chica Ecological Reserve to the south (fig. 17.24). The channel into the harbor from the open ocean is through Anaheim Bay, which also provides tidal flushing for the lagoon and canals.

In 1981 the Orange County Board of Supervisors approved a land-use plan concept with emphasis on marsh restoration and

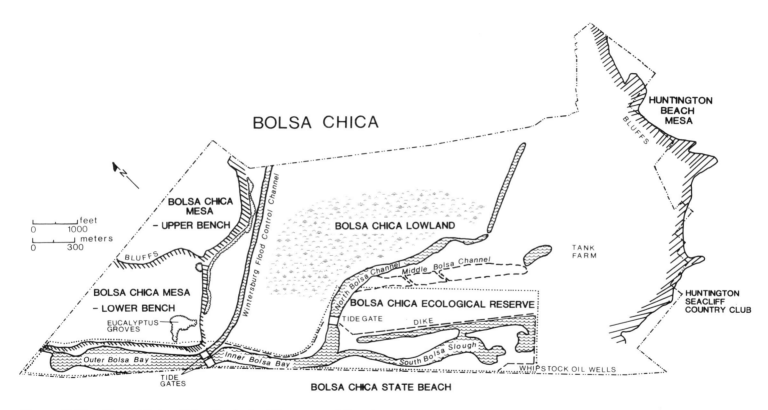

Figure 17.23. Map of Bolsa Chica Ecological Reserve and Beach State Park. See figure 17.2 for aerial view of the same area in 1952.

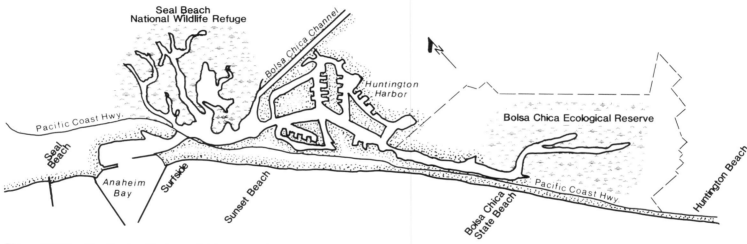

Figure 17.24. Huntington Harbour divides natural wetlands of Bolsa Chica into 2 parts. See figure 17.2 for aerial perspective of harbor and lagoon.

public, visitor-serving recreation. The plan includes a navigable ocean entrance that will provide another access channel for recreational boating into the Huntington Harbour area. There will be a 600-acre salt marsh system; parks connected by equestrian, pedestrian, and bicycle trails; 5,700 residential units along the northern edge of the marsh; and commercial areas. The physical and environmental problems connected with such a project are legion and range from increasing tidal currents throughout the lagoon and erosion of its banks to blocking littoral drift at the proposed ocean entrance. The Corps of Engineers estimates drift rates of 276,000 cubic yards per year along the beach at Bolsa Chica, one of the higher values measured along the southern California coast. What will be the impact on downcoast beaches if these jetties are built without providing for sand bypassing? Haven't we learned from our mistakes at Santa Barbara and Redondo Beach? A great deal of research and model testing, including flood control, will have to be performed by responsible agencies in order to allow us to live with this part of the California coast. This marsh was created during the latter part of the ice ages when the Santa Ana River flowed in what is now called Bolsa Gap. It took nature thousands of years to build the marshlands, and once the marsh is destroyed or seriously damaged it will not be possible to reverse the process (see table 17.2).

Bolsa Chica Beach State Park and Huntington Beach are well exposed to winter storm waves. The coastal plain behind Bolsa Chica Beach is so low that wave overwash and inundation of the park, Pacific Coast Highway (U.S. 101), the oil production facili-

Table 17.2 Longshore transport of sand in the Los Angeles–Orange County area

Location	Direction	Rate (Cubic yds/yr)
El Segundo	downcoast	165,000
Redondo Beach	downcoast	30,000
Redondo Beach to Malaga Cove	upcoast	52,000
Surfside to Sunset Beach	downcoast	275,000
Newport Beach	downcoast	300,000
Newport Pier to Newport Jetty	downcoast	72,000
Newport Jetty to Dana Point	downcoast	0
San Juan Capistrano	downcoast	100,000

Note: A large dump truck can hold about 10 cubic yards, so dividing annual rates by 10 gives an idea of how many truckloads of sand are being carried yearly along the beach.

ties, and the lagoon occurred during 1983 storms (fig. 17.25). High tides and storm surge did major damage to 182 homes and minor damage to 280 others. Several hundred mobile homes along flood-control channels were also damaged. A large part of the $14 million damage was due to high tides and their effect upon flood runoff in confined channels. Information on the elevations protected from inundation by floods of various magnitudes is available from Orange County and Huntington Beach disaster agencies and should be consulted before purchase of property or structures in these areas. Similarly, design storm-

Figure 17.25. Damage to Pacific Coast Highway (arrow), Bolsa Chica Beach State Park, and oil production facilities in storms of February 1983. Photo by Ed Vacile, City of Huntington Beach Fire Department.

Figure 17.26. Storm-damaged section of Seal Beach Pier. Heavy surf overtopped several sections of the pier producing heavy damage. Photo courtesy Huntington Beach Fire Department.

wave heights used in the past for piers and offshore structures should be reevaluated in light of the damage during the winter of 1983 (fig. 17.26).

Newport Beach (fig. 17.21)

During late summer and fall of each year, Antarctic storms and hurricanes off Baja California bring potentially destructive waves to southern California. Newport Beach is seriously affected because of its south-facing shoreline and the lack of protective offshore islands. Erosion at the end of summer is a fact of life for its residents, and in 1965 the situation became critical when the

beach eroded 165 feet before being stabilized by sandbags just 5 feet from property lines. In August and September of 1968 several homes were undermined, and the same sandbags used to protect the homes in 1965 were exposed in the cut eroded by wave action. Most of the erosion was in an area locally known as West Newport, between the Newport Pier and the Santa Ana River jetties. About 500 feet offshore from the pier in water 25 feet deep is the Newport Submarine Canyon, formed by the ancestral Santa Ana River when sea level was much lower than today (fig. 17.27).

Newport Beach is assumed to be the southern end of the San Pedro littoral cell. The submarine canyon provides an exit path-

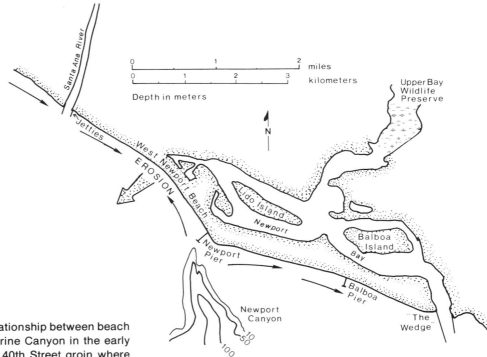

Figure 17.27. Newport Beach showing relationship between beach features, littoral drift, and Newport Submarine Canyon in the early 1970s. Extreme erosion took place at the 40th Street groin where strong rip currents are created by convergence of longshore currents on groin barrier.

way for sand moving southward in the littoral system. Measurements by the U.S. Army Corps of Engineers indicate the rate of littoral drift there is about 300,000 cubic yards per year. Curiously enough, only about 75,000 cubic yards of sand accumulates each year at the north entrance jetty to Newport Bay ("The Wedge"), and is the end of the line for the littoral cell. Thus it appears that the remainder of the sand, almost a quarter of a million cubic yards, is flushed down the canyon just upcoast. The Corps of Engineers wanted to reduce the sand loss to the canyon, and a variety of plans were proposed; the most outrageous was a parallel breakwater just north of the canyon head to arrest the sand and then recycle it by pumping to the river jetties for another round trip. Other investigators found that under almost all conditions waves were refracted in a fanlike manner at the canyon head to create longshore currents moving northward on the north side of the canyon and southward on the southside (fig. 17.27). After considerable research it was concluded that when flood control structures were built on the Santa Ana River, coarse sand was trapped, leaving only finer material to be carried to the beach. This finer sand is not stable under the present wave climate, so the beaches eroded. The problem was not one of too little sand so much as sand of the wrong size. As a result of a detailed coastal study by the University of Southern California and input from other agencies, researchers, and interested parties, the Corps of Engineers responded by artificially rebuilding the beach with sand deposited in the Santa Ana River during the 1969 floods and by constructing 6 groins to hold the

Figure 17.28. Groin field constructed to hold artificially replenished sand at Newport Beach. Some sand is provided naturally by the Santa Ana River during heavy rains. Photo by Bernard Pipkin.

sand (fig. 17.28). This has been a tremendously successful beach replenishment effort and is an example of people and technology living with the California coast, albeit at considerable expense.

Corona del Mar to Dana Point (figs. 17.21 and 17.31)

This 13-mile stretch of coastline is entirely cliffed, with many charming pocket beaches and some of the most beautiful scenery in the state. Laguna Beach, the largest community in this area, is famous for its Festival of the Arts. In contrast to the Palos Verdes Peninsula, this section is more irregular and scalloped, and the cliffs are generally lower. This is due to the difference in rock types that support the cliffs and affect the slope and shape that the shoreline will take. From Corona del Mar to Abalone Point the shore is relatively straight. It is composed of layered shales that resist erosion rather uniformly. From Abalone Point to Laguna Beach the cliffs are supported by hard volcanic rocks that form rugged points, whereas softer shales form the bays. From Laguna Beach to Dana Point the cliffs are supported by coarse sandstones that vary widely in resistance to wave attack. Some of the highest cliffs along this coastline (almost 200 feet) occur at Dana Point (fig. 17.29). Large landslides are also common (Bluebird Canyon, for example) and must be considered in the planning of any future developments.

A striking feature of this coastline is almost 4 miles of open space between Laguna Beach and Corona del Mar. This is part of the 77,000-acre Irvine Ranch put together from Spanish land grants by James Irvine. Development of the ranch has been planned in great detail and calls for special treatment of 10,000 acres along the coast. The state of California wishes to acquire over 3,000 acres of this property for parkland.

With the development of Dana Point Harbor, demand for coastal property south of Laguna Beach has greatly increased. The cliffs here are active, that is, wave erosion tends to over-steepen the cliffs and make them unstable (fig. 17.30). Average rates of cliff retreat are somewhat misleading because the different rock types present erode at different rates (see chapter 3). Thus, as along much of the California shoreline, the local geology and human activities at each site must be carefully evaluated in order to live in harmony with the coastal zone.

Figure 17.29.   Dana Point and Salt Creek as they looked in 1952. Photo by Robert Stevenson, Office Naval Research.

Figure 17.30.   Massive landslides just north of Dana Point. View looking toward San Clemente in 1952. Photo by Robert Stevenson, Office of Naval Research.

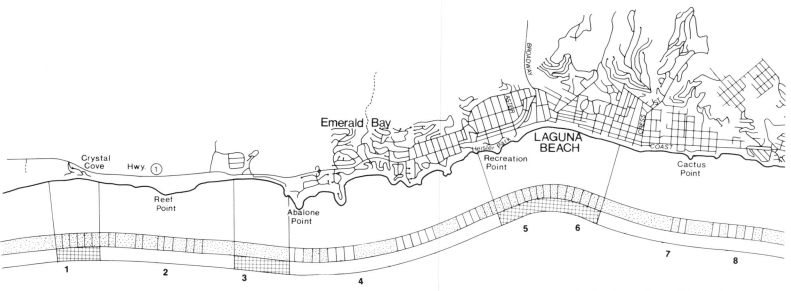

1   House threatened under severe wave
    conditions.
2   Landslides common in cliffs.
3   Eroding cliffs. Mobile home park
    threatened under severe wave conditions.
4   Resistant volcanic rocks form points. Bays
    form in weaker shales.

5   Some damage due to erosion of cliffs at
    Heisler Park.
6   Occasional wave damage to buildings on
    beach. Discontinuous low seawalls.
7   Sandstones of variable resistance to
    erosion form bluffs.

8   Structures at base of bluff subject to wave
    damage.
9   Buildings in low areas subject to wave
    damage.
10 Numerous landslides now removed for
    new projects at Laguna Niguel.
11 Rocky point with offshore rocks and reef
    backed by high, vegetated rocky cliff.
    Active slides at top and base of cliff.
    Recent rockfall along base of cliff.

Figure 17.31.   Site analysis: Crystal Cove to Doheny State Beach.

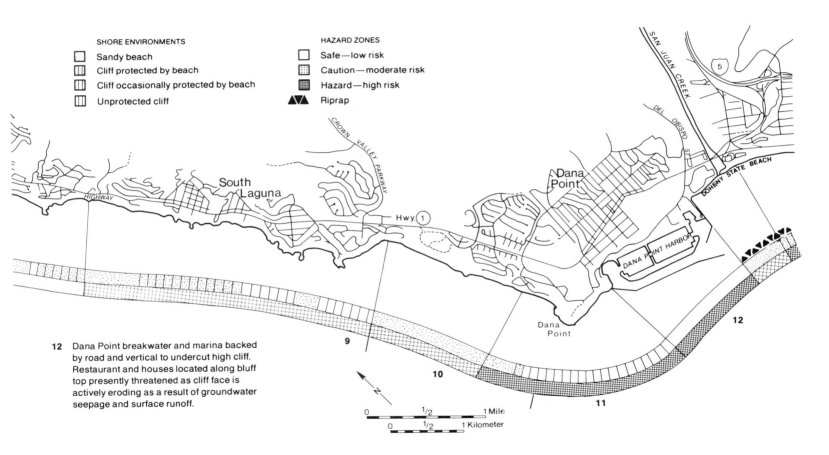

SHORE ENVIRONMENTS

- Sandy beach
- Cliff protected by beach
- Cliff occasionally protected by beach
- Unprotected cliff

HAZARD ZONES

- Safe—low risk
- Caution—moderate risk
- Hazard—high risk
- Riprap

SAN JUAN CREEK

DEL OBISPO ST.

DOHENY STATE BEACH

CROWN VALLEY PARKWAY

South Laguna

Dana Point

Hwy. ①

HIGHWAY

DANA POINT HARBOR

Dana Point

9

10

11

12

N

0   ½   1 Mile

0   ½   1 Kilometer

12   Dana Point breakwater and marina backed by road and vertical to undercut high cliff. Restaurant and houses located along bluff top presently threatened as cliff face is actively eroding as a result of groundwater seepage and surface runoff.

18. Dana Point to the Mexican border

Gerald Kuhn and Francis P. Shepard

During the past 2 decades there has been a tremendous boom of land development along the cliffs at the seaward margins of the coastal terraces of San Diego County, particularly to the north of La Jolla. Until the enactment of the California Coastal Zone Conservation initiative in late 1972, local agencies frequently permitted construction of single-family residences and multiple-unit residential structures within only a few feet of bluffs and on the beaches themselves. Developers justified this practice by stating that none of the seacliffs were retreating at an appreciable rate. What was overlooked was that the cited low retreat rates were usually based on only the experience of the last 25 or 30 years, an unusually quiescent time, characterized by low rainfall and few coastal storms capable of producing heavy surf. Earlier studies of coastal erosion in the area, which provided a far less optimistic picture of the stability of the seacliffs, were ignored or discounted.

In 1889, during a wet, stormy period that lasted from 1884 to about 1893, the U.S. Coast and Geodetic Survey, while conducting topographic surveys along the coast of San Diego County, noted that "new erosion during each winter storm is the characteristic feature of this coast." It was also recorded somewhat later that the cliffs north of La Jolla were actively retreating during the wet and stormy years of the 1920s. The bluffs near Scripps Institution of Oceanography retreated 10 to 20 feet between 1923 and 1930.

There are good indications that the sedimentary cliffs have retreated episodically due to large rock falls, although rock cliffs in many places showed no indications of having had appreciable retreat since photographic records began some 50 years earlier.

The stormy period with unusually high rainfall that began in early 1978 has changed the picture considerably, especially on the beaches and along the bluffs from north of Del Mar to Oceanside, where much building has taken place on the low terrace close to the bluff in recent years.

(Much of the information presented in the following pages is the result of previous research by the authors as well as other scientists who are not directly referenced. For additional information, the reader should consult the original publications listed in the bibliography at the end of the book.)

Dana Point, San Juan Capistrano, and San Clemente (figs. 17.29 and 18.2)

This coastal sketch begins at Dana Point to the north and extends south to the Mexican border (fig. 18.1). Dana Point itself is a rock headland comprised of sedimentary rocks. Directly to the south of the point lies San Juan Creek, which is important because it still contributes sediment to the shore during floods; its headwaters have not been cut off entirely by the construction of dams inland. South of the river is Doheny State Beach, located

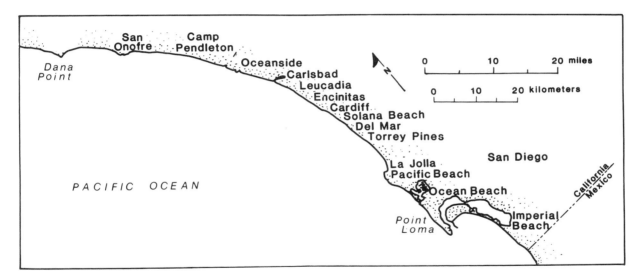

Figure 18.1.   Location map of coastline between Dana Point and the Mexican border.

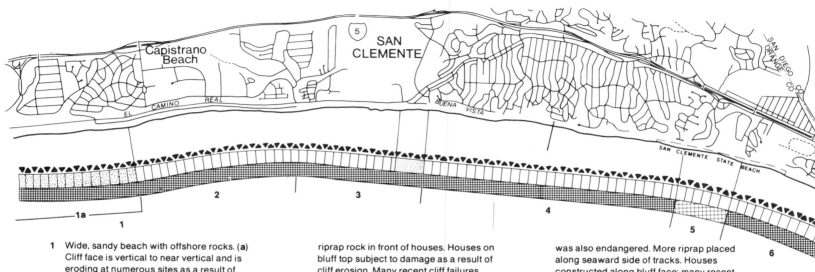

1
1a
2
3
4
5
6

1 Wide, sandy beach with offshore rocks. (**a**) Cliff face is vertical to near vertical and is eroding at numerous sites as a result of groundwater flow. Buildings are subject to danger as a result of cliff collapse.

2 Narrow sandy beach backed by houses at beach level (fronted by riprap rock revetment) backed by high coastal bluffs. Many homes have low wooden or concrete block seawalls. Houses on beach road and railroad subject to damage during high wave conditions as waves break directly on riprap rock in front of houses. Houses on bluff top subject to damage as a result of cliff erosion. Many recent cliff failures visible.

3 Narrow sandy beach backed by low wooden seawall, mobile home park, railroad, highway, and high eroding coastal bluff. Three sand-filled Longard tubes placed in front of timber wall collapsed. Seawall overtopped and mobile homes sustained severe damage during winter storms of January-March 1983. Railroad was also endangered. More riprap placed along seaward side of tracks. Houses constructed along bluff face; many recent slides and groundwater seepage visible at many sites.

4 Narrow sandy beach backed by park facilities, railroad, and high coastal bluffs with houses and apartments built along rim. Groundwater seepage, storm drain collapse, recent cliff failure visible along bluff face. Rock riprap seawall semi-protects railroad. Winter storms of 1983 damaged park facilities. Houses located along bluff top subject to damage as a result of landslides and cliff collapse.

Figure 18.2. Site analysis: Capistrano Beach through San Onofre State Park.

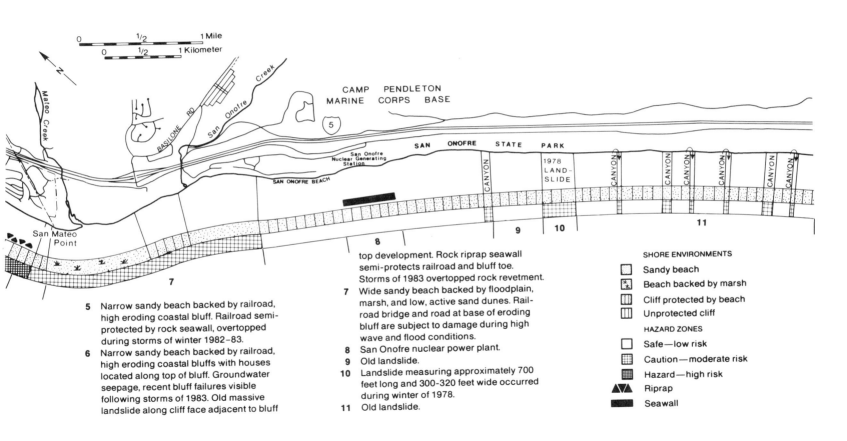

CAMP PENDLETON
MARINE CORPS BASE

SAN ONOFRE STATE PARK

San Onofre
Nuclear Generating
Station

SAN ONOFRE BEACH

1978
LAND-
SLIDE

San Mateo
Point

CANYON 9   10   CANYON   CANYON   CANYON   CANYON   11   CANYON

8

7

SHORE ENVIRONMENTS

Sandy beach

Beach backed by marsh

Cliff protected by beach

Unprotected cliff

HAZARD ZONES

Safe—low risk

Caution—moderate risk

Hazard—high risk

Riprap

Seawall

5   Narrow sandy beach backed by railroad, high eroding coastal bluff. Railroad semi-protected by rock seawall, overtopped during storms of winter 1982–83.

6   Narrow sandy beach backed by railroad, high eroding coastal bluffs with houses located along top of bluff. Groundwater seepage, recent bluff failures visible following storms of 1983. Old massive landslide along cliff face adjacent to bluff

top development. Rock riprap seawall semi-protects railroad and bluff toe. Storms of 1983 overtopped rock revetment.

7   Wide sandy beach backed by floodplain, marsh, and low, active sand dunes. Railroad bridge and road at base of eroding bluff are subject to damage during high wave and flood conditions.

8   San Onofre nuclear power plant.

9   Old landslide.

10   Landslide measuring approximately 700 feet long and 300-320 feet wide occurred during winter of 1978.

11   Old landslide.

in the San Juan Capistrano area. In recent years, the cliffs of this area have been retreating actively as a result of weakening by groundwater and erosion due to surface runoff. Groundwater levels along the cliffs have shown a steady rise in recent years commensurate with urbanization along and inland of the seacliff. This results from excessive lawn watering, an attempt to grow nonnative vegetation, and the input from septic tanks and cesspools. It has been estimated that landscape irrigation alone is the equivalent of 50 to 60 inches of additional rainfall each year.

The extensive watering in the coastal areas has had at least 3 important effects: a slow but steady rise in the water table that has progressively weakened the cliff material; lubrication of surfaces along which slides and block falls are initiated; and in some instances increased dissolution in the underlying rocks.

Directly to the south is the city of San Clemente, located on a terrace that has experienced landslides, in many cases initiated by bluff-top construction and related groundwater problems. During the storms of January–March 1983 many of the structures located on the beach seaward of the cliffs were severely damaged. These uplifted terraces extend south all the way down to San Mateo Creek near the San Onofre nuclear power plant. Former President Nixon's summer White House was located on the terrace near San Mateo Point.

San Onofre and Camp Pendleton (figs. 18.2 and 18.3)

Extending south from San Onofre for some 14 miles, the coast

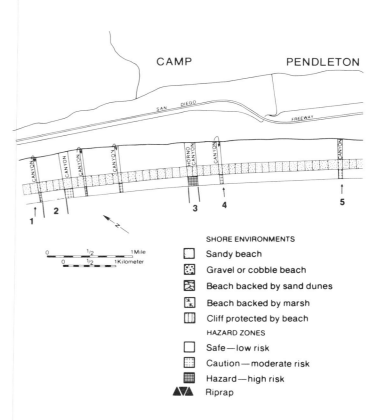

SHORE ENVIRONMENTS
☐ Sandy beach
▦ Gravel or cobble beach
▨ Beach backed by sand dunes
▥ Beach backed by marsh
▥ Cliff protected by beach

HAZARD ZONES
☐ Safe—low risk
▨ Caution—moderate risk
▦ Hazard—high risk
▲▼▲ Riprap

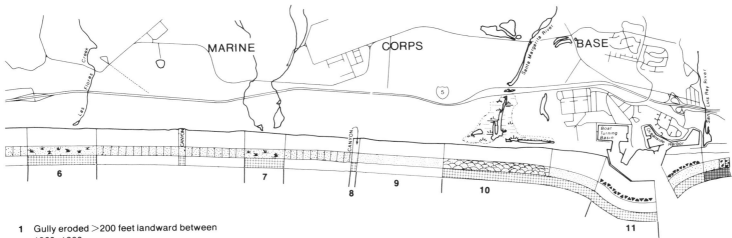

1. Gully eroded >200 feet landward between 1968–1980.
2. Landslide measuring 1700 feet in length and 350 feet in width occurred in June 1941.
3. Horno Canyon landslide.
4. Dead Dog Canyon eroded approximately 560 feet headward during very wet years between 1932–1977. This same canyon eroded another 100 feet during a two-week period during January–February 1978, and approximately 100 feet in February 1980 alone.

5. Canyon exhibits recent headward erosion. Frequent active landslides along top and base of bluff along canyon walls.
6. Area subject to damage during high wave and flood conditions.
7. Area subject to damage during high waves and flood conditions.
8. Area subject to damage during high waves and flood conditions.

9. Frequent small slides visible along rim, face, and toe of bluff, especially during storm years of 1978, 1980, and 1983.
10. Area subject to severe erosion and damage during high wave conditions.
11. Harbors act as a sediment trap, and more than 12 million cubic yards of sedimentary material was dredged between 1942 and 1981. Harbor subject to great damage from southerly storm "Chubasco."

Figure 18.3. Site analysis: Camp Pendleton Marine Corps Base.

consists of broad, sedimentary fans built out from the local mountains. Most of the region is included in the Camp Pendleton Marine Corps Reservation. This area is unique in having extensive beaches that buffer the low cliffs fronting this sloping terrace. In fact, it is about the only area in San Diego County where large beaches, such as existed very commonly in the past, are still present. These beaches are largely due to the indirect effects of highway construction in this area. The concentration of runoff in the underlying culverts cuts through these loosely consolidated sediments during heavy rains eroding small canyons and quickly producing large volumes of sediment. In one case a canyon was lengthened over 200 feet in 24 hours and contributed about 50,000 cubic yards of material to the adjacent beach (fig. 18.4). The eroded sand is initially carried to the beach and then moved offshore by large winter waves to form large sandbars. Under former conditions this sand might have been transported southward and added to the beaches of the Oceanside area. Unfortunately, the general longshore currents in the area, which had been primarily north to south, have changed during the past few years and are now from south to north or more variable. As a result, these sandbars have been moved landward to create very wide beaches where there used to be narrow beaches. This coast now has some of the best beaches in southern California.

At the southern edge of this area the Santa Margarita and San Luis Rey rivers enter the coast. During past floods these rivers carried considerable sediment to the coast, much of which

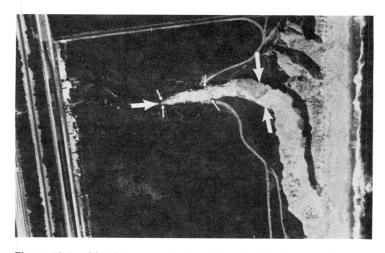

Figure 18.4. March 1980 vertical aerial photo showing development of 460-foot canyon perpendicular to shoreline. It was extended 235 feet on 20 February 1980, within 24 hours as indicated by small arrows. Photo by U.S. Navy.

was ultimately deposited in the Oceanside area to the south; the recent shift in currents has changed all of this, however.

Oceanside and Carlsbad (fig. 18.5)

Two harbors have also been excavated at the southern end of Camp Pendleton. One is a turning basin for the Marine Corps, and the other, directly north of Oceanside, is a harbor for pleasure boats. The jetties at the harbor mouths have cut off the sand moving southward, which nourished the wide beaches in the Carlsbad and Oceanside areas. Although the sediment from each of the harbors has, in the past, been carried by dredge pipe to the beaches farther south, the shift in littoral currents has changed the area significantly. Problems have developed at Oceanside. The formerly wide beach, which was of great importance to tourism, has virtually disappeared. The beach has had temporary additions from periodic dredging, but most of the dredge spoils have been carried away during storms. The general northerly littoral drift that has developed in recent years has carried much of the beach material northward back to the pleasure boat harbor, leaving behind a cobble-strewn beach that extends for many miles. The large cobbles have been picked up by the great storm waves of the past few years and thrown against the low 30-foot terrace and the ocean-front houses (fig. 18.6). This has produced considerable erosion and damage along the northern part of the city where construction has occurred directly on a former beach.

Recent storms have also resulted in the loss of the piers that have been built along the shore.

South Carlsbad has one of the longest cobble beaches in southern California (fig. 18.7) which was extensively mined for abrasives between 1912 and 1945. The beach has grown significantly since 1980 when large storm swells moved great quantities of cobbles from offshore onto the beach, at times even covering the low-lying coastal land.

Leucadia, Encinitas, Cardiff-by-the-Sea, and Solana Beach

From Batiquitos Lagoon in South Carlsbad the bluffs rise abruptly to a height of about 90–110 feet and consist of older beach and dune ridges. The towns of Leucadia, Encinitas, Cardiff-by-the-Sea, and Solana Beach were constructed on the beach ridges. In a few sites the cliffs are severely undercut. In Leucadia large caves have formed both at the base and higher up on the cliff as a result of groundwater solution along fault zones.

Construction along most of the bluff tops occurred during the decade prior to 1980. Very little erosion apparently took place along this section of the coast between 1950 and 1970. Between 1978 and 1983, however, there has been extensive localized retreat primarily during the storms of January through February 1983. In Leucadia small buildings and walls have collapsed onto the beach during the past few years. Moonlight State Beach directly to the south has experienced severe erosion during recent

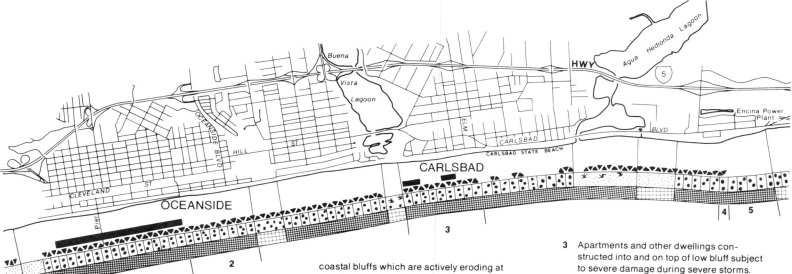

1 Oceanside and Carlsbad: Pier, road, and buildings located on the beach are subject to damage during periods of high surf. Winter storms in 1978, 1980, and 1983 caused extensive damage when sections of Oceanside Pier and coastal road collapsed, and high waves threw cobbles at the buildings along the shore. Beach development backed by 30 foot high coastal bluffs which are actively eroding at certain sites and are threatening bluff-top houses and apartments. Concrete walls and riprap extensive. Some riprap-protected cliffs retreated 15-20 feet in 1983 alone as waves overtopped seawalls.

2 Damage occurred at selected sites during the winters of 1978, 1980, and 1983 as beach cobbles were thrown as "projectiles" at buildings along the shore. Several buildings were severely damaged in 1980 and were subsequently demolished and replaced by even larger buildings.

3 Apartments and other dwellings constructed into and on top of low bluff subject to severe damage during severe storms. Damage occurred at selected sites along this area during storms of 1978, 1980, and 1983.

4 Up to 26 feet of coastal retreat in the bluff forming terrace deposits and 6-8 feet of retreat in more resistant cliff material between 7 and 9 August 1983.

5 Cliff failures are a result of both marine (waves and high tides) and subaerial (groundwater-induced) processes. As many as 10 blockfalls occurred between 1978 and 1983.

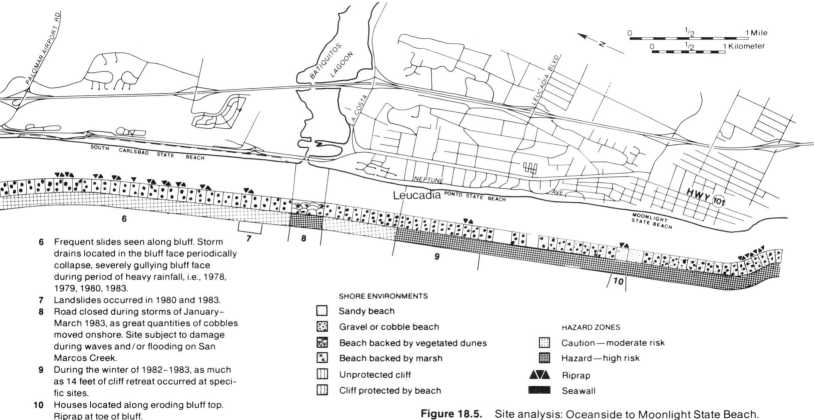

6 Frequent slides seen along bluff. Storm drains located in the bluff face periodically collapse, severely gullying bluff face during period of heavy rainfall, i.e., 1978, 1979, 1980, 1983.

7 Landslides occurred in 1980 and 1983.

8 Road closed during storms of January–March 1983, as great quantities of cobbles moved onshore. Site subject to damage during waves and/or flooding on San Marcos Creek.

9 During the winter of 1982–1983, as much as 14 feet of cliff retreat occurred at specific sites.

10 Houses located along eroding bluff top. Riprap at toe of bluff.

SHORE ENVIRONMENTS

☐ Sandy beach

▨ Gravel or cobble beach

▨ Beach backed by vegetated dunes

▨ Beach backed by marsh

▥ Unprotected cliff

▥ Cliff protected by beach

HAZARD ZONES

▦ Caution—moderate risk

▦ Hazard—high risk

🔺 Riprap

▬ Seawall

Figure 18.5. Site analysis: Oceanside to Moonlight State Beach.

Figure 18.6. February 1980 view of damaged beach cottage in South Oceanside. Note windows were shattered and roof partially collapsed as a result of cobbles being thrown by waves less than 6 feet high. Photo by G. Kuhn.

Figure 18.7. 1983 photo looking north from mouth of Bataquitos Lagoon along extensive cobble beach at south Carlsbad. Photo by G. Kuhn.

years, and the sand that now covers the beach front has been trucked in.

The next 2 miles of coast to the south is actively eroding as a result of elevated groundwater conditions. Since 1967 the groundwater table in this area has been monitored by the Self Realization Fellowship in Encinitas. A marked rise in the water table began in 1973–74 commensurate with construction directly on and inland of the bluff top. Many stairways, both private and public, have collapsed to the beach in this area as a result of the cliff failure (fig. 18.8). One particular area that has had extensive

erosion since the 1930s is the Self Realization Fellowship property. Landslide problems here date well back into the 1890s. This organization constructed their temple very close to the cliff margin in 1938 (fig. 18.9), and following storms in the early 1940s the building collapsed and slid toward the beach (fig. 18.10). During the late 1950s the seaward lanes of the major highway to the south, which was constructed along these cliffs, collapsed for a distance of over 300 feet as a result of groundwater solution.

The beach ridges continue along this section of coast as far as the valley of San Elijo Lagoon, where the highway gradually descends to a sandbar that crosses the mouth of the lagoon. Many

Figure 18.9. 1938 oblique aerial view of Self Realization Fellowship temple at Encinitas. Photo by Self Realization Fellowship.

Figure 18.8. April 1983 view of public stairway to beach. Stairs were crushed by a mass that separated from the cliff in North Leucadia. Photo by G. Kuhn.

restaurants have been constructed directly on this barrier bar in recent years. During the winter storms of 1982–83 restaurants on the seaward side of the highway were severely damaged (fig. 18.11). Both beach cobbles and blocks of protective riprap up to a half ton in size, emplaced to protect the structures, were thrown through the buildings. In effect, the rocks were actually launched as they were rolled backward on the cobbles and then catapulted through the buildings and onto the highway. Although temporary repairs have been made to allow the restaurants to continue operation, these repairs may be of only short-term benefit.

Cliffs rise very abruptly at the south end of Cardiff-by-the-Sea. The elevated beach ridges continue south from this point as

Figure 18.10. View of temple collapse in the early 1940s following significant storms accompanied by saturation of the bluff top. Photo by Self Realization Fellowship.

far as the San Dieguito River near Del Mar. Between 1970 and 1977, which was the terminal part of a long, calm climatic period, the bluffs of this area were subject to considerable erosion at specific sites. Storms from 1978 to 1982 resulted in even more erosion, necessitating the construction of seawalls and concrete revetments at certain locations. Seacliffs in this area retreated as much as 6 to 10 feet during the late January and February storms of 1983.

Following massive bluff-top construction in south Solana Beach in the early 1970s, it was discovered that buildings were located on top of an old, easily eroded river channel deposit (fig. 18.12). The base of the cliff at the south end of this site retreated

10 feet in 2 episodes between 1972 and 1978 before the larger storms of 1980 and 1983. A massive concrete seawall was built to protect the area before the most recent storm years, and this has, to date, prevented significant further erosion.

Del Mar and vicinity (fig. 18.14)

The city of Del Mar is partly constructed on the former delta of the San Dieguito River and on adjacent terraces and the canyons

Figure 18.11. April 1983 view looking north along the beach at Cardiff. The restaurants were severely damaged during the storms of January–February 1983 when beach cobbles and riprap were thrown by large waves into buildings and onto the coast highway. Photo by G. Kuhn.

that cut them. A rather large river valley, which drains Lake Hodges, is subject to occasional flooding. A beach and delta exist that have been extensively urbanized, and large waves or floods pose very real threats to the survival of these homes. Large waves in January, February, and March of 1983 seriously threatened the area, and the beach homes were preserved only by the furious dumping of riprap and sand bags, and the use of other remedial methods by homeowners (fig. 18.13). Even so, damage was extensive before the storms were over. The city of Del Mar is built on 2 terraces deeply cut by small valleys. Around the turn of the century, a more gentle grade was dug into the seaward terrace for the railroad. The steep, landward side of this

Figure 18.13.   Riprap being placed along the beach at Del Mar in an attempt to slow erosion by large storm waves during the winter of 1983. Photo by G. Kuhn.

Figure 18.12.   1974 aerial photo of condominium construction along the bluffs in south Solana Beach. Photo by B & A Engineering.

cut is subject to occasional slumping, which presents a severe hazard. On the seaward side the slope leading to the beach is very unstable, and on several occasions trains have fallen off the cliff, resulting in loss of life and severe damage to the railroad. In recent years the groundwater table has risen as a result of extensive development inland, causing cliff failure even during the summer months.

To the south of Del Mar, Sorrento, another large valley, comes in from the southeast. This valley has been built up as an industrial park, but there has been no attempt to prepare for potential floods. At the mouth of the valley is a small estuary called Pena-

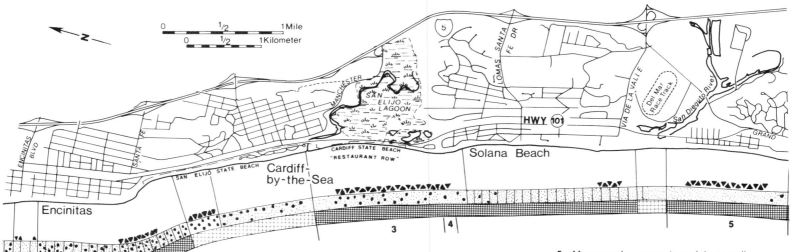

1. **Leucadia to Del Mar:** Apartments, houses, and condominiums located along bluff top. Sandstone and siltstone cliffs collapse as a result of groundwater-induced landslides, as well as undercutting by storm waves. Narrow sandy beaches or bars backed by highway along lagoon areas are subject to damage during high wave conditions or flooding by streams. Many wooden, concrete, and riprap walls located in front of beach development. Storms during 1978, 1980, and 1983 damaged many of the houses, lifeguard towers, and restaurants.

2. **Self-Realization Fellowship property.** History of large landslides.

3. During the storms of January–March 1983, several of the largest restaurants built on this fragile barrier bar were severely damaged.

4. Numerous slides in the bluff face

5. Many wooden, concrete, and riprap walls located in front of beach houses. Storms during winter 1978, 1980, and 1983 damaged many of the houses, lifeguard towers, and restaurants.

6. Numerous trains have been derailed along this section of coast during significant storm years. A recent one was January 1, 1941, when a train collapsed to the beach between 7th and 8th streets. During the very dry summers of 1979–1980, numerous sections of cliff collapsed as a result of inland sources of groundwater.

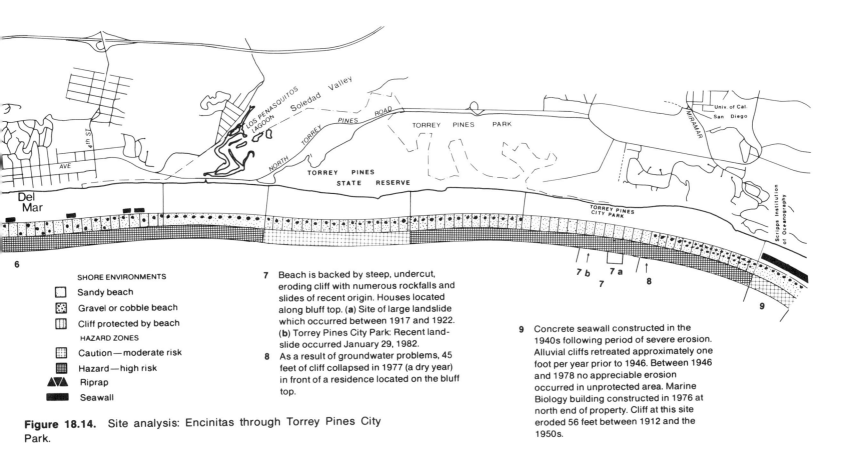

SHORE ENVIRONMENTS

☐ Sandy beach

▨ Gravel or cobble beach

▥ Cliff protected by beach

HAZARD ZONES

▦ Caution—moderate risk

▦ Hazard—high risk

▲▼ Riprap

▬ Seawall

7 Beach is backed by steep, undercut, eroding cliff with numerous rockfalls and slides of recent origin. Houses located along bluff top. (a) Site of large landslide which occurred between 1917 and 1922. (b) Torrey Pines City Park: Recent landslide occurred January 29, 1982.

8 As a result of groundwater problems, 45 feet of cliff collapsed in 1977 (a dry year) in front of a residence located on the bluff top.

9 Concrete seawall constructed in the 1940s following period of severe erosion. Alluvial cliffs retreated approximately one foot per year prior to 1946. Between 1946 and 1978 no appreciable erosion occurred in unprotected area. Marine Biology building constructed in 1976 at north end of property. Cliff at this site eroded 56 feet between 1912 and the 1950s.

Figure 18.14. Site analysis: Encinitas through Torrey Pines City Park.

squitos Lagoon, ordinarily cut off from the sea by a gravel bar. The bar formerly rose some 30 feet above sea level and provided a large source of cobbles for use in city pavements and various industrial purposes around the turn of the century. During small floods, this barrier had to be opened artificially at the north end to allow the river water to flow out to sea.

North La Jolla (fig. 18.14 and 18.15)

Directly south of Sorrento Valley is Torrey Pines State Park, a high-cliffed area with remarkable gorges. Somewhat south of the park, the sandstone and siltstone cliffs are the highest of any along this southern California coast, rising to as high as 350 feet with vertical faces at certain localities. One of the valleys that enters the coast was accessible to the public from the beach prior to 1940, but a great storm in December of that year removed most of the talus, making beach access to this valley impossible for several decades.

The cliffs have been subject to extensive landslides (fig. 18.16). The most recent failure was in 1982 at Blacks Beach (fig. 18.17); as the sandstone blocks from the cliff were broken up by the waves, they provided a large temporary source of beach sand. It is fortunate that the slide did not occur when the popular beach was packed with people as it commonly is during summer months. Farther south, another slide in 1980 came right up to the edge of a home and carried away an entire yard. Much of this cliff area is quite unstable.

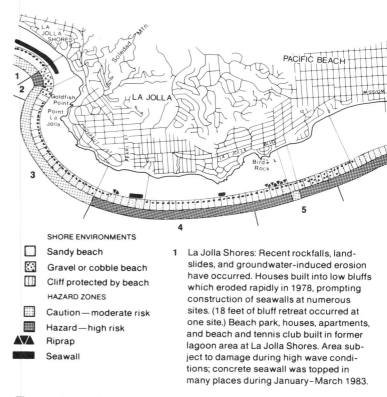

SHORE ENVIRONMENTS

☐	Sandy beach
▦	Gravel or cobble beach
▥	Cliff protected by beach

HAZARD ZONES

▦	Caution—moderate risk
▩	Hazard—high risk
▲▼▲	Riprap
■	Seawall

1   La Jolla Shores: Recent rockfalls, landslides, and groundwater-induced erosion have occurred. Houses built into low bluffs which eroded rapidly in 1978, prompting construction of seawalls at numerous sites. (18 feet of bluff retreat occurred at one site.) Beach park, houses, apartments, and beach and tennis club built in former lagoon area at La Jolla Shores. Area subject to damage during high wave conditions; concrete seawall was topped in many places during January–March 1983.

Figure 18.15.   Site analysis: La Jolla Shores to Point Loma.

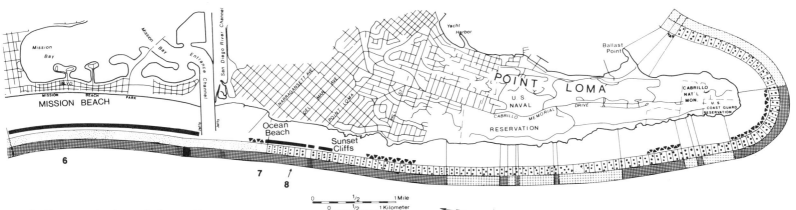

2 Numerous deeply penetrating caves underlie coastal roads and buildings. Cracks developed in overlying road during winter of 1983.

3 La Jolla: Very narrow sand and pocket beaches backed by low and undercut bluff. Bluff backed by coastal road with development east of road in La Jolla. To south, houses and apartments located along bluff top backshore of beach subject to damage during high wave conditions. Concrete seawalls and riprap found at several sites. Rockfalls and groundwater seepage also occur.

4 Buildings and roads along this area subject to damage during high wave conditions and especially during periods of severe sea storms.

5 Houses, apartments, and park facilities along bluff top. Groundwater seepage along cliff face, and several sections of cliff base locally undercut. Erosion of bluff face and top occurs as a result of surface runoff at certain street ends. Houses, apartments, and park facilities subject to damage as a result of cliff erosion during high wave conditions.

6 Pacific Beach and Mission Beach: Artificially nourished sandy beach supplied by infrequent dredging of Mission Bay. In Pacific Beach, extensive urban development occurs on low, undercut bluff and is endangered by erosion during high wave conditions. Along Mission Beach, a con-

crete seawall protects Esplanade, hotels, motels, apartments, and houses. Area subject to damage from severe flooding and/or extreme high wave conditions. Numerous sites were damaged in 1983 as storm waves broke over the sea wall.

7 Ocean Beach: Artificially nourished beach between south jetty of Mission Bay Entrance Channel and headland backed by commercial and residential area on low coastal plain. Houses destroyed at this location in storms of the 1920s. Ocean Beach Pier and buildings subject to damage during large storms.

8 Section of apartments undercut and overhanging.

Figure 18.16. 1949 view of Torrey Pines landslide, which measures approximately 1,700 feet along the cliff. Photo from F. Shepard Collection.

The U.S. Fish and Wildlife Service buildings located just south of this area were actually built on top of a landslide, which has been moving very slowly since construction in the early 1960s (fig. 18.18). Seaward portions of the buildings actually moved .5 inch seaward between 1968 and 1973, although it was also noted that much of the measured movement was settling related to rather distant earthquakes. One wonders what the fate of the building complex would be during a local earthquake of some severity, which is a real possibility. Interestingly enough, the U.S. Geological Survey occupied part of this building 20 years ago,

Figure 18.17. 1982 landslide at Blacks Beach north of Scripps Institution of Oceanography (SIO). Note recent slide material on the beach and the fresh vertical scarp behind the slide. Photo by Ron McConnaughey.

Figure 18.18. March 1978 aerial view of National Marine Fisheries building. Note funicular tram built at precarious position north of building and rockfall directly in front of it. An old landslide was discovered to be under the building after it was constructed. Photo by G. Kuhn.

and at that time estimated the structure's life expectancy at 20 years due to the landslide threat.

The cliffs farther south diminish rapidly down to bluffs where the original Scripps Institution of Oceanography was located (fig. 18.19). In recent years construction has occurred along and on top of these low cliffs. Between 1946 and 1977, during a 30-year period of somewhat calm climatic conditions when almost no destructive storms occurred, there was a broad beach from Scripps Institution south through La Jolla Shores. A recent climatic shift and acceleration of erosion, which began in 1978, led to the construction of seawalls in an attempt to arrest the undermining of houses and buildings that were so unwisely built at the top of the low bluffs. Most of this area is now at least temporarily protected by seawalls and is relatively safe; only in one small zone is there likely to be any serious trouble in the near future.

Farther south is La Jolla Shores, a low, flat area that had formerly been an estuary cut off from the sea by a barrier. The estuary has been filled in both by sediment washed from adjacent slopes (fig. 18.20) and dumping of fill for residential development. The lowest portions are still prone to flooding during very heavy rains, in which case there would be very serious damage to the homes and commercial district that virtually cover this area.

The beach along the La Jolla Shores area is of special interest. The breakers are commonly much larger in the northern portions than in the south, where most bathers congregate during the summer. This is the result of the wave divergence, or decrease in energy, over the head of the La Jolla Submarine Canyon, and the wave convergence, or increase in energy, farther north approaching the Scripps Institution of Oceanography.

South of La Jolla Shores we encounter the seaward continuation of Soledad Mountain. The relatively low cliffs just south of the La Jolla Beach and Tennis Club consist of sandstone and are bordered seaward by a wide, wave-cut terrace that at low tide exposes a series of small, resistant ridges produced by erosion of the rock. This is an amazing tidal terrace and is unique for its width along the San Diego County coast. Although rock terraces exist in other places, they are not so well exposed at low tide.

The sandstones with their caves on the cliffed north side of

Figure 18.19. 1910 view of original Scripps building and water tower, Scripps Institution of Oceanography. Note absence of talus at base of cliffs, indicating active wave erosion. Photo from Scripps Institution of Oceanography Archives.

Figure 18.20. The unmodified lagoon shoreward of La Jolla Submarine Canyon as it appeared in 1930. Photo from F. Shepard Collection.

Point La Jolla are remarkable, too. The lack of a wave-cut terrace at the base of the sandstone cliffs suggests that the caves are the result of a process other than wave erosion alone. It does not seem probable that wave erosion by itself could have extended so deeply into this sandstone. It is more likely that some type of groundwater solution of the sandstone cement is involved. Thus we have a similar situation to that discovered in the caves under Solana Beach to the north, where there appear to be caverns deep below the surface that may connect with the sea caves found along the shore, the latter being clearly related to wave erosion.

Erosion of the bluffs at La Jolla is an interesting problem. Examination of the geological formation at the west end of the bluffs, and comparison with photographs taken about 1900, shows a conspicuous lack of erosion. A single arch at Goldfish Point has shown very little change since 1908. The same is true of some massive boulders a little to the east, near La Jolla Cove, which are exposed to large waves. West of the cliffs, however, old photographs of the cove show decided changes in the well-known arches that existed in the early part of the century and were partially reinforced with cement in later years. They finally collapsed, the last one falling in 1978.

An interesting site to watch for seasonal changes is Boomer Beach, located just south of Alligator Head. It consists of a steep, coarse-sanded beach during summer months, but with the first winter storm the sand is washed away to expose underlying boulders. The sand is apparently carried partly to the south,

Figure 18.21.   1873 view of Cathedral Rock Arch, La Jolla. Photo from F. Shepard Collection.

forming narrow beach fringes, and partly carried seaward. However, with the onset of southerly waves in summer, the sand again returns to Boomer Beach, and the narrow fringes to the south disappear.

Most of the relatively straight, rocky coast extending toward the Children's Pool had little erosion since the 1930s until the storms in January through March 1983, when many large cracks developed. A beautiful arch called Cathedral Rock once existed on the west side of the Children's Pool (fig. 18.21). This arch collapsed in 1906, but the buttresses of the old arch were still standing. Every remnant of the arch is now gone, and other rocks as well have since gradually disappeared. Many sections of the partially submerged rock point have also been removed.

South of La Jolla, the cliffs and terraces are of much softer rock. At Bird Rock, a stack stands above the tidal terrace that had shown little erosion as documented through photographs during the past few decades up until January 1983. The low cliffs are now actively eroding, although the amount of retreat has not been established.

Pacific Beach and Mission Beach; San Diego Bay and River (figs. 18.15 and 18.22)

Pacific and Mission beaches rest on the former delta of the San Diego River. There is one slightly elevated bench at Crown Point, which is an extension of the La Jolla Marine terrace. The community of Mission Beach to the south is constructed on a sandy barrier bar and in recent years had been protected from wave erosion by seawalls. During a storm period from 23 December 1940 to 7 January 1941, however, extremely large waves threw concrete benches through the buildings along the boardwalk.

The southern part of Mission Beach is a sandspit built across Mission Bay. The San Diego River occupies a rather broad valley extending east and west. Where it enters Mission Bay it has deposited a considerable amount of material, forming a delta. The mouth of the river has been greatly modified by human

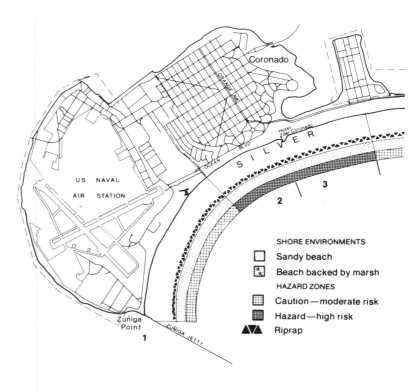

Figure 18.22. Site analysis: Zuñiga Point through California-Mexico border.

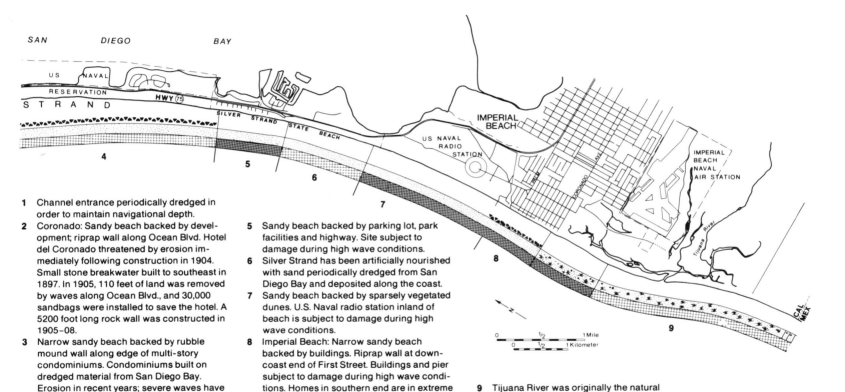

SAN   DIEGO   BAY

US   NAVAL
RESERVATION
S T R A N D   HWY 75
SILVER STRAND STATE BEACH

IMPERIAL
BEACH

US NAVAL
RADIO
STATION

IMPERIAL
BEACH
NAVAL
AIR STATION

Tijuana River

CAL
MEX

0   ½   1 Mile
0   ½   1 Kilometer

1 Channel entrance periodically dredged in order to maintain navigational depth.

2 Coronado: Sandy beach backed by development; riprap wall along Ocean Blvd. Hotel del Coronado threatened by erosion immediately following construction in 1904. Small stone breakwater built to southeast in 1897. In 1905, 110 feet of land was removed by waves along Ocean Blvd., and 30,000 sandbags were installed to save the hotel. A 5200 foot long rock wall was constructed in 1905–08.

3 Narrow sandy beach backed by rubble mound wall along edge of multi-story condominiums. Condominiums built on dredged material from San Diego Bay. Erosion in recent years; severe waves have outflanked wall at up- and downcoast ends.

4 Sandy beach backed by low active dunes landward of beach face backed by highway.

5 Sandy beach backed by parking lot, park facilities and highway. Site subject to damage during high wave conditions.

6 Silver Strand has been artificially nourished with sand periodically dredged from San Diego Bay and deposited along the coast.

7 Sandy beach backed by sparsely vegetated dunes. U.S. Naval radio station inland of beach is subject to damage during high wave conditions.

8 Imperial Beach: Narrow sandy beach backed by buildings. Riprap wall at downcoast end of First Street. Buildings and pier subject to damage during high wave conditions. Homes in southern end are in extreme danger in winter. Area was badly damaged in 1952–53 and again in 1978, 1980, and 1983.

9 Tijuana River was originally the natural source of sediment during floods which nourished the Silver Strand. River is now dammed.

activity in recent years. A flood channel has been built across the beach to carry floodwaters from Mission Bay to the sea. It is doubtful whether this outlet could carry the volume of flood-water that could sweep down the valley in a repetition of the great floods of 1862, 1884, 1889, 1891, 1916, and subsequent smaller floods. Three rock jetties built across the mouth of Mission Bay have cut off the supply of sand from Mission Beach to Ocean Beach (fig. 18.23), and as a result the cliffs at Ocean Beach

Figure 18.23. 1968 aerial photo following construction of jetties at the mouth of San Diego River and entrance to Mission Bay. The area north of the jetties was extensively dredged to create a yacht basin and recreational facilities. Photo by U.S. Army.

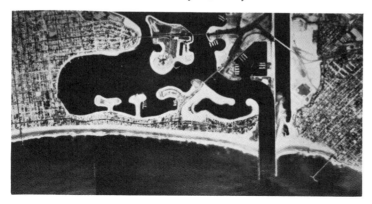

have receded considerably. Sand has been introduced in recent years from north of the jetties to prevent serious erosion of that community's ocean front.

The oldest city on the west coast is San Diego. San Diego Harbor is largely kept open by tidal currents, with the exception of a sandbar that has to be periodically dredged, and is the best-protected natural harbor on the southern California coast. Mission Bay, formerly called False Bay, was deep enough up until 1810 to allow even relatively deep-draft vessels to enter.

During the early nineteenth century southern California rivers have changed their courses periodically as the result of numerous great floods. Prior to 1821 the San Diego River entered San Diego Harbor most of the time. In the fall of 1821, however, a flood changed the river channel in one night, and the greater volume of the flow was diverted into what was then known as False Bay, leaving only a small stream still flowing into the harbor. This flood was remarkable in that no rain was reported along the coast at the time. The river was later observed to flow into San Diego Harbor in 1849, and a survey map of 1859 shows this to have been the case once again. The federal government diverted the flow of the river into Mission Bay and built an earthen levee extending from near Old Town to Point Loma in the fall of 1853 because of the river's large depositional rate during floods. Later that year heavy rains caused the river to change course once again, washing out part of the levee and resuming its old course into the harbor. The great flood in 1862, appropriately called the "Noachian Deluge," inundated San Diego; houses in the lower

Old Town section of Mission Valley were flooded when severe coastal winds from the south backed the water from the bay into the river. The levee was then reconstructed in 1876, and no further diversions into San Diego Bay have occurred. Since then a considerable volume of sediment has been added to the San Diego River delta in Mission Bay by occasional floods.

The Mission Bay/San Diego River jetties were built in 1948 at a time when the shore of the bay was subject to alternating periods of recession and advance. By February 1951 the river levees had been connected to the jetties, and all tidal flow was confined to a new channel. Since the river discharges only during flooding, the middle channel was soon completely filled with sediment.

Sunset Cliffs and Point Loma (fig. 18.15)

The community of Ocean Beach extends south into Sunset Cliffs and is located at the north end of Point Loma. The northern part of Ocean Beach is laid out on a remnant of a former delta of the San Diego River. Following floods that came down through Mission Bay, the beach in this area was subject to periods of buildup, or accretion. Since the construction of the San Diego River jetties, however, the beach has had to be maintained by dredged spoils from the channels.

Seacliff retreat measured at Sunset Cliffs in 1973 amounted to 3 feet in the preceeding 75 years, or about 2 inches per year. Where a sea cave roof collapsed, however, local retreat could be much greater. More rapid retreat in recent years has endangered the road along the cliff, necessitating the dumping of concrete debris at several locations, which inadvertently accelerated erosion, especially during the storms of January to March 1983.

The U.S. Army Corps of Engineers has investigated the coastal erosion at Sunset Cliffs more recently and determined that as much as 38 feet of landward retreat occurred at the toe of Del Mar Avenue and 40 feet of erosion took place along the top of the bluff during the period from 1962 to 1976 (fig. 18.24). The majority of the erosion at this site at that time was the result of surface runoff, overwatering of the bluff top, pedestrian traffic, and burrowing by animals. Erosion has occurred in many places back to the coastal road, and city authorities have dumped concrete debris and asphalt at these locations in an attempt to slow erosion. In March 1968, as a result of an earthquake, a section of Sunset Cliffs separated from the cliff and slumped, endangering the coast road. During the storms of January to March 1983 many sections of the cliffs were undercut and began to collapse.

Coronado Beach and Imperial Beach (fig. 18.22)

Coronado was initially a combination of two islands with a swampy area between them. The swamp is now entirely filled with sediment brought in by the U.S. Navy when its large base was established to the south. There is a long spit extending from the embayment of the Sweetwater and Tijuana rivers, and inside this an extensive delta has built up much of the land, extending

Figure 18.24. 1976 view looking south along the bluffs at Del Mar Street, Sunset Cliffs. Vertical line indicates 1954 property line. The bluff top retreated 40 feet and the cliff base retreated 38 feet between 1962 and 1976. From 1952 to 1976 total retreat at the top of the bluff was approximately 75 feet. Photo by D. Pain.

as far south as the Mexican border.

The Hotel Del Coronado was constructed in 1888 at the northern end of the Coronado sand spit. In 1893 construction began on the 7,500-foot-long Zuniga Shoal jetty, which was designed to stabilize the entrance to San Diego Harbor. The jetty was finally completed in 1904. Immediately following construction of the hotel, erosion problems developed on the spit. A small stone breakwater was built to the southeast in 1897 to offer protection from wave erosion; it was soon damaged and subsequently repaired.

Erosion of the beach just west of Spanish Bight occurred in 1900 following extension of the Zuniga Jetty. Erosion continued during the next extension of the jetty in 1903–4, which took it farther east of Spanish Bight. Beginning in January 1905 severe storms caused erosion both north and south of the Hotel Del Coronado. Waves focused on the hotel area, necessitating installation of 30,000 200-pound sandbags north and south of the hotel. Storms in 1905 continued to cause serious wave erosion and a total of 110 feet of land was removed by the waves along Ocean Boulevard (fig. 18.25). Between 1905 and 1908 a 5,200-foot-long seawall was constructed from the hotel west along Ocean Boulevard.

In recent years over 30 million cubic yards of sand have been dredged from San Diego harbor. Sixteen million cubic yards were deposited to the south of the hotel, and 1 million of that was deposited to the north, which greatly widened the beach. Subsequently, construction of huge condominium apartments has dominated the spit directly south of the hotel.

Imperial Beach erosion has been an increasing problem since 1953, during the relative drought periods (preceding the floods of 1978 to 1980). Sand sources were cut off from the beaches during this dry period because dams had been built on the Tijuana River. Approximately 660,000 cubic yards of sand would normally have reached the beach each year if it had not been entrapped by these dams. As a result, the beach has had to be artificially nourished by dredging.

During the early part of the winter of 1983 the beaches south of the border were severely eroded, and the sediment apparently moved north, accreting in the Imperial Beach area. Since construction of the Rodriguez Flood Control Dam on the Tijuana River in Mexico, the beach cliff for at least 3 miles south of the border has eroded markedly. Much construction along the shore, including the highway, is disappearing as this is written (fig. 18.26). It is obvious that the longshore currents in the area are quite complicated and have a pronounced effect on beach stability.

Figure 18.25.   March 1905 view of Hotel del Coronado. Note severe erosion of beach cliff. Approximately 30,000 sandbags had been placed north and south of the hotel to save it from wave destruction. Photo from F. Shepard Collection.

Figure 18.26.   1983 view along the beach at Tijuana. Note the severe erosion of the beach and road. Photo by C. Everts.

19. Afterword

**After all has been said and done,
a lot more has been said than done**

Our intent in the previous chapters has been to provide a readable and understandable perspective on the hazards and risks associated with living directly on the California coastline. As scientists we are not against coastal development, but instead we are trying to discourage construction and rebuilding in unsafe areas. The coastline of California has many areas that are stable and protected from erosion and wave inundation. On the other hand, there are countless other locations that are either subject to periodic wave attack or are unstable or rapidly eroding. We have attempted to delineate these areas and describe the hazards and risks involved with them.

Damage along the coastline normally occurs seasonally, and frequently during major storms and high tides. You must keep in mind that many coastal areas that appear high and dry during the summer months can be quickly threatened during a winter storm.

Protective efforts are very expensive, for each type of structure and coastal terrain has its drawbacks or liabilities. History indicates that most structures should be considered temporary.

Look at the situation carefully before investing your life savings in an ocean-front house or lot that may have no future. If in doubt, hire a professional geologist with experience in the coastal zone or check with staff geologists in the local city or county. Also ask local residents about conditions in the area during winter storms.

The coastal storms of recent years have been tragic; we must learn something from them. Always keep in mind that the Pacific Ocean is a very powerful force to oppose, and most of us cannot long afford to protect an ocean-front home in an area of periodic inundation or active coastal retreat.

Appendix A. Glossary

alluvial: referring to earth material transported and deposited by rivers and streams.

backshore: upper shore zone beyond the reach of ordinary waves or tides.

barrier beach: a single elongate sand ridge rising slightly above the high tide level and extending generally parallel with the coast, but separated from it by a lagoon.

beach: the gently sloping shore of a body of water that is washed by waves or tides, especially the parts covered by sand or pebbles.

beach compartment: see littoral cell.

bed: a deposit of material.

berm: a nearly horizontal portion of the beach or backshore formed by the deposit of material by wave action. Some beaches have no berms, others have one or several.

blowout: various saucer-, cup-, or trough-shaped hollows formed by wind erosion on a dune.

breaker: a wave breaking on the shore, over a reef, etc.

breakwater: a structure protecting a harbor, shore area, anchorage, or basin from waves.

budget: see sand budget.

bulkhead: a structure that may be constructed of steel, wood, or concrete, which usually serves as both a barrier to wave attack and also as a retaining wall.

coast: a strip of land of indefinite width (up to several miles) that extends from the shoreline inland to the first major change in terrain features.

coastal zone: coastal waters and lands that exert a measurable influence on the uses of the sea and its ecology.

coastline: the line that forms the boundary between the land and the water.

conglomerate: a rock containing rounded fragments of gravel or pebbles cemented together.

continental shelf: the gently sloping, shallow, submerged edge of the continent extending from the shore to a marked increase in the bottom slope toward deep water.

colluvium: a general term applied to loose and incoherent deposits, usually at the foot of a slope or cliff and brought there chiefly by gravity.

creep: the imperceptibly slow, more or less continuous downward and outward movement of slope-forming soil or rock.

dune: see sand dune.

dynamic equilibrium: a condition that exists along some coastlines where neither erosion nor buildup is occurring, but the beach is continually being shaped by wave action.

earthflow: a slow downslope flow of earth lubricated with water.

erosion: the group of processes whereby earth or rock material is loosened or dissolved and removed from any part of the earth's surface. It includes the processes of weathering, solution, abrasion, and transportation.

fault: a fracture or fracture zone along which there has been displacement of the sides relative to one another parallel to the fracture.

filter cloth: a type of strong permeable plastic cloth that is used behind seawalls or under revetments to reduce or minimize sand removal by scour.

flood plain: that portion of a river valley, adjacent to the river channel, that is built by sediments deposited by the river and is covered with water when the river overflows its banks at flood stages.

fold: a bend or flexure in rock.

foredune: see primary dune.

foreshore: lower shore zone, between ordinary low and high water levels.

fracture: breaks in rock due to intense folding or faulting.

glacial deposits: sediments deposited by glaciers.

groin: an elongate structure usually constructed perpendicular to the coast for building or protecting a beach by trapping littoral drift.

groundwater: subsurface water.

gunnite: concrete that is sprayed in a slurry form onto some framework or structure to which it will adhere and harden.

headland: any projection of the land into the sea, generally applied to a cape or promontory of some boldness and elevation.

hindcast: calculating prior wave conditions from historic wind charts, often used as a guide for expected future waves.

hooked bay (log spiral bay): an embayment with a form like a hook, formed behind a protective headland.

jetty: an elongate structure extending into a body of water to direct and confine a stream or tidal flow to a selected channel. Jetties are built in pairs to help protect or stabilize a channel for navigation.

joint: fracture in a rock, along which no appreciable movement has occurred. See fault.

land division: the legal process of dividing a parcel of land into 2 or more smaller parcels.

landslide: (1) the perceptible downward sliding or falling of a relatively dry mass of earth, rock, or mixture of the two. (2) earth and rock that becomes loosened from a hillside by moisture or snow, and slides or falls down the slope.

littoral cell: a self-contained section of coast consisting of 3 elements: (1) a source of beach sand, (2) littoral drift that moves the sand downcoast, and (3) a sink for the sand.

littoral drift: the movement of beach sand along the coast due to wave action.

longshore current: a current generated by waves approaching the coast at an angle that then moves parallel to the beach within the surf zone.

marine terrace: an elevated, seaward-sloping, wave-cut bench or platform exposed by uplift along the coast. Several terraces commonly exist at different elevations.

mass movement: downslope movement of a portion of the land surface as in creep, landslide, or slip.

Mean higher high water (MHHW): average height of the higher high tides over a 19-year period.

Mean high water (MHW): average height of the high tides over a 19-year period.

Mean lower low water (MLLW): frequently abbreviated lower low water. Average height of the lower low tides over a 19-year period.

Mean low water (MLW): average height of low tide over a 19-year period.

Mean sea level: average height of the surface of the sea for all stages of the tide over a 19-year period, usually determined from hourly height readings. Also known as national geodetic vertical datum (NGVD).

melange: a heterogeneous mixture or medley of rock materials.

metamorphic rock: includes all those rocks that have formed in the solid state in response to pronounced changes of temperature, pressure, and chemical environment.

mudstone: a general group of sedimentary rocks that includes clay, silt, siltstone, claystone, and shale.

nearshore: in beach terminology, a zone of indefinite width extending seaward from the shoreline well beyond the breaker zone.

perimeter foundation: a common foundation type in which the structure is supported on a low concrete wall or footing extending around the outside edge of the building.

platform: see wave-cut platform.

pocket beach: a small beach formed between 2 points or headlands, often at the mouth of a coastal stream.

primary dune: the first or frontal dune, directly landward of the beach, that is periodically eroded and then built up again.

reflection: see wave reflection.

refraction: see wave refraction.

retaining wall: a low wall used to support or retain an earth embankment or an area of fill.

revetment: a facing of rock built along the shoreline to protect a bluff, embankment, or structure from wave attack. The rock is layered with smaller core stone and larger cap stone to minimize sand scour by wave action.

riprap: a wall or facing of large (1–5 ton) rocks stacked along the shoreline to protect the cliffs, bluffs, dunes, or structures from wave attack.

rockfall: the relatively free falling of a detached segment of bedrock of any size from a cliff, steep slope, cave, or arch.

runoff: the discharge of water through surface streams, or the quantity of water discharged through surface streams, usually expressed in units of volume.

runup: see wave runup.

sand budget: an accounting of the sand along a particular stretch of coast: the sources, sinks, and rates of movement, or the supply and loss.

sand dune: a mound, ridge, or hill of loose sand, heaped up by the wind. See also primary and secondary dunes.

sandstone: a cemented or otherwise compacted sediment composed primarily of sand.

scour: see wave scour.

sea (waves): waves caused by wind at the place and time of observation, normally some distance offshore.

seacliff: a cliff formed by wave action.

seastack: small, steep-sided rocky projection above sea level near the coast.

seawall: a solid structure built along the coastline to prevent erosion and other damage by wave action. Seawalls are normally more massive than other types of shoreline engineering structures.

secondary dune: the second line of dunes along a coastline, landward of the primary dune, and normally permanently vegetated.

sedimentary rock: rock formed by the accumulation of sediment deposited by water or by air. The sediment may consist of rock fragments or particles of various sizes (conglomerate, sandstone, shale) and of the remains or products of animals or plants (certain limestones and coal).

setback: an exclusion zone adjacent to some hazardous or sensitive feature (an eroding seacliff, for example) in which no building or structures are allowed.

shoal: (1) to become shallow gradually. (2) a part of the area covered by water, in the sea, a lake, or a river, when the depth is little.

shoreline: the line of intersection of the sea with the land. The region immediately to the landward of the shoreline is called the coast.

siltstone: a fine-grained rock composed of silt-sized particles.

slab foundation: a common type of foundation in which the structure or building is supported on a relatively thin (usually 6 inches) steel reinforced layer of concrete.

slump: the downward movement of material of any size, moving as a unit or as several subsidiary units, characterized by rotational motion.

soil creep: see creep.

steel sheet pile bulkhead: a protective structure consisting of interlocking steel plates that are usually driven into the sand or subsurface.

storm surge: the rise above normal water or tidal level on the open coast due to the action of wind on the water surface.

submarine canyon: a steep-sided submarine valley commonly crossing the continental shelf and slope.

surf zone: the area between the outermost breaker and the limit of wave uprush.

swell: wind-generated waves that have left the storm area and have advanced into regions of weaker winds or calm.

terrace: see marine terrace.

tidal range: the difference between the level of water at high tide and low tide.

tidal wave: see tsunami.

tide: the periodic rise and fall of oceans and bodies of water connecting them, caused chiefly by the gravitational attraction of the sun and moon.

tombolo: a sandbar connecting an island or rock with the mainland. Morro Rock is an example of a tombolo.

tsunami: a very large sea wave produced by a submarine earthquake or volcanic eruption. Commonly misnamed a tidal wave.

uplift: elevation of any extensive part of the earth's surface relative to some other parts.

wave: an oscillatory movement in a body of water manifested by an alternate rise and fall of the surface.

wave-cut platform: a marine-cut terrace, plain of marine abrasion, shore platform, wave-cut plain, or wave-cut terrace.

wave reflection: the wave that is returned seaward after a wave impinges on a steep beach, cliff, or other barrier.

wave refraction: the process by which the direction of a train of waves moving in shallow water at an angle to the contours is changed.

wave runup: the distance or extent that water from a breaking wave will extend up a beach or structure.

wave scour: erosion or sand removal produced by wave action.

wave shadow: an area protected from direct wave attack by a natural or engineered structure, for example, an island or a breakwater.

Appendix B. Useful references

Beaches And Beach Processes

Beach Processes and Sedimentation, by P. D. Komar. Englewood Cliffs, N.J.: Prentice-Hall, 1976.

Coastal Data Information Program, 1980–83, by Richard Seymour. U.S. Army Corps of Engineers Monthly Report, No. 52-82.

Field Guide to Beaches, by J. H. Hoyt. American Geological Institute, Earth Science Curriculum Project Pamphlet Series PS-7, 1971.

The Movement of Beach Sand, by J. C. Ingle. Amsterdam: Elsevier, 1966.

Submarine Geology, by F. P. Shepard. New York: Harper and Row, 1963.

Waves and Beaches, by Willard Bascom. Garden City, N.Y.: Doubleday, 1964, 1982.

Coastal Construction

Coastal Design: A Guide for Planners, Developers, and Home-owners, by Orrin H. Pilkey, Jr., Orrin H. Pilkey, Sr., Walter D. Pilkey, and William J. Neal. New York: Van Nostrand Reinhold, 1983.

Construction Materials for Coastal Structures, by Moffatt and Nichol, engineers. Fort Belvoir, Va.: Coastal Engineering Research Center, U.S. Army Corps of Engineers, 1983.

Design and Construction Manual for Residential Buildings in Coastal High Hazard Areas, by Dames and Moore for HUD on behalf of the Federal Emergency Management Agency. Washington, D.C.: U.S. Government Printing Office, 1981.

Elevated Residential Structures, prepared by the American Institute of Architects Foundation for the Federal Emergency Management Agency. Washington, D.C.: U.S. Government Printing Office, 1984.

Flood Emergency and Residential Repair Handbook, by the National Association of Homebuilders Research Advisory Board of the National Academy of Science. Washington, D.C.: U.S. Government Printing Office, 1980.

Guidelines for Beachfront Construction with Special Reference to the Coastal Construction Setback Line, by C. A. Collier and others. Gainesville, Fla.: Marine Advisory Program, Coastal Engineering Laboratory, University of Florida, 1977.

Houses Can Resist Hurricanes, by the U.S. Forest Service. Madison, Wis.: Forest Products Laboratory, Forest Service, U.S. Department of Agriculture, 1965.

Hurricane-Resistant Construction for Homes, by T. L. Walton, Jr. Gainesville, Fla.: Marine Advisory Program, Coastal Engineering Laboratory, University of Florida, 1976.

Interim Guidelines for Building Occupant Protection from Tornadoes and Extreme Winds, and *Tornado Protection— Selecting and Designing Safe Areas in Buildings*, by Delbart B. Ward. Washington, D.C.: Civil Defense Preparedness Agency, the Pentagon.

Masonry Design Manual, by the Masonry Institute of America. Los Angeles, Calif.: Masonry Institute of America.

Peace of Mind in Earthquake Country: How to Save Your Home and Life, by Peter Yanev. San Francisco, Calif.: Chronicle Books, 1974.

Protecting Mobile Homes from High Winds, by the Civil Defense Preparedness Agency. Washington, D.C.: U.S. Government Printing Office, 1974.

Standard Details for One-Story Concrete Block Residences, by the Masonry Institute of America. Los Angeles, Calif.: Masonry Institute of America.

Wind-Resistant Design Concepts for Residences, by Delbart B. Ward. Washington, D.C.: Civil Defense Preparedness Agency, the Pentagon.

Coastal Hazards and Erosion, Geology, and Land Use

"Analysis of Historical Shoreline Changes," by V. Goldsmith and others, in *Coastal Zone '78* 4(1978): 2819–36, New York: American Society of Civil Engineers.

Assessment and Atlas of Shoreline Erosion Along the California Coast, by John S. Hable and George A. Armstrong. California Department of Navigation and Ocean Development, 1977.

The Beaches Are Moving, by Wallace Kaufman and Orrin Pilkey, Jr. Durham, N.C.: Duke University Press, 1984.

The Coastal Almanac, by P. L. Ringold and J. Clark. W. H. Freeman, 1980.

The Coastal Challenge, by Douglas L. Inman and Birchard M. Brush. *Science* 181, 6 July 1973: 20–32.

Coastal Mapping Book, Melvin Y. Ellis, ed. U.S. Geological Survey, March 1976.

Erosion and Sediment Control Handbook, by Perry Amimoto. U. S. Environmental Protection Agency, Water Planning Division, EPA 440/3-78-003, 1978.

Geologic Hazards, Resources, and Environmental Planning by Gary B. Griggs and John A. Gilchrist. Belmont, Calif.: Duxbury Press, 1983.

"Landslides—The Descent of Man," by Raymond Pestrong, *California Geology* 29, 7, July 1976: 147–51.

A Manual for Researching Historical Coastal Erosion, by Kim Fulton. California Sea Grant College Program Report No. T-CSGCP-003, 1981.

National Shoreline Study—California Regional Inventory, U.S. Army Corps of Engineers. U.S. Army Corps Report to Congress, House Document 93-121, 1971.

Nature to be Commanded, by G. D. Robinson and A. Spieker. U.S. Geological Survey Professional Paper 950, 1978.

Our Changing Coastlines, by F. P. Shepard and H. R. Wanless. New York: McGraw-Hill, 1971.

Peace of Mind in Earthquake Country: How to Save Your Home and Life, by Peter Yanev. San Francisco: Chronicle Books, 1974.

A Summary of Man's Impact on the California Coastal Zone, by Douglas L. Inman. California Department of Navigation and Ocean Development (now Boating and Waterways), 1976.

Conservation and Planning

"Coastal Processes and Long-Range Planning," by Richard M. Brush and Douglas L. Inman, in *8th Annual Conference and Exposition* Reprints, Washington, D.C.: Marine Technological Society, 11–13 September 1972, 215–21.

Design with Nature, by Ian McHarg. Garden City, N.Y.: Doubleday, 1971.

Natural Hazard Management in Coastal Areas, by Gilbert F. White and others. Washington D.C.: U.S. Department of Commerce, National Oceanic and Atmospheric Administration, Office of Coastal Zone Management, 1976.

Planning for an Eroding Shoreline. San Francisco: California Coastal Commission, 1978.

Urban Landslides: Targets for Land Use Planning in California, by F. Beach Leighton. Geological Society of America Special Paper 174, 1976: 37–60.

Geographic References

Northern California Coast—Oregon to San Francisco

"Beach Erosion Control Study Ocean Beach, San Francisco, California," by G. W. Domurat and others, in *Shore and Beach*, October 1979: 20–32.

"Coastal Faulting and Erosion Hazards in Humboldt County, Northern California," by Derek J. Rust, in *Papers, Ocean Studies Symposium, November 1982*. Asilomar, Calif.: California Coastal Commission, 1984, 821–42.

"Coastal Zone Geology Near Gualala, California," by John W. Williams and Trinda L. Bedrossian. *California Geology*, California Division of Mines and Geology, February 1977, 27–34.

"Coastal Zone Geology Near Mendocino, California," by John W. Williams and Trinda L. Bedrossian. *California Geology*, California Division of Mines and Geology, October 1976, 232–37.

Exploring the North Coast from the Golden Gate to the Oregon Border, by Mike Hayden. San Francisco: Chronicle Books, 1976.

Exploring Point Reyes, by Phil Arnot and Elvira Monroe. San Carlos, Calif.: Wide World, revised 1979.

Exploring Redwood National Park, by Robert J. Dolezal. Eureka, Calif.: Interface California Corp., 1974.

Feasibility Study, Groin Reconstruction and Bulkhead Replacement Bolinas Beach, California, Report prepared by

Woodward-Clyde Consultants and the Bolinas Beach and Cliff Association for the Bolinas Community Public Utilities District, 1982.

Geology for Planning Eureka and Fields Landing 7.5 Quadrangles, Humboldt County, California, by R. T. Kilbourne and others. California Division of Mines and Geology OFR 80-9 SF, 1980.

Geology for Planning in Sonoma County, by M. E. Huffman and C. F. Armstrong. California Division of Mines and Geology Special Report 120, 1980.

Geology for Planning on the Sonoma County Coast Between the Russian and Gualala Rivers, by M. E. Huffman. California Division of Mines and Geology Report, 1972.

Geology for Planning in Western Marin County, California, by David L. Wagner. San Francisco: California Division of Mines and Geology Report, 1977.

Guidebook to the Northern California Coast, by Mike Hayden. Los Angeles: Ward Ritchie Press, 1970.

Investigation of Coastline Retreat at Shelter Cove, California, by D. C. Tuttle (Humboldt County Public Works). Sea Grant Report, 1982.

Investigation of Methods for Determining Coastal Bluff Erosion: Historical Section, Gold Bluffs to the Little River, Humboldt Co., by D. C. Tuttle (Humboldt County Public Works). Sea Grant Report, 1981.

The Mendocino Coast, by Barbara D. Muller. Mendocino, Calif.: Community Land Trust, Inc., 1981.

Sand Dune Control Benefits Everybody: The Bodega Bay Story. Pamphlet by the U.S. Dept. of Agriculture Soil Conservation Service, 1967.

Central California Coast—San Francisco to Point Conception

Coastal Processes in Southern Monterey Bay: Coastal Erosion, Sand Sources, and Sand Mining, by B. Allyaud. Central Coast Regional Commission, 1978.

Combing the Coast: San Francisco through Big Sur, by Ruth A. Jackson. San Francisco: Chronicle Books, 1977.

"Effects of the Santa Cruz Harbor on Coastal Processes of Northern Monterey Bay, California," by Gary B. Griggs and Rogers E. Johnson. *Environmental Geology* 1, 5 (1976): 299–312.

"Coastline Erosion: Santa Cruz County, California," Gary B. Griggs and Rogers E. Johnson. *California Geology* 32, 4 (April 1979): 67–76.

"Impact of 1983 Storms on the Coastline of Northern Monterey Bay," Gary B. Griggs and Rogers E. Johnson. *California Geology* 36, 8 (August 1983): 163–74.

Monterey Bay Area: Natural History and Cultural Imprints, B. L. Gordon. Pacific Grove, Calif.: Boxwood Press, 1979.

Natural History of Año Nuevo, edited by Bernie Laboeuf and Stephanie Kaza. Pacific Grove, Calif.: Boxwood Press, 1981.

Nature Walks on the San Luis Coast, by Harold Wieman. San Luis Obispo, Calif.: Padre Publications, 1980.

"Rates of Coastal Bluff Retreat, Pismo Beach, California," by D. O. Asquith, in *Coastal Zone '83* 9 (1983): 1195–1207, New York: American Society of Civil Engineers.

Southern California Coast—Point Conception to the Mexican Border

"Accelerated Beach-Cliff Erosion Related to Unusual Storms in Southern California," by G. G. Kuhn and F. P. Shepard. *California Geology* 32, 3 (1979): 58–59.

"Coastal Erosion in San Diego County," by G. G. Kuhn and F. P. Shepard, in *Coastal Zone '80* 6 (1980): 1899–1918, New York: American Society of Civil Engineers.

Coastal Landslides in Southern California, by Barnard Pipkin and Michael Ploessel. Los Angeles: University of Southern California Sea Grant publication no. USC-SG-3-72, 1972.

Coastal Southern California, by R. P. Sharp. Dubuque, Iowa: Kendall/Hunt, 1978.

Coastal Zone Geology and Related Sea Cliff Erosion: San Dieguito River to San Elijo Lagoon, San Diego County, by G. G. Kuhn. San Diego County Integrated Planning Organization Contract No. 11596-0800E, prepared for San Diego County Board of Supervisors, 1977.

"Dams and Beach Sand Supply in Southern California," *Papers in Marine Geology*, by R. M. Norris. R. L. Miller, ed. New York: Macmillan, 1964, pp. 154–71.

Detailed Studies of Seacliffs, Beaches, and Coastal Valleys of San Diego County with some Amazing Histories and some Horrifying Implications, by G. G. Kuhn and F. P. Shepard. Berkeley: University of California Press, 1984.

"Greatly Accelerated Man-Induced Coastal Erosion and New Sources of Beach Sand, San Onofre State Park and Camp Pendleton, Northern San Diego County, California," by G. G. Kuhn, E. D. Baker, and F. P. Shepard. *Shore and Beach* 48, 4 (1980): 9–13.

"Impact of a Low Impermeable Groin on Shorezone Geometry," by A. R. Orme, *Geoscience and Man* 18 (1978): 81–95.

Littoral Processes and the Development of Shorelines, by D. L. Inman and J. D. Frautschy, in *Santa Barbara Specialty Conference on Coastal Engineering* (1966), pp. 511–36.

"Is Southern California Ready for a Wet Period?," by William J. Ganus, *Environment Southwest* 472 (Winter 1976): 9–13.

The Sea Off Southern California: A Modern Habitat of Petroleum, by K. O. Emery. New York: John Wiley and Sons, 1960.

"Sea Cliff Erosion at Sunset Cliffs, San Diego," by M. P. Kennedy, *California Geology* 26 (1973): 27–31.

"Sea Cliff Retreat Near Santa Barbara, California," by Robert M. Norris. *Mineral Information Service* 21, 6 (June 1968): 87–91, California Division of Mines and Geology.

"Destructive High Waves Along the Southern California Coast," by G. F. McEwen, *Shore and Beach* 3, 2 (April 1935): 61–64.

"Should Southern California Build Defenses Against Violent Storms Resulting in Lowland Flooding as Discovered in Records of the Past Century?," by G. G. Kuhn and F. P.

Shepard, *Shore and Beach* 49, 4 (1981): 2–10.

Soil Slips, Debris Flows, and Rainstorms in the Santa Monica Mountains and Vicinity, Southern California, by R. H. Campbell. U.S. Geological Survey Professional Paper 851, 1975.

"Temporal Variability of a Summer Shorezone," by A. R. Orme, in *Space and Time in Geomorphology*, C. E. Thorn, ed. London: George Allen Unwin, 1982, pp. 285–313.

"Variable Sediment Flux and Beach Management Ventura County, California," by A. R. Orme and A. J. Brown, in *Coastal Zone '83* 9 (1983): 2328–42, New York: American Society of Civil Engineers.

Legislation and Regulations

California Coastal Plan. San Francisco: California Coastal Zone Conservation Commissions, 1975.

"Legal Battles on the California Coast: A Review of the Rules," by G. Bowden, *Coastal Zone Management Journal* 2, 3 (1975): 273–96.

"The Law of the Coast in a Clamshell: Part III, the California Approach," by Peter H. F. Graber in *Shore and Beach* 49 (1981).

Recreation

Beachcomber's Guide to the Pacific Coast, Menlo Park, Calif.: Sunset Books, 1966.

California Coastal Access Guide, by the California Coastal Commission. Berkeley: University of California Press, 1981.

California State Parks. Menlo Park, Calif.: Sunset Books, 1972.

Edge of a Continent, by D. G. Kelley, Palo Alto, Calif.: American West Publishing Company, 1971.

West Coast Beaches: A Complete Guide, by Sarah Dixon and Peter Dixon. New York: E. P. Dutton, 1979.

Explore, a series of pamphlets for the Year of the Coast, U. S. Army Corps of Engineers, San Francisco, 1980.

Shoreline Engineering and Protection

"The Case For/Against Coastal Protection," by Gary B. Griggs. Ocean Studies Symposium, California Coastal Commission, 1982.

"Emergency Protection of Eroding Shores," by Jon T. Moore, in *Coastal Zone '78* 4 (1978): 2897–910, New York: American Society of Civil Engineers.

"Low-Cost Shore Protection," by Billy Edge and Adrian Combe, in *Coastal Zone '78* 4 (1978): 2925–47, New York: American Society of Civil Engineers.

Low Cost Shore Protection: A Property Owner's Guide, U.S. Army Corps of Engineers, 1981.

Low Cost Shore Protection: A Guide for Local Government Officials, U.S. Army Corps of Engineers, 1981.

Low Cost Shore Protection: A Guide for Engineers and Contractors, U.S. Army Corps of Engineers, 1981.

Shore and Beach, journal of the American Shore and Beach Preservation Association, P.O. Drawer 2087, Wilmington, NC 28401.

Shore Protection in California. California Department of Navigation and Ocean Development (now Boating and Waterways), 1972.

Shore Protection Manual (3 volumes). Washington, D.C.: U.S. Army Coastal Engineering Research Center, U.S. Government Printing Office, 1973.

Storms and Storm Damage

Extreme Waves in California During Winter 1983, by R. J. Seymour. California Department of Boating and Waterways, April 1983.

Preliminary Report on January 1983 Coastal Storm Damage, by Mary Lou Swisher. San Francisco: California Coastal Commission, 1983.

"Selected Coastal Storm Damage in California, Winter of 1977–78," by G. W. Domurat. *Shore and Beach* (July 1978): 15–20.

Wave Damage Along the California Coast, 1977–1978 by Steve Howe. San Francisco: California Coastal Commission, 1978.

Appendix C. Sources for coastal information

Federal government offices

Army Corps of Engineers
Coastal Engineering Research Center
256 Kingman Building
Fort Belvoir, VA 22060

These regional offices assist state and local governments in shoreline protection.

Army Corps of Engineers
Public Affairs Office
Los Angeles District
300 N. Los Angeles Street
Los Angeles, CA 90012
(213) 688-5320

Army Corps of Engineers
Public Affairs Office
San Francisco District
211 Main Street
San Francisco, CA 94105
(415) 556-0594

Army Corps of Engineers
Public Affairs Office
South Pacific Division
630 Sansome Street
San Francisco, CA 94111
(415) 556-5630

For disaster and flood insurance information.

Federal Emergency Management Agency
Region IX
Presidio of San Francisco
Building 105
San Francisco, CA 94129
(415) 556-9840

Small Business Administration
General Information
211 Main Street
San Francisco, CA 94105
(415) 556-4530

For information on national forests.

Forest Service
Office of Information
630 Sansome Street
San Francisco, CA 94111

For weather information

National Oceanic and Atmospheric Administration
(NOAA)
National Weather Records Center
Environmental Data Service
Federal Building
Asheville, NC 28807

For information on geology and geologic hazards in California.

U.S. Geological Survey
Associate Chief Geologist
MS 919
345 Middlefield Road
Menlo Park, CA 94025
For public inquiries, call (415) 323–8111, extension 2817

California state offices

For information on beach erosion and stabilization studies, shore protection programs, and development of small craft harbors.

California Department of Boating and Waterways
(formerly Department of Navigation and Ocean Development
—DNOD)
1629 S Street
Sacramento, CA 95814
Information (916) 445–2615

For information regarding development and development regulations within the coastal zone.

California Coastal Commission
Headquarters
631 Howard Street
San Francisco, CA 94105

Contact the regional offices for local information.

California Coastal Commission
Regional Office
350 E Street, 4th Floor
Eureka, CA 95501

California Coastal Commission
Regional Office
710 Ocean Street, Room 310
Santa Cruz, CA 95060

California Coastal Commission
Regional Office
735 State Street
Balboa Building, Suite 232
Santa Barbara, CA 93101

California Coastal Commission
Regional Office
245 W. Broadway, Suite 380
Long Beach, CA 90802

California Coastal Commission
Regional Office
6154 Mission Gorge Road
San Diego, CA 92120

For information on state parks and beaches.

California Department of Parks and Recreation
Administrative Offices
P.O. Box 2390
Sacramento, CA 95811

For information on the geologic processes and hazards of California.

California Division of Mines and Geology
Division Headquarters
1416 Ninth Street, Room 1314
Sacramento, CA 95814
(916) 445-1825
UC California Sea Grant Marine Advisory Programs
Department of Animal Physiology
University of California
Davis, CA 95616
(916) 752-3342

For regional information contact the local marine advisor.

Humboldt–Del Norte counties
Marine Advisory-Extension Service
Humboldt State University
Arcata, CA 95521
(707) 443-8369

Marin–Sonoma–Mendocino counties
2555 Mendocino Avenue
Room 100-P
Santa Rosa, CA 95401
(707) 527-2621 Sonoma
(415) 479-1100, extension 2374, Marin
(707) 468-4495, extension 276 or 277, Mendocino

San Francisco–San Mateo counties
P.O. Box 34066
Cow Palace, South Hall
San Francisco, CA 94134
(415) 586-4115

Santa Cruz–Monterey counties
Cooperative Extension
1432 Freedom Blvd.
Watsonville, CA 95076
(408) 724-4734 Santa Cruz
(408) 758-4637 Monterey

San Luis Obispo–Santa Barbara–Ventura counties
P.O. Box 126
140 E. Carrillo Street
Santa Barbara, CA 93102
(805) 963-4269 Santa Barbara
(805) 659-0990 Ventura
(805) 543-1550, extension 241, San Luis Obispo

Los Angeles–Orange counties
100 West Water Street
Wilmington, CA 90744
(213) 830-6328

San Diego County
1140 N. Harbor Drive, Room 11
San Diego, CA 92101
(714) 234-4033
(714) 565-5110 (messages)

For general information on the coastal zone, including the location of shoreline structures, dredging activities, and shoreline boundaries.

California State Lands Commission
State Lands Division
1807 13th Street
Sacramento, CA 95814

California Water Resources Control Board
Office of Public Affairs
2125 19th Street
Sacramento, CA 95818

California Department of Transportation
(Caltrans)
Division of Administrative Services
General and Technical Services Branch
1120 N Street
Sacramento, CA 95814

For specific information regarding local ordinances and permit procedures, contact the planning department of the appropriate county or city government.

Index

List of contributors

Chapters 1, 3, 4, and 5 were written by Gary B. Griggs and Lauret E. Savoy of the Department of Earth Sciences at the University of California at Santa Cruz.

Chapter 2 was written by Gary Griggs and Lauret Savoy, Gerald Kuhn and Francis Shepard of the Scripps Institution of Oceanography at La Jolla, and Donald Disraeli of the Department of Geography at the University of California at Santa Barbara.

Chapter 6 was written by James Pepper of the Department of Environmental Studies at the University of California at Santa Cruz.

Chapter 7 was written by Lauret Savoy (for the coast from the Oregon border to Redwood National Park) and by Derek J. Rust of the Department of Geology at Humboldt State University (for the coast from Big Lagoon to Shelter Cove).

Chapters 8 and 9 were written by Lauret Savoy.

Chapter 10 was written by Kim Fulton of the Department of Earth Sciences at the University of California at Santa Cruz and Lauret Savoy.

Chapter 11 was written by Kenneth R. Lajoie and Scott A. Mathieson of the U.S. Geological Survey in Menlo Park, California.

Chapter 12 was written by Gary Griggs.

Chapter 13 was written by Jef Parsons of the Department of Earth Sciences at the University of California at Santa Cruz.

Chapter 14 was written by Amalie Jo Brown, Kevin Mulligan, and Vatche Tchakerian of the Department of Geography at the University of California at Los Angeles (for the coast from Point San Luis to Point Conception), and by Thomas Rockwell of the Department of Geology at San Diego State University and Don Johnson (for the coast from Point Conception to Gaviota State Beach).

Chapter 15 was written by Robert M. Norris of the Department of Geological Sciences at the University of California at Santa Barbara.

Chapter 16 was written by Antony Orme and Linda O'Hirok of the Department of Geography at the University of California at Los Angeles.

Chapter 17 was written by Bernard Pipkin of the Department of Geological Sciences at the University of Southern California.

Chapter 18 was written by Gerald Kuhn and Francis P. Shephard of the Scripps Institution of Oceanography at La Jolla, California.